Sue Pemberton

Series Editor: Julian Gilbey

Cambridge International
AS & A Level Mathematics:

# Pure Mathematics 1
## Coursebook

CAMBRIDGE
UNIVERSITY PRESS

# CAMBRIDGE
## UNIVERSITY PRESS

University Printing House, Cambridge CB2 8BS, United Kingdom

One Liberty Plaza, 20th Floor, New York, NY 10006, USA

477 Williamstown Road, Port Melbourne, VIC 3207, Australia

314–321, 3rd Floor, Plot 3, Splendor Forum, Jasola District Centre, New Delhi – 110025, India

79 Anson Road, #06–04/06, Singapore 079906

Cambridge University Press is part of the University of Cambridge.

It furthers the University's mission by disseminating knowledge in the pursuit of education, learning and research at the highest international levels of excellence.

www.cambridge.org
Information on this title: www.cambridge.org/9781108407144

First published 2018

20  19  18  17  16  15  14  13  12  11  10 9 8 7 6 5 4

Printed in Malaysia by Vivar Printing

*A catalogue record for this publication is available from the British Library*

ISBN 978-1-108-40714-4 Paperback
ISBN 978-1-108-56289-8 Paperback + Cambridge Online Mathematics, 2 years
ISBN 978-1-108-46220-4 Cambridge Online Mathematics, 2 years

Cambridge University Press has no responsibility for the persistence or accuracy of URLs for external or third-party internet websites referred to in this publication, and does not guarantee that any content on such websites is, or will remain, accurate or appropriate. Information regarding prices, travel timetables, and other factual information given in this work is correct at the time of first printing but Cambridge University Press does not guarantee the accuracy of such information thereafter.

® *IGCSE is a registered trademark*

Past exam paper questions throughout are reproduced by permission of Cambridge Assessment International Education. Cambridge Assessment International Education bears no responsibility for the example answers to questions taken from its past question papers which are contained in this publication.

The questions, example answers, marks awarded and/or comments that appear in this book were written by the author(s). In examination, the way marks would be awarded to answers like these may be different.

....................................................................................................................................................

# Contents

iv

v

# Series introduction

Cambridge International AS & A Level Mathematics can be a life-changing course. On the one hand, it is a facilitating subject: there are many university courses that either require an A Level or equivalent qualification in mathematics or prefer applicants who have it. On the other hand, it will help you to learn to think more precisely and logically, while also encouraging creativity. Doing mathematics can be like doing art: just as an artist needs to master her tools (use of the paintbrush, for example) and understand theoretical ideas (perspective, colour wheels and so on), so does a mathematician (using tools such as algebra and calculus, which you will learn about in this course). But this is only the technical side: the joy in art comes through creativity, when the artist uses her tools to express ideas in novel ways. Mathematics is very similar: the tools are needed, but the deep joy in the subject comes through solving problems.

You might wonder what a mathematical 'problem' is. This is a very good question, and many people have offered different answers. You might like to write down your own thoughts on this question, and reflect on how they change as you progress through this course. One possible idea is that a mathematical problem is a mathematical question that you do not immediately know how to answer. (If you do know how to answer it immediately, then we might call it an 'exercise' instead.) Such a problem will take time to answer: you may have to try different approaches, using different tools or ideas, on your own or with others, until you finally discover a way into it. This may take minutes, hours, days or weeks to achieve, and your sense of achievement may well grow with the effort it has taken.

In addition to the mathematical tools that you will learn in this course, the problem-solving skills that you will develop will also help you throughout life, whatever you end up doing. It is very common to be faced with problems, be it in science, engineering, mathematics, accountancy, law or beyond, and having the confidence to systematically work your way through them will be very useful.

This series of Cambridge International AS & A Level Mathematics coursebooks, written for the Cambridge Assessment International Education syllabus for examination from 2020, will support you both to learn the mathematics required for these examinations and to develop your mathematical problem-solving skills. The new examinations may well include more unfamiliar questions than in the past, and having these skills will allow you to approach such questions with curiosity and confidence.

In addition to problem solving, there are two other key concepts that Cambridge Assessment International Education have introduced in this syllabus: namely communication and mathematical modelling. These appear in various forms throughout the coursebooks.

Communication in speech, writing and drawing lies at the heart of what it is to be human, and this is no less true in mathematics. While there is a temptation to think of mathematics as only existing in a dry, written form in textbooks, nothing could be further from the truth: mathematical communication comes in many forms, and discussing mathematical ideas with colleagues is a major part of every mathematician's working life. As you study this course, you will work on many problems. Exploring them or struggling with them together with a classmate will help you both to develop your understanding and thinking, as well as improving your (mathematical) communication skills. And being able to convince someone that your reasoning is correct, initially verbally and then in writing, forms the heart of the mathematical skill of 'proof'.

Mathematical modelling is where mathematics meets the 'real world'. There are many situations where people need to make predictions or to understand what is happening in the world, and mathematics frequently provides tools to assist with this. Mathematicians will look at the real world situation and attempt to capture the key aspects of it in the form of equations, thereby building a model of reality. They will use this model to make predictions, and where possible test these against reality. If necessary, they will then attempt to improve the model in order to make better predictions. Examples include weather prediction and climate change modelling, forensic science (to understand what happened at an accident or crime scene), modelling population change in the human, animal and plant kingdoms, modelling aircraft and ship behaviour, modelling financial markets and many others. In this course, we will be developing tools which are vital for modelling many of these situations.

To support you in your learning, these coursebooks have a variety of new features, for example:

- Explore activities: These activities are designed to offer problems for classroom use. They require thought and deliberation: some introduce a new idea, others will extend your thinking, while others can support consolidation. The activities are often best approached by working in small groups and then sharing your ideas with each other and the class, as they are not generally routine in nature. This is one of the ways in which you can develop problem-solving skills and confidence in handling unfamiliar questions.
- Questions labelled as **P**, **M** or **PS**: These are questions with a particular emphasis on 'Proof', 'Modelling' or 'Problem solving'. They are designed to support you in preparing for the new style of examination. They may or may not be harder than other questions in the exercise.
- The language of the explanatory sections makes much more use of the words 'we', 'us' and 'our' than in previous coursebooks. This language invites and encourages you to be an active participant rather than an observer, simply following instructions ('you do this, then you do that'). It is also the way that professional mathematicians usually write about mathematics. The new examinations may well present you with unfamiliar questions, and if you are used to being active in your mathematics, you will stand a better chance of being able to successfully handle such challenges.

At various points in the books, there are also web links to relevant Underground Mathematics resources, which can be found on the free **undergroundmathematics.org** website. Underground Mathematics has the aim of producing engaging, rich materials for all students of Cambridge International AS & A Level Mathematics and similar qualifications. These high-quality resources have the potential to simultaneously develop your mathematical thinking skills and your fluency in techniques, so we do encourage you to make good use of them.

We wish you every success as you embark on this course.

Julian Gilbey
London, 2018

# How to use this book

Throughout this book you will notice particular features that are designed to help your learning. This section provides a brief overview of these features.

**In this chapter you will learn how to:**
- use the expansion of $(a + b)^n$, where $n$ is a positive integer
- recognise arithmetic and geometric progressions
- use the formulae for the $n$th term and for the sum of the first $n$ terms to solve problems involving arithmetic or geometric progressions
- use the condition for the convergence of a geometric progression, and the formula for the sum to infinity of a convergent geometric progression.

**Learning objectives** indicate the important concepts within each chapter and help you to navigate through the coursebook.

**KEY POINT 1.4**

If we multiply or divide both sides of an inequality by a negative number then the inequality sign must be reversed.

**Key point** boxes contain a summary of the most important methods, facts and formulae.

**PREREQUISITE KNOWLEDGE**

| Where it comes from | What you should be able to do | Check your skills |
|---|---|---|
| IGCSE / O Level Mathematics | Solve quadratic equations by factorising. | 1 Solve. <br> a $x^2 + x - 12 = 0$ <br> b $x^2 - 6x + 9 = 0$ <br> c $3x^2 - 17x - 6 = 0$ |
| IGCSE / O Level Mathematics | Solve linear inequalities. | 2 Solve. <br> a $5x - 8 > 2$ <br> b $3 - 2x \leqslant 7$ |
| IGCSE / O Level Mathematics | Solve simultaneous linear equations. | 3 Solve. <br> a $2x + 3y = 13$ <br> $7x - 5y = -1$ <br> b $2x - 7y = 31$ <br> $3x + 5y = -31$ |
| IGCSE / O Level Additional Mathematics | Carry out simple manipulation of surds. | 4 Simplify. <br> a $\sqrt{20}$ <br> b $(\sqrt{5})^2$ <br> c $\dfrac{8}{\sqrt{2}}$ |

**Prerequisite knowledge** exercises identify prior learning that you need to have covered before starting the chapter. Try the questions to identify any areas that you need to review before continuing with the chapter.

completing the square

**Key terms** are important terms in the topic that you are learning. They are highlighted in orange bold. The **glossary** contains clear definitions of these key terms.

**WORKED EXAMPLE 3.12**

Find the centre and the radius of the circle $x^2 + y^2 + 10x - 8y - 40 = 0$.

**Answer**

We answer this question by first completing the square

| | |
|---|---|
| $x^2 + 10x + y^2 - 8y - 40 = 0$ | Complete the square. |
| $(x + 5)^2 - 5^2 + (y - 4)^2 - 4^2 - 40 = 0$ | Collect constant terms together. |
| $(x + 5)^2 + (y - 4)^2 = 81$ | Compare with $(x - a)^2 + (y - b)^2 = r^2$. |

$a = -5 \qquad b = 4 \qquad r^2 = 81$

Centre $= (-5, 4)$ and radius $= 9$.

**EXPLORE 6.2**

Consider the expansion:
$$(1 + x)^5 = 1 + 5x + 10x^2 + 10x^3 + 5x^4 + x^5$$

The coefficients are: 1    5    10    10    5    1

Find the $nCr$ function on your calculator. On some calculators this may be $_nC_r$ or $\binom{n}{r}$.

1 Use your calculator to find the value of:
$$\binom{5}{0}, \binom{5}{1}, \binom{5}{2}, \binom{5}{3}, \binom{5}{4} \text{ and } \binom{5}{5}.$$

**Explore** boxes contain enrichment activities for extension work. These activities promote group work and peer-to-peer discussion, and are intended to deepen your understanding of a concept. (Answers to the Explore questions are provided in the Teacher's Resource.)

**Worked examples** provide step-by-step approaches to answering questions. The left side shows a fully worked solution, while the right side contains a commentary explaining each step in the working.

**TIP**

It is important to remember to show appropriate calculations in coordinate geometry questions. Answers from scale drawings are not accepted.

**Tip** boxes contain helpful guidance about calculating or checking your answers.

**REWIND**

In Section 2.5 we learnt about the inverse of a function. Here we will look at the particular case of the inverse of a trigonometric function.

**FAST FORWARD**

In the Pure Mathematics 2 and 3 Coursebook, Chapter 7, you will learn how to expand these expressions for any real value of $n$.

**Rewind** and **Fast forward** boxes direct you to related learning. **Rewind** boxes refer to earlier learning, in case you need to revise a topic. **Fast forward** boxes refer to topics that you will cover at a later stage, in case you would like to extend your study.

**DID YOU KNOW?**

A geographical coordinate system is used to describe the location of any point on the Earth's surface. The coordinates used are longitude and latitude. 'Horizontal' circles and 'vertical' circles form the 'grid'. The horizontal circles are perpendicular to the axis of rotation of the Earth and are known as lines of latitude. The vertical circles pass through the North and South poles and are known as lines of longitude.

**Did you know?** boxes contain interesting facts showing how Mathematics relates to the wider world.

**Checklist of learning and understanding**

**Binomial expansions**

Binomial coefficients, denoted by $^nC_r$ or $\binom{n}{r}$, can be found by using:

● Pascal's triangle

● the formulae $\binom{n}{r} = \dfrac{n!}{r!(n-r)!}$ or $\binom{n}{r} = \dfrac{n \times (n-1) \times (n-2) \times \cdots \times (n-r+1)}{r \times (r-1) \times (r-2) \times \cdots \times 3 \times 2 \times 1}$

At the end of each chapter there is a **Checklist of learning and understanding**.

The checklist contains a summary of the concepts that were covered in the chapter. You can use this to quickly check that you have covered the main topics.

**CROSS-TOPIC REVIEW EXERCISE 1**

1  A car of mass 1500 kg is on a straight horizontal road. The car accelerates from $20\,\text{m s}^{-1}$ to $24\,\text{m s}^{-1}$ in 10 s. The car has a constant driving force and there is resistance of 100 N. Find the size of the driving force.  **[4]**

2  A particle starts from rest at a point $X$ and moves in a straight line until 40 s later it reaches a point $Y$, which is 145 m from $X$. For $0\,\text{s} < t < 5\,\text{s}$ the particle accelerates at $0.8\,\text{m s}^{-2}$. For $5\,\text{s} < t < 30\,\text{s}$ it remains at constant velocity. For $30\,\text{s} < t < 40\,\text{s}$ it decelerates at a constant rate, but does not come to rest.

**Cross-topic review exercises** appear after several chapters, and cover topics from across the preceding chapters.

**E**

**Extension** material goes beyond the syllabus. It is highlighted by a red line to the left of the text.

**WEB LINK**

Try the *Sequences* and *Counting and Binomial* resources on the Underground Mathematics website.

**Web link** boxes contain links to useful resources on the internet.

Throughout each chapter there are multiple exercises containing practice questions. The questions are coded:

**PS** These questions focus on problem-solving.

**P** These questions focus on proofs.

**M** These questions focus on modelling.

You should not use a calculator for these questions.

You can use a calculator for these questions.

These questions are taken from past examination papers.

The **End-of-chapter review** contains exam-style questions covering all topics in the chapter. You can use this to check your understanding of the topics you have covered. The number of marks gives an indication of how long you should be spending on the question. You should spend more time on questions with higher mark allocations; questions with only one or two marks should not need you to spend time doing complicated calculations or writing long explanations.

**END-OF-CHAPTER REVIEW EXERCISE 7**

1  Differentiate $\dfrac{3x^5 - 7}{4x}$ with respect to $x$.  **[3]**

2  Find the gradient of the curve $y = \dfrac{8}{4x - 5}$ at the point where $x = 2$.  **[3]**

3  A curve has equation $y = 3x^3 - 3x^2 + x - 7$. Show that the gradient of the curve is never negative.  **[3]**

4  The equation of a curve is $y = (3 - 5x)^3 - 2x$. Find $\dfrac{dy}{dx}$ and $\dfrac{d^2y}{dx^2}$.  **[3]**

5  Find the gradient of the curve $y = \dfrac{15}{x^2 - 2x}$ at the point where $x = 5$.  **[4]**

# Acknowledgements

*The authors and publishers acknowledge the following sources of copyright material and are grateful for the permissions granted. While every effort has been made, it has not always been possible to identify the sources of all the material used, or to trace all copyright holders. If any omissions are brought to our notice, we will be happy to include the appropriate acknowledgements on reprinting.*

Past examination questions throughout are reproduced by permission of Cambridge Assessment International Education.

The following questions are used by permission of the Underground Mathematics website: Exercise 1F Question 9, Exercise 3C Question 16, Exercise 3E Questions 6 and 7, Exercise 4B Question 10.

*Thanks to the following for permission to reproduce images:*

Cover image iStock/Getty Images

Inside (*in order of appearance*) English Heritage/Heritage Images/Getty Images, Sean Russell/Getty Images, Gopinath Duraisamy/EyeEm/Getty Images, Frank Fell/robertharding/Getty Images, Fred Icke/EyeEm/Getty Images, Ralph Grunewald/Getty Images, Gustavo Miranda Holley/Getty Images, shannonstent/Getty Images, wragg/Getty Images, Dimitrios Pikros/EyeEm/Getty Images

x

# Chapter 1
# Quadratics

**In this chapter you will learn how to:**

- carry out the process of completing the square for a quadratic polynomial $ax^2 + bx + c$ and use a completed square form
- find the discriminant of a quadratic polynomial $ax^2 + bx + c$ and use the discriminant
- solve quadratic equations, and quadratic inequalities, in one unknown
- solve by substitution a pair of simultaneous equations of which one is linear and one is quadratic
- recognise and solve equations in $x$ that are quadratic in some function of $x$
- understand the relationship between a graph of a quadratic function and its associated algebraic equation, and use the relationship between points of intersection of graphs and solutions of equations.

| PREREQUISITE KNOWLEDGE | | |
|---|---|---|
| **Where it comes from** | **What you should be able to do** | **Check your skills** |
| IGCSE® / O Level Mathematics | Solve quadratic equations by factorising. | 1 Solve:<br>  a  $x^2 + x - 12 = 0$<br>  b  $x^2 - 6x + 9 = 0$<br>  c  $3x^2 - 17x - 6 = 0$ |
| IGCSE / O Level Mathematics | Solve linear inequalities. | 2 Solve:<br>  a  $5x - 8 > 2$<br>  b  $3 - 2x \leqslant 7$ |
| IGCSE / O Level Mathematics | Solve simultaneous linear equations. | 3 Solve:<br>  a  $2x + 3y = 13$<br>     $7x - 5y = -1$<br>  b  $2x - 7y = 31$<br>     $3x + 5y = -31$ |
| IGCSE / O Level Additional Mathematics | Carry out simple manipulation of surds. | 4 Simplify:<br>  a  $\sqrt{20}$<br>  b  $(\sqrt{5})^2$<br>  c  $\dfrac{8}{\sqrt{2}}$ |

## Why do we study quadratics?

At IGCSE / O Level, you will have learnt about straight-line graphs and their properties. They arise in the world around you. For example, a cell phone contract might involve a fixed monthly charge and then a certain cost per minute for calls: the monthly cost, $y$, is then given as $y = mx + c$, where $c$ is the fixed monthly charge, $m$ is the cost per minute and $x$ is the number of minutes used.

Quadratic functions are of the form $y = ax^2 + bx + c$ (where $a \neq 0$) and they have interesting properties that make them behave very differently from linear functions. A quadratic function has a maximum or a minimum value, and its graph has interesting symmetry. Studying quadratics offers a route into thinking about more complicated functions such as $y = 7x^5 - 4x^4 + x^2 + x + 3$.

You will have plotted graphs of quadratics such as $y = 10 - x^2$ before starting your A Level course. These are most familiar as the shape of the path of a ball as it travels through the air (called its *trajectory*). Discovering that the trajectory is a quadratic was one of Galileo's major successes in the early 17th century. He also discovered that the vertical motion of a ball thrown straight upwards can be modelled by a quadratic, as you will learn if you go on to study the Mechanics component.

**WEB LINK**

Try the *Quadratics* resource on the Underground Mathematics website (www.underground mathematics.org).

## 1.1 Solving quadratic equations by factorisation

You already know the factorisation method and the quadratic formula method to solve quadratic equations algebraically.

This section consolidates and builds on your previous work on solving quadratic equations by factorisation.

---

**EXPLORE 1.1**

$$2x^2 + 3x - 5 = (x - 1)(x - 2)$$

This is Rosa's solution to the previous equation:

| | |
|---|---|
| Factorise the left-hand side: | $(x - 1)(2x + 5) = (x - 1)(x - 2)$ |
| Divide both sides by $(x - 1)$: | $2x + 5 = x - 2$ |
| Rearrange: | $x = -7$ |

Discuss her solution with your classmates and explain why her solution is not fully correct.

Now solve the equation correctly.

---

**WORKED EXAMPLE 1.1**

Solve:

**a** $6x^2 + 5 = 17x$          **b** $9x^2 - 39x - 30 = 0$

**Answer**

**a**
$$6x^2 + 5 = 17x$$    Write in the form $ax^2 + bx + c = 0$.

$$6x^2 - 17x + 5 = 0$$    Factorise.

$$(2x - 5)(3x - 1) = 0$$    Use the fact that if $pq = 0$, then $p = 0$ or $q = 0$.

$$2x - 5 = 0 \quad \text{or} \quad 3x - 1 = 0$$    Solve.

$$x = \frac{5}{2} \quad \text{or} \quad x = \frac{1}{3}$$

**b**
$$9x^2 - 39x - 30 = 0$$    Divide both sides by the common factor of 3.

$$3x^2 - 13x - 10 = 0$$    Factorise.

$$(3x + 2)(x - 5) = 0$$

$$3x + 2 = 0 \quad \text{or} \quad x - 5 = 0$$    Solve.

$$x = -\frac{2}{3} \quad \text{or} \quad x = 5$$

> **TIP**
>
> Divide by a common factor first, if possible.

**WORKED EXAMPLE 1.2**

Solve $\dfrac{21}{2x} - \dfrac{2}{x+3} = 1$.

**Answer**

$$\dfrac{21}{2x} - \dfrac{2}{x+3} = 1$$
Multiply both sides by $2x(x+3)$.

$$21(x+3) - 4x = 2x(x+3)$$
Expand brackets and rearrange.

$$2x^2 - 11x - 63 = 0$$
Factorise.

$$(2x+7)(x-9) = 0$$

$$2x+7 = 0 \quad \text{or} \quad x-9 = 0$$
Solve.

$$x = -\dfrac{7}{2} \quad \text{or} \quad x = 9$$

**WORKED EXAMPLE 1.3**

Solve $\dfrac{3x^2 + 26x + 35}{x^2 + 8} = 0$.

**Answer**

$$\dfrac{3x^2 + 26x + 35}{x^2 + 8} = 0$$
Multiply both sides by $x^2 + 8$.

$$3x^2 + 26x + 35 = 0$$
Factorise.

$$(3x+5)(x+7) = 0$$

$$3x+5 = 0 \quad \text{or} \quad x+7 = 0$$
Solve.

$$x = -\dfrac{5}{3} \quad \text{or} \quad x = -7$$

**WORKED EXAMPLE 1.4**

A rectangle has sides of length $x$ cm and $(6x - 7)$ cm.

The area of the rectangle is $90$ cm$^2$.

Find the lengths of the sides of the rectangle.

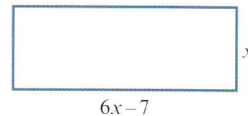

**Answer**

Area $= x(6x - 7) = 6x^2 - 7x = 90$      Rearrange.

$$6x^2 - 7x - 90 = 0$$      Factorise.

$$(2x - 9)(3x + 10) = 0$$

$$2x - 9 = 0 \quad \text{or} \quad 3x + 10 = 0$$      Solve.

$$x = \frac{9}{2} \quad \text{or} \quad x = -\frac{10}{3}$$      Length is a positive quantity, so $x = 4\frac{1}{2}$.

When $x = 4\frac{1}{2}$, $6x - 7 = 20$.

The rectangle has sides of length $4\frac{1}{2}$ cm and $20$ cm.

**EXPLORE 1.2**

| **A** $\quad 4^{(2x^2 + x - 6)} = 1$ | **B** $\quad (x^2 - 3x + 1)^6 = 1$ | **C** $\quad (x^2 - 3x + 1)^{(2x^2 + x - 6)} = 1$ |
|---|---|---|

**1**   Discuss with your classmates how you would solve each of these equations.

**2**   Solve:

    **a**   equation **A**        **b**   equation **B**        **c**   equation **C**

**3**   State how many values of $x$ satisfy:

    **a**   equation **A**        **b**   equation **B**        **c**   equation **C**

**4**   Discuss your results.

**TIP**

Remember to check each of your answers.

**EXERCISE 1A**

1 Solve by factorisation.

a $x^2 + 3x - 10 = 0$  
b $x^2 - 7x + 12 = 0$  
c $x^2 - 6x - 16 = 0$

d $5x^2 + 19x + 12 = 0$  
e $20 - 7x = 6x^2$  
f $x(10x - 13) = 3$

2 Solve:

a $x - \dfrac{6}{x-5} = 0$  
b $\dfrac{2}{x} + \dfrac{3}{x+2} = 1$

c $\dfrac{5x+1}{4} - \dfrac{2x-1}{2} = x^2$  
d $\dfrac{5}{x+3} + \dfrac{3x}{x+4} = 2$

e $\dfrac{3}{x+1} + \dfrac{1}{x(x+1)} = 2$  
f $\dfrac{3}{x+2} + \dfrac{1}{x-1} = \dfrac{1}{(x+1)(x+2)}$

3 Solve:

a $\dfrac{3x^2 + x - 10}{x^2 - 7x + 6} = 0$  
b $\dfrac{x^2 + x - 6}{x^2 + 5} = 0$  
c $\dfrac{x^2 - 9}{7x + 10} = 0$

d $\dfrac{x^2 - 2x - 8}{x^2 + 7x + 10} = 0$  
e $\dfrac{6x^2 + x - 2}{x^2 + 7x + 4} = 0$  
f $\dfrac{2x^2 + 9x - 5}{x^4 + 1} = 0$

4 Find the real solutions of the following equations.

a $8^{(x^2 + 2x - 15)} = 1$  
b $4^{(2x^2 - 11x + 15)} = 1$  
c $2^{(x^2 - 4x + 6)} = 8$

d $3^{(2x^2 + 9x + 2)} = \dfrac{1}{9}$  
e $(x^2 + 2x - 14)^5 = 1$  
f $(x^2 - 7x + 11)^8 = 1$

5 The diagram shows a right-angled triangle with sides $2x$ cm, $(2x + 1)$ cm and 29 cm.

a Show that $2x^2 + x - 210 = 0$.

b Find the lengths of the sides of the triangle.

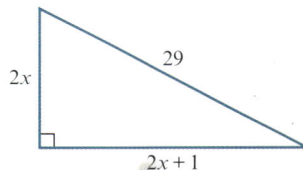

> **TIP**
>
> Check that your answers satisfy the original equation.

6 The area of the trapezium is $35.75\,\text{cm}^2$.  
Find the value of $x$.

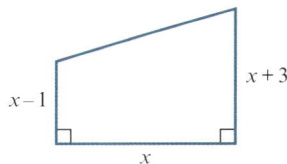

> **WEB LINK**
>
> Try the *Factorisable quadratics* resource on the Underground Mathematics website.

**PS** 7 Solve $(x^2 - 11x + 29)^{(6x^2 + x - 2)} = 1$.

## 1.2 Completing the square

Another method we can use for solving quadratic equations is **completing the square**.

The method of completing the square aims to rewrite a quadratic expression using only one occurrence of the variable, making it an easier expression to work with.

If we expand the expressions $(x + d)^2$ and $(x - d)^2$, we obtain the results:

$$(x + d)^2 = x^2 + 2dx + d^2 \quad \text{and} \quad (x - d)^2 = x^2 - 2dx + d^2$$

Rearranging these gives the following important results:

> **KEY POINT 1.1**
>
> $x^2 + 2dx = (x + d)^2 - d^2$  and  $x^2 - 2dx = (x - d)^2 - d^2$

To complete the square for $x^2 + 10x$, we can use the first of the previous results as follows:

$$10 \div 2 = 5$$
$$x^2 + 10x = (x + 5)^2 - 5^2$$
$$x^2 + 10x = (x + 5)^2 - 25$$

To complete the square for $x^2 + 8x - 7$, we again use the first result applied to the $x^2 + 8x$ part, as follows:

$$8 \div 2 = 4$$
$$x^2 + 8x - 7 = (x + 4)^2 - 4^2 - 7$$
$$x^2 + 8x - 7 = (x + 4)^2 - 23$$

To complete the square for $2x^2 - 12x + 5$, we must first take a factor of 2 out of the first two terms, so:

$$2x^2 - 12x + 5 = 2(x^2 - 6x) + 5$$
$$6 \div 2 = 3$$
$$x^2 - 6x = (x - 3)^2 - 3^2, \text{ giving}$$
$$2x^2 - 12x + 5 = 2\left[(x - 3)^2 - 9\right] + 5 = 2(x - 3)^2 - 13$$

We can also use an algebraic method for completing the square, as shown in Worked example 1.5.

> **WORKED EXAMPLE 1.5**
>
> Express $2x^2 - 12x + 3$ in the form $p(x - q)^2 + r$, where $p$, $q$ and $r$ are constants to be found.
>
> **Answer**
>
> $$2x^2 - 12x + 3 = p(x - q)^2 + r$$
>
> Expanding the brackets and simplifying gives:
>
> $$2x^2 - 12x + 3 = px^2 - 2pqx + pq^2 + r$$
>
> Comparing coefficients of $x^2$, coefficients of $x$ and the constant gives
>
> $$2 = p \text{--------} (1) \qquad -12 = -2pq \text{--------} (2) \qquad 3 = pq^2 + r \text{------} (3)$$
>
> Substituting $p = 2$ in equation (2) gives $q = 3$
>
> Substituting $p = 2$ and $q = 3$ in equation (3) therefore gives $r = -15$
>
> $$2x^2 - 12x + 3 = 2(x - 3)^2 - 15$$

**WORKED EXAMPLE 1.6**

Express $4x^2 + 20x + 5$ in the form $(ax + b)^2 + c$, where $a$, $b$ and $c$ are constants to be found.

**Answer**

$$4x^2 + 20x + 5 = (ax + b)^2 + c$$

Expanding the brackets and simplifying gives:

$$4x^2 + 20x + 5 = a^2x^2 + 2abx + b^2 + c$$

Comparing coefficients of $x^2$, coefficients of $x$ and the constant gives

$$4 = a^2 \text{ ------(1)} \quad 20 = 2ab \text{ ------ (2)} \quad 5 = b^2 + c \text{ ------ (3)}$$

Equation (1) gives $a = \pm 2$.

Substituting $a = 2$ into equation (2) gives $b = 5$.

Substituting $b = 5$ into equation (3) gives $c = -20$.

$$4x^2 + 20x + 5 = (2x + 5)^2 - 20$$

**Alternatively:**

Substituting $a = -2$ into equation (2) gives $b = -5$.

Substituting $b = -5$ into equation (3) gives $c = -20$.

$$4x^2 + 20x + 5 = (-2x - 5)^2 - 20 = (2x + 5)^2 - 20$$

**WORKED EXAMPLE 1.7**

Use completing the square to solve the equation $\dfrac{5}{x + 2} + \dfrac{3}{x - 5} = 1$.

Leave your answers in surd form.

**Answer**

$$\frac{5}{x + 2} + \frac{3}{x - 5} = 1 \qquad \text{Multiply both sides by } (x + 2)(x - 5).$$

$$5(x - 5) + 3(x + 2) = (x + 2)(x - 5) \qquad \text{Expand brackets and collect terms.}$$

$$x^2 - 11x + 9 = 0 \qquad \text{Complete the square.}$$

$$\left(x - \frac{11}{2}\right)^2 - \left(\frac{11}{2}\right)^2 + 9 = 0$$

$$\left(x - \frac{11}{2}\right)^2 = \frac{85}{4}$$

$$x - \frac{11}{2} = \pm\sqrt{\frac{85}{4}}$$

$$x = \frac{11}{2} \pm \frac{\sqrt{85}}{2}$$

$$x = \frac{1}{2}(11 \pm \sqrt{85})$$

**EXERCISE 1B**

1 Express each of the following in the form $(x + a)^2 + b$.

    **a**   $x^2 - 6x$        **b**   $x^2 + 8x$        **c**   $x^2 - 3x$        **d**   $x^2 + 15x$

    **e**   $x^2 + 4x + 8$      **f**   $x^2 - 4x - 8$      **g**   $x^2 + 7x + 1$      **h**   $x^2 - 3x + 4$

2 Express each of the following in the form $a(x + b)^2 + c$.

    **a**   $2x^2 - 12x + 19$    **b**   $3x^2 - 12x - 1$     **c**   $2x^2 + 5x - 1$     **d**   $2x^2 + 7x + 5$

3 Express each of the following in the form $a - (x + b)^2$.

    **a**   $4x - x^2$        **b**   $8x - x^2$        **c**   $4 - 3x - x^2$      **d**   $9 + 5x - x^2$

4 Express each of the following in the form $p - q(x + r)^2$.

    **a**   $7 - 8x - 2x^2$     **b**   $3 - 12x - 2x^2$    **c**   $13 + 4x - 2x^2$    **d**   $2 + 5x - 3x^2$

5 Express each of the following in the form $(ax + b)^2 + c$.

    **a**   $9x^2 - 6x - 3$     **b**   $4x^2 + 20x + 30$   **c**   $25x^2 + 40x - 4$   **d**   $9x^2 - 42x + 61$

6 Solve by completing the square.

    **a**   $x^2 + 8x - 9 = 0$        **b**   $x^2 + 4x - 12 = 0$       **c**   $x^2 - 2x - 35 = 0$

    **d**   $x^2 - 9x + 14 = 0$      **e**   $x^2 + 3x - 18 = 0$       **f**   $x^2 + 9x - 10 = 0$

7 Solve by completing the square. Leave your answers in surd form.

    **a**   $x^2 + 4x - 7 = 0$        **b**   $x^2 - 10x + 2 = 0$      **c**   $x^2 + 8x - 1 = 0$

    **d**   $2x^2 - 4x - 5 = 0$      **e**   $2x^2 + 6x + 3 = 0$      **f**   $2x^2 - 8x - 3 = 0$

8 Solve $\dfrac{5}{x + 2} + \dfrac{3}{x - 4} = 2$. Leave your answers in surd form.

**PS** 9 The diagram shows a right-angled triangle with sides $x$ m, $(2x + 5)$ m and $10$ m.

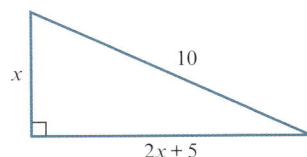

    Find the value of $x$. Leave your answer in surd form.

**PS** 10 Find the real solutions of the equation $(3x^2 + 5x - 7)^4 = 1$.

**PS** 11 The path of a projectile is given by the equation $y = (\sqrt{3})x - \dfrac{49x^2}{9000}$, where $x$ and $y$ are measured in metres.

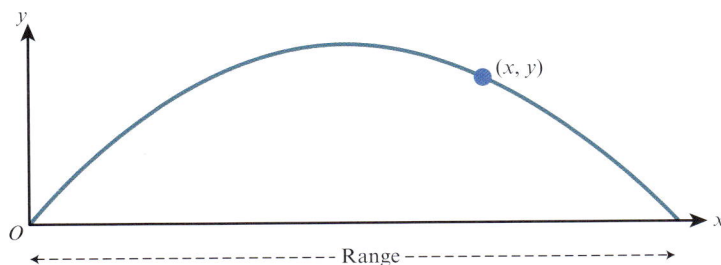

> **TIP**
>
> You will learn how to derive formulae such as this if you go on to study Further Mathematics.

    **a**   Find the range of this projectile.

    **b**   Find the maximum height reached by this projectile.

## 1.3 The quadratic formula

We can solve quadratic equations using the **quadratic formula**.

If $ax^2 + bx + c = 0$, where $a$, $b$ and $c$ are constants and $a \neq 0$, then

> ### KEY POINT 1.2
>
> $$x = \frac{-b \pm \sqrt{b^2 - 4ac}}{2a}$$

The quadratic formula can be proved by completing the square for the equation $ax^2 + bx + c = 0$:

$$ax^2 + bx + c = 0 \qquad \text{Divide both sides by } a.$$

$$x^2 + \frac{b}{a}x + \frac{c}{a} = 0 \qquad \text{Complete the square.}$$

$$\left(x + \frac{b}{2a}\right)^2 - \left(\frac{b}{2a}\right)^2 + \frac{c}{a} = 0 \qquad \text{Rearrange the equation.}$$

$$\left(x + \frac{b}{2a}\right)^2 = \frac{b^2}{4a^2} - \frac{c}{a} \qquad \text{Write the right-hand side as a single fraction.}$$

$$\left(x + \frac{b}{2a}\right)^2 = \frac{b^2 - 4ac}{4a^2} \qquad \text{Find the square root of both sides.}$$

$$x + \frac{b}{2a} = \pm\frac{\sqrt{b^2 - 4ac}}{2a} \qquad \text{Subtract } \frac{b}{2a} \text{ from both sides.}$$

$$x = -\frac{b}{2a} \pm \frac{\sqrt{b^2 - 4ac}}{2a} \qquad \text{Write the right-hand side as a single fraction.}$$

$$x = \frac{-b \pm \sqrt{b^2 - 4ac}}{2a}$$

### WORKED EXAMPLE 1.8

Solve the equation $6x^2 - 3x - 2 = 0$.

Write your answers correct to 3 significant figures.

**Answer**

Using $a = 6$, $b = -3$ and $c = -2$ in the quadratic formula gives:

$$x = \frac{-(-3) \pm \sqrt{(-3)^2 - 4 \times 6 \times (-2)}}{2 \times 6}$$

$$x = \frac{3 + \sqrt{57}}{12} \text{ or } x = \frac{3 - \sqrt{57}}{12}$$

$$x = 0.879 \quad \text{or } x = -0.379 \text{ (to 3 significant figures)}$$

**EXERCISE 1C**

1   Solve using the quadratic formula. Give your answer correct to 2 decimal places.

   **a**   $x^2 - 10x - 3 = 0$       **b**   $x^2 + 6x + 4 = 0$       **c**   $x^2 + 3x - 5 = 0$

   **d**   $2x^2 + 5x - 6 = 0$       **e**   $4x^2 + 7x + 2 = 0$       **f**   $5x^2 + 7x - 2 = 0$

2   A rectangle has sides of length $x$ cm and $(3x - 2)$ cm.

   The area of the rectangle is $63\,\text{cm}^2$.

   Find the value of $x$, correct to 3 significant figures.

3   Rectangle A has sides of length $x$ cm and $(2x - 4)$ cm.

   Rectangle B has sides of length $(x + 1)$ cm and $(5 - x)$ cm.

   Rectangle A and rectangle B have the same area.

   Find the value of $x$, correct to 3 significant figures.

4   Solve the equation $\dfrac{5}{x - 3} + \dfrac{2}{x + 1} = 1$.

   Give your answers correct to 3 significant figures.

5   Solve the quadratic equation $ax^2 - bx + c = 0$, giving your answers in terms
   of $a$, $b$ and $c$.

   How do the solutions of this equation relate to the solutions of
   the equation $ax^2 + bx + c = 0$?

> **⊕ WEB LINK**
>
> Try the *Quadratic
> solving sorter* resource
> on the Underground
> Mathematics website.

11

## 1.4 Solving simultaneous equations (one linear and one quadratic)

In this section, we shall learn how to solve simultaneous equations where one equation is
linear and the second equation is quadratic.

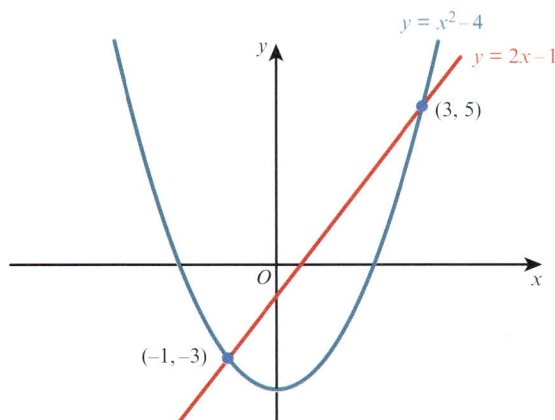

The diagram shows the graphs of $y = x^2 - 4$ and $y = 2x - 1$.

The coordinates of the points of intersection of the two graphs are $(-1, -3)$ and $(3, 5)$.

It follows that $x = -1$, $y = -3$ and $x = 3$, $y = 5$ are the solutions of the simultaneous
equations $y = x^2 - 4$ and $y = 2x - 1$.

The solutions can also be found algebraically:

$$y = x^2 - 4 \quad \text{--------- (1)}$$
$$y = 2x - 1 \quad \text{--------- (2)}$$

Substitute for $y$ from equation (2) into equation (1):

$$2x - 1 = x^2 - 4 \qquad \text{Rearrange.}$$
$$x^2 - 2x - 3 = 0 \qquad \text{Factorise.}$$
$$(x + 1)(x - 3) = 0$$
$$x = -1 \quad \text{or} \quad x = 3$$

Substituting $x = -1$ into equation (2) gives $y = -2 - 1 = -3$.

Substituting $x = 3$ into equation (2) gives $y = 6 - 1 = 5$.

The solutions are: $x = -1$, $y = -3$ and $x = 3$, $y = 5$.

In general, an equation in $x$ and $y$ is called *quadratic* if it has the form $ax^2 + bxy + cy^2 + dx + ey + f = 0$, where at least one of $a$, $b$ and $c$ is non-zero.

Our technique for solving one linear and one quadratic equation will work for these more general quadratics, too. (The graph of a general quadratic function such as this is called a *conic*.)

---

**WORKED EXAMPLE 1.9**

Solve the simultaneous equations.

$$2x + 2y = 7$$
$$x^2 - 4y^2 = 8$$

**Answer**

$$2x + 2y = 7 \quad \text{-------- (1)}$$
$$x^2 - 4y^2 = 8 \quad \text{-------- (2)}$$

From equation (1), $x = \dfrac{7 - 2y}{2}$.

Substitute for $x$ in equation (2):

$$\left(\frac{7 - 2y}{2}\right)^2 - 4y^2 = 8 \qquad\qquad \text{Expand brackets.}$$

$$\frac{49 - 28y + 4y^2}{4} - 4y^2 = 8 \qquad\qquad \text{Multiply both sides by 4.}$$

$$49 - 28y + 4y^2 - 16y^2 = 32 \qquad\qquad \text{Rearrange.}$$

$$12y^2 + 28y - 17 = 0 \qquad\qquad \text{Factorise.}$$

$$(6y + 17)(2y - 1) = 0$$

$$y = -\frac{17}{6} \quad \text{or} \quad y = \frac{1}{2}$$

Substituting $y = -\dfrac{17}{6}$ in equation (1) gives $x = \dfrac{19}{3}$.

Substituting $y = \dfrac{1}{2}$ into equation (1) gives $x = 3$.

The solutions are: $x = \dfrac{19}{3}$, $y = -\dfrac{17}{6}$ and $x = 3$, $y = \dfrac{1}{2}$.

**Alternative method:**

From equation (1), $2y = 7 - 2x$.

Substitute for $2y$ in equation (2):

$x^2 - (7 - 2x)^2 = 8$ .................................... Expand brackets.

$x^2 - 49 + 28x - 4x^2 = 8$ .................................... Rearrange.

$3x^2 - 28x + 57 = 0$ .................................... Factorise.

$(3x - 19)(x - 3) = 0$

$x = \dfrac{19}{3}$ or $x = 3$

The solutions are: $x = \dfrac{19}{3}$, $y = -\dfrac{17}{6}$ and $x = 3$, $y = \dfrac{1}{2}$.

## EXERCISE 1D

1 Solve the simultaneous equations.

a $y = 6 - x$
  $y = x^2$

b $x + 4y = 6$
  $x^2 + 2xy = 8$

c $3y = x + 10$
  $x^2 + y^2 = 100$

d $y = 3x - 1$
  $8x^2 - 2xy = 4$

e $x - 2y = 6$
  $x^2 - 4xy = 20$

f $4x - 3y = 5$
  $x^2 + 3xy = 10$

g $2x + y = 8$
  $xy = 8$

h $2y - x = 5$
  $2x^2 - 3y^2 = 15$

i $x + 2y = 6$
  $x^2 + y^2 + 4xy = 24$

j $5x - 2y = 23$
  $x^2 - 5xy + y^2 = 1$

k $x - 4y = 2$
  $xy = 12$

l $2x - y = 14$
  $y^2 = 8x + 4$

m $2x + 3y + 19 = 0$
  $2x^2 + 3y = 5$

n $x + 2y = 5$
  $x^2 + y^2 = 10$

o $x - 12y = 30$
  $2y^2 - xy = 20$

2 The sum of two numbers is 26. The product of the two numbers is 153.

a What are the two numbers?

b If instead the product is 150 (and the sum is still 26), what would the two numbers now be?

3 The perimeter of a rectangle is 15.8 cm and its area is 13.5 cm². Find the lengths of the sides of the rectangle.

4 The sum of the perimeters of two squares is 50 cm and the sum of the areas is 93.25 cm$^2$.

Find the side length of each square.

5 The sum of the circumferences of two circles is $36\pi$ cm and the sum of the areas is $170\pi$ cm$^2$.

Find the radius of each circle.

6 A cuboid has sides of length 5 cm, $x$ cm and $y$ cm. Given that $x + y = 20.5$ and the volume of the cuboid is 360 cm$^3$, find the value of $x$ and the value of $y$.

7 The diagram shows a solid formed by joining a hemisphere, of radius $r$ cm, to a cylinder, of radius $r$ cm and height $h$ cm.

The total height of the solid is 18 cm and the surface area is $205\pi$ cm$^2$.

Find the value of $r$ and the value of $h$.

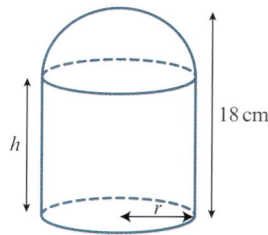

18 cm

$h$

$r$

> **TIP**
>
> The surface area, $A$, of a sphere with radius $r$ is $A = 4\pi r^2$.

8 The line $y = 2 - x$ cuts the curve $5x^2 - y^2 = 20$ at the points $A$ and $B$.

a Find the coordinates of the points $A$ and $B$.

b Find the length of the line $AB$.

9 The line $2x + 5y = 1$ meets the curve $x^2 + 5xy - 4y^2 + 10 = 0$ at the points $A$ and $B$.

a Find the coordinates of the points $A$ and $B$.

b Find the midpoint of the line $AB$.

10 The line $7x + 2y = -20$ intersects the curve $x^2 + y^2 + 4x + 6y - 40 = 0$ at the points $A$ and $B$. Find the length of the line $AB$.

11 The line $7y - x = 25$ cuts the curve $x^2 + y^2 = 25$ at the points $A$ and $B$.

Find the equation of the perpendicular bisector of the line $AB$.

12 The straight line $y = x + 1$ intersects the curve $x^2 - y = 5$ at the points $A$ and $B$.

Given that $A$ lies below the $x$-axis and the point $P$ lies on $AB$ such that $AP : PB = 4 : 1$, find the coordinates of $P$.

13 The line $x - 2y = 1$ intersects the curve $x + y^2 = 9$ at two points, $A$ and $B$.

Find the equation of the perpendicular bisector of the line $AB$.

**PS** 14 a Split 10 into two parts so that the difference between the squares of the parts is 60.

b Split $N$ into two parts so that the difference between the squares of the parts is $D$.

> **WEB LINK**
>
> Try the *Elliptical crossings* resource on the Underground Mathematics website.

## 1.5 Solving more complex quadratic equations

You may be asked to solve an equation that is quadratic in some function of $x$.

**WORKED EXAMPLE 1.10**

Solve the equation $4x^4 - 37x^2 + 9 = 0$.

**Answer**

**Method 1:** Substitution method

$4x^4 - 37x^2 + 9 = 0$

Let $y = x^2$.

$4y^2 - 37y + 9 = 0$        Substitute $x^2$ for $y$.

$(4y - 1)(y - 9) = 0$

$y = \dfrac{1}{4}$ or $y = 9$

$x^2 = \dfrac{1}{4}$ or $x^2 = 9$

$x = \pm\dfrac{1}{2}$ or $x = \pm 3$

**Method 2:** Factorise directly

$4x^4 - 37x^2 + 9 = 0$

$(4x^2 - 1)(x^2 - 9) = 0$

$x^2 = \dfrac{1}{4}$ or $x^2 = 9$

$x = \pm\dfrac{1}{2}$ or $x = \pm 3$

**WORKED EXAMPLE 1.11**

Solve the equation $x - 4\sqrt{x} - 12 = 0$.

**Answer**

$x - 4\sqrt{x} - 12 = 0$

Let $y = \sqrt{x}$.

$y^2 - 4y - 12 = 0$

$(y - 6)(y + 2) = 0$

$y = 6$ or $y = -2$        Substitute $\sqrt{x}$ for $y$.

$\sqrt{x} = 6$ or $\sqrt{x} = -2$        $\sqrt{x} = -2$ has no solutions as $\sqrt{x}$ is never negative.

$\therefore x = 36$

**WORKED EXAMPLE 1.12**

Solve the equation $3(9^x) - 28(3^x) + 9 = 0$.

**Answer**

$3(3^x)^2 - 28(3^x) + 9 = 0$ .............................. Let $y = 3^x$.

$3y^2 - 28y + 9 = 0$

$(3y - 1)(y - 9) = 0$

$y = \dfrac{1}{3}$ or $y = 9$ .............................. Substitute $3^x$ for $y$.

$3^x = \dfrac{1}{3}$ or $3^x = 9$ .............................. $\dfrac{1}{3} = 3^{-1}$ and $9 = 3^2$.

$x = -1$ or $x = 2$

**EXERCISE 1E**

1   Find the real values of $x$ that satisfy the following equations.

    a   $x^4 - 13x^2 + 36 = 0$                 b   $x^6 - 7x^3 - 8 = 0$                c   $x^4 - 6x^2 + 5 = 0$

    d   $2x^4 - 11x^2 + 5 = 0$             e   $3x^4 + x^2 - 4 = 0$                f   $8x^6 - 9x^3 + 1 = 0$

    g   $x^4 + 2x^2 - 15 = 0$              h   $x^4 + 9x^2 + 14 = 0$            i   $x^8 - 15x^4 - 16 = 0$

    j   $32x^{10} - 31x^5 - 1 = 0$         k   $\dfrac{9}{x^4} + \dfrac{5}{x^2} = 4$              l   $\dfrac{8}{x^6} + \dfrac{7}{x^3} = 1$

2   Solve:

    a   $2x - 9\sqrt{x} + 10 = 0$         b   $\sqrt{x}(\sqrt{x} + 1) = 6$          c   $6x - 17\sqrt{x} + 5 = 0$

    d   $10x + \sqrt{x} - 2 = 0$           e   $8x + 5 = 14\sqrt{x}$             f   $3\sqrt{x} + \dfrac{5}{\sqrt{x}} = 16$

3   The curve $y = 2\sqrt{x}$ and the line $3y = x + 8$ intersect at the points $A$ and $B$.

    a   Write down an equation satisfied by the $x$-coordinates of $A$ and $B$.

    b   Solve your equation in part **a** and, hence, find the coordinates of $A$ and $B$.

    c   Find the length of the line $AB$.

**PS**   4   The graph shows $y = ax + b\sqrt{x} + c$ for $x \geqslant 0$. The graph crosses the $x$-axis at the points $(1, 0)$ and $\left(\dfrac{49}{4}, 0\right)$ and it meets the $y$-axis at the point $(0, 7)$. Find the value of $a$, the value of $b$ and the value of $c$.

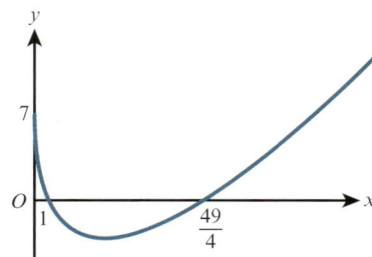

**PS**  5  The graph shows $y = a(2^{2x}) + b(2^x) + c$.
The graph crosses the axes at the points $(2, 0)$, $(4, 0)$ and $(0, 90)$.
Find the value of $a$, the value of $b$ and the value of $c$.

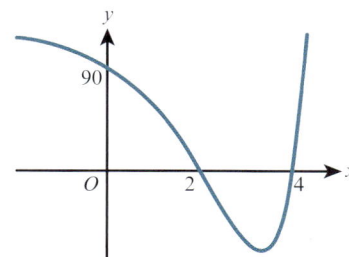

## 1.6 Maximum and minimum values of a quadratic function

The general form of a quadratic function is $f(x) = ax^2 + bx + c$, where $a$, $b$ and $c$ are constants and $a \neq 0$.

The shape of the graph of the function $f(x) = ax^2 + bx + c$ is called a **parabola**.
The orientation of the parabola depends on the value of $a$, the coefficient of $x^2$.

> **TIP**
>
> A point where the gradient is zero is called a **stationary point** or a **turning point**.

If $a > 0$, the curve has a **minimum point** that occurs at the lowest point of the curve.

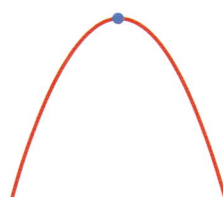

If $a < 0$, the curve has a **maximum point** that occurs at the highest point of the curve.

In the case of a parabola, we also call this point the **vertex** of the parabola.

Every parabola has a line of symmetry that passes through the vertex.

One important skill that we will develop during this course is 'graph sketching'.

A sketch graph needs to show the key features and behaviour of a function.

When we sketch the graph of a quadratic function, the key features are:

- the general shape of the graph
- the axis intercepts
- the coordinates of the vertex.

Depending on the context we should show some or all of these.

The skills you developed earlier in this chapter should enable you to draw a clear sketch graph for any quadratic function.

> **WEB LINK**
>
> Try the *Quadratic symmetry* resource on the Underground Mathematics website for a further explanation of this.

17

18

> **i** **DID YOU KNOW?**
>
> If we rotate a parabola about its axis of symmetry, we obtain a three-dimensional shape called a *paraboloid*. Satellite dishes are paraboloid shapes. They have the special property that light rays are reflected to meet at a single point, if they are parallel to the axis of symmetry of the dish. This single point is called the *focus* of the satellite dish. A receiver at the focus of the paraboloid then picks up all the information entering the dish.

**WORKED EXAMPLE 1.13**

For the function $f(x) = x^2 - 3x - 4$:

**a** Find the axes crossing points for the graph of $y = f(x)$.

**b** Sketch the graph of $y = f(x)$ and find the coordinates of the vertex.

**Answer**

**a** $y = x^2 - 3x - 4$

When $x = 0$, $y = -4$

When $y = 0$, $x^2 - 3x - 4 = 0$

$$(x + 1)(x - 4) = 0$$

$$x = -1 \text{ or } x = 4$$

Axes crossing points are: $(0, -4), (-1, 0)$ and $(4, 0)$.

**b** The line of symmetry cuts the $x$-axis midway between the axis intercepts of $-1$ and $4$.

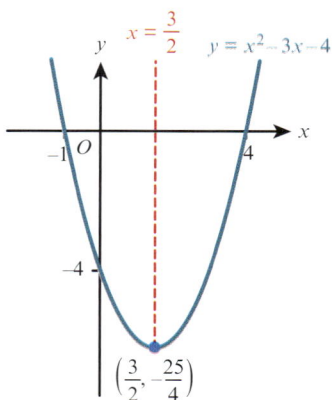

Hence, the line of symmetry is $x = \dfrac{3}{2}$.

When $x = \dfrac{3}{2}$, $y = \left(\dfrac{3}{2}\right)^2 - 3\left(\dfrac{3}{2}\right) - 4$

$$y = -\dfrac{25}{4}$$

Since $a > 0$, the curve is U-shaped.

Minimum point $= \left(\dfrac{3}{2}, -\dfrac{25}{4}\right)$

> **TIP**
>
> Write your answer in fraction form.

Completing the square is an alternative method that can be used to help sketch the graph of a quadratic function.

Completing the square for $x^2 - 3x - 4$ gives:

$$x^2 - 3x - 4 = \left(x - \frac{3}{2}\right)^2 - \left(\frac{3}{2}\right)^2 - 4$$

$$= \left(x - \frac{3}{2}\right)^2 - \frac{25}{4}$$

> This part of the expression is a square so it will be at least zero. The smallest value it can be is 0. This occurs when $x = \frac{3}{2}$.

The minimum value of $\left(x - \frac{3}{2}\right)^2 - \frac{25}{4}$ is $-\frac{25}{4}$ and this minimum occurs when $x = \frac{3}{2}$.

So the function $f(x) = x^2 - 3x - 4$ has a minimum point at $\left(\frac{3}{2}, -\frac{25}{4}\right)$.

The line of symmetry is $x = \frac{3}{2}$.

---

**KEY POINT 1.3**

If $f(x) = ax^2 + bx + c$ is written in the form $f(x) = a(x - h)^2 + k$, then:

- the line of symmetry is $x = h = -\dfrac{b}{2a}$
- if $a > 0$, there is a minimum point at $(h, k)$
- if $a < 0$, there is a maximum point at $(h, k)$.

**WORKED EXAMPLE 1.14**

Sketch the graph of $y = 16x - 7 - 4x^2$.

**Answer**

Completing the square gives:
$16x - 7 - 4x^2 = 9 - 4(x - 2)^2$

> This part of the expression is a square so $(x - 2)^2 \geqslant 0$. The smallest value it can be is 0. This occurs when $x = 2$. Since this is being *subtracted* from 9, the whole expression is *greatest* when $x = 2$.

The maximum value of $9 - 4(x - 2)^2$ is 9 and this maximum occurs when $x = 2$.

So the function $f(x) = 16x - 7 - 4x^2$ has a maximum point at $(2, 9)$.

The line of symmetry is $x = 2$.

When $x = 0$, $y = -7$

When $y = 0$, $9 - 4(x - 2)^2 = 0$

$$(x - 2)^2 = \frac{9}{4}$$

$$x - 2 = \pm \frac{3}{2}$$

$$x = 3\tfrac{1}{2} \text{ or } x = \frac{1}{2}$$

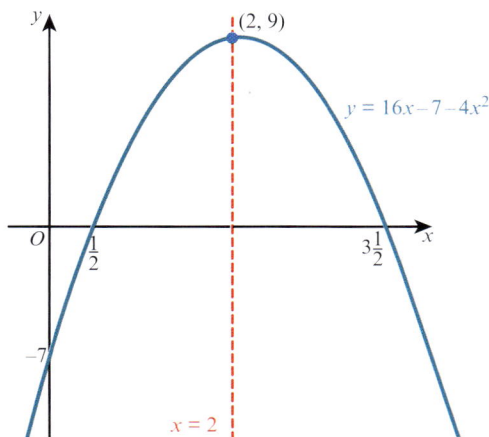

## EXERCISE 1F

**1** Use the symmetry of each quadratic function to find the maximum or minimum points.

Sketch each graph, showing all axes crossing points.

    **a** $y = x^2 - 6x + 8$         **b** $y = x^2 + 5x - 14$         **c** $y = 2x^2 + 7x - 15$         **d** $y = 12 + x - x^2$

**2**   **a** Express $2x^2 - 8x + 5$ in the form $a(x + b)^2 + c$, where $a$, $b$ and $c$ are integers.

    **b** Write down the equation of the line of symmetry for the graph of $y = 2x^2 - 8x + 1$.

**3**   **a** Express $7 + 5x - x^2$ in the form $a - (x + b)^2$, where $a$, and $b$ are constants.

    **b** Find the coordinates of the turning point of the curve $y = 7 + 5x - x^2$, stating whether it is a maximum or a minimum point.

**4**   **a** Express $2x^2 + 9x + 4$ in the form $a(x + b)^2 + c$, where $a$, $b$ and $c$ are constants.

    **b** Write down the coordinates of the vertex of the curve $y = 2x^2 + 9x + 4$, and state whether this is a maximum or a minimum point.

**5** Find the minimum value of $x^2 - 7x + 8$ and the corresponding value of $x$.

**6**   **a** Write $1 + x - 2x^2$ in the form $p - 2(x - q)^2$.

    **b** Sketch the graph of $y = 1 + x - 2x^2$.

**7** Prove that the graph of $y = 4x^2 + 2x + 5$ does not intersect the $x$-axis.

**PS**   **8** Find the equations of parabolas A, B and C.

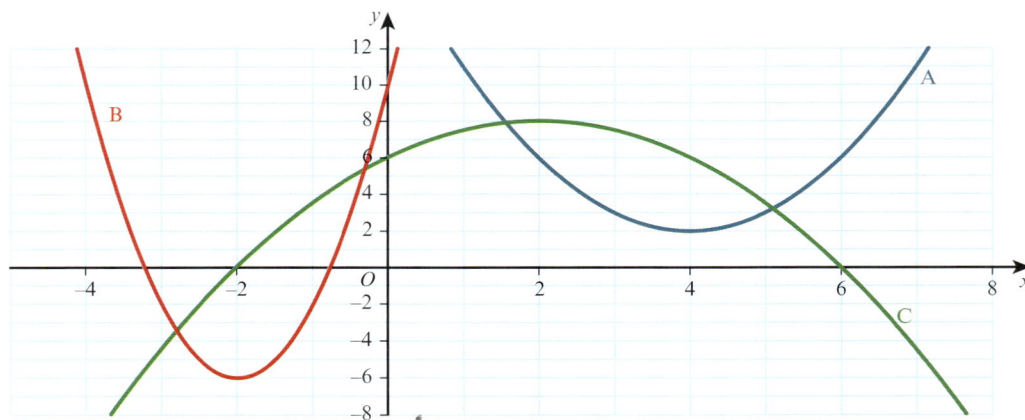

**PS** 9 The diagram shows eight parabolas.

The equations of two of the parabolas are $y = x^2 - 6x + 13$ and $y = -x^2 - 6x - 5$.

a Identify these two parabolas and find the equation of each of the other parabolas.

b Use graphing software to create your own parabola pattern.

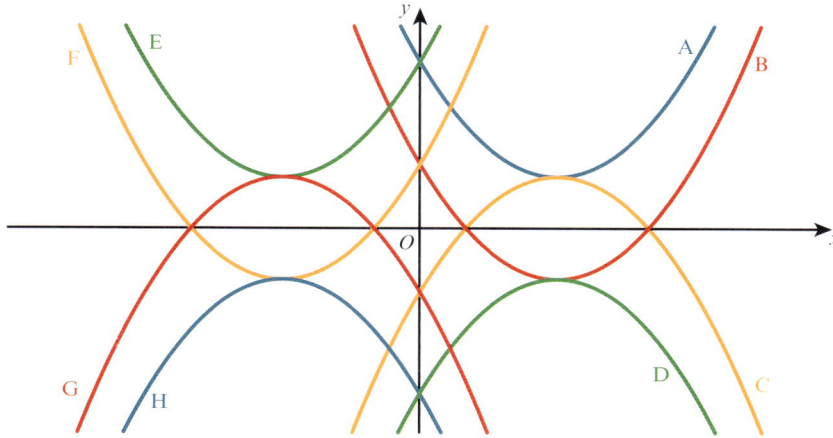

[This question is an adaptation of *Which parabola?* on the Underground Mathematics website and was developed from an original idea from NRICH.]

**PS** 10 A parabola passes through the points $(0, -24)$, $(-2, 0)$ and $(4, 0)$.

Find the equation of the parabola.

**PS** 11 A parabola passes through the points $(-2, -3)$, $(2, 9)$ and $(6, 5)$.

Find the equation of the parabola.

**P** 12 Prove that any quadratic that has its vertex at $(p, q)$ has an equation of the form $y = ax^2 - 2apx + ap^2 + q$ for some non-zero real number $a$.

## 1.7 Solving quadratic inequalities

We already know how to solve linear inequalities.

The following text shows two examples.

| | |
|---|---|
| Solve $2(x + 7) < -4$. | Expand brackets. |
| $2x + 14 < -4$ | Subtract 14 from both sides. |
| $2x < -18$ | Divide both sides by 2. |
| $x < -9$ | |

| | |
|---|---|
| Solve $11 - 2x \geqslant 5$. | Subtract 11 from both sides. |
| $-2x \geqslant -6$ | Divide both sides by $-2$. |
| $x \leqslant 3$ | |

The second of the previous examples uses the important rule that:

> ### 🔍 KEY POINT 1.4
>
> If we multiply or divide both sides of an inequality by a negative number, then the inequality sign must be reversed.

Quadratic inequalities can be solved by sketching a graph and considering when the graph is above or below the $x$-axis.

### WORKED EXAMPLE 1.15

Solve $x^2 - 5x - 14 > 0$.

**Answer**

Sketch the graph of $y = x^2 - 5x - 14$.

When $y = 0$, $x^2 - 5x - 14 = 0$

$\qquad (x + 2)(x - 7) = 0$

$\qquad x = -2$ or $x = 7$

So the $x$-axis crossing points are $-2$ and $7$.

For $x^2 - 5x - 14 > 0$ we need to find the range of values of $x$ for which the curve is positive (above the $x$-axis).

The solution is $x < -2$ or $x > 7$.

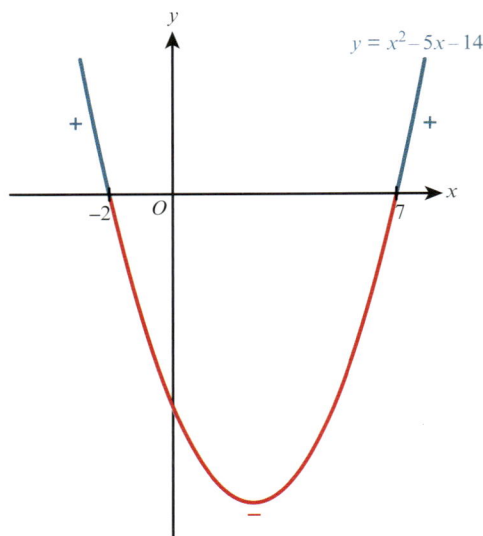

$y = x^2 - 5x - 14$

> ### 💡 TIP
>
> For the sketch graph, you only need to identify which way up the graph is and where the $x$-intercepts are: you do not need to find the vertex or the $y$-intercept.

### WORKED EXAMPLE 1.16

Solve $2x^2 + 3x \leqslant 27$.

**Answer**

Rearranging: $2x^2 + 3x - 27 \leqslant 0$

Sketch the graph of $y = 2x^2 + 3x - 27$.

When $y = 0$, $2x^2 + 3x - 27 = 0$

$\qquad (2x + 9)(x - 3) = 0$

$\qquad x = -4\frac{1}{2}$ or $x = 3$

So the $x$-axis intercepts are $-4\frac{1}{2}$ and $3$.

For $2x^2 + 3x - 27 \leqslant 0$ we need to find the range of values of $x$ for which the curve is either zero or negative (below the $x$-axis).

The solution is $-4\frac{1}{2} \leqslant x \leqslant 3$.

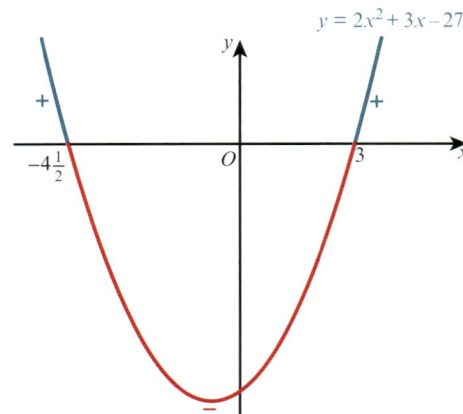

$y = 2x^2 + 3x - 27$

**EXPLORE 1.3**

Ivan is asked to solve the inequality $\dfrac{2x-4}{x} \geqslant 7$.

This is his solution:

Multiply both sides by $x$: $\qquad\qquad 2x - 4 \geqslant 7x$

Subtract $2x$ from both sides: $\qquad\quad -4 \geqslant 5x$

Divide both sides by 5: $\qquad\qquad\quad x \leqslant -\dfrac{4}{5}$

Anika checks to see if $x = -1$ satisfies the original inequality.

She writes:

When $x = -1$: $(2(-1) - 4) \div (-1) = 6$

Hence, $x = -1$ is a value of $x$ that does not satisfy the original inequality.

So Ivan's solution must be incorrect!

Discuss Ivan's solution with your classmates and explain Ivan's error.

How could Ivan have approached this problem to obtain a correct solution?

**EXERCISE 1G**

1  Solve:

   a  $x(x - 3) \leqslant 0$
   
   b  $(x - 3)(x + 2) > 0$
   
   c  $(x - 6)(x - 4) \leqslant 0$

   d  $(2x + 3)(x - 2) < 0$
   
   e  $(5 - x)(x + 6) \geqslant 0$
   
   f  $(1 - 3x)(2x + 1) < 0$

2  Solve:

   a  $x^2 - 25 \geqslant 0$
   
   b  $x^2 + 7x + 10 \leqslant 0$
   
   c  $x^2 + 6x - 7 > 0$

   d  $14x^2 + 17x - 6 \leqslant 0$
   
   e  $6x^2 - 23x + 20 < 0$
   
   f  $4 - 7x - 2x^2 < 0$

3  Solve:

   a  $x^2 < 36 - 5x$
   
   b  $15x < x^2 + 56$
   
   c  $x(x + 10) \leqslant 12 - x$

   d  $x^2 + 4x < 3(x + 2)$
   
   e  $(x + 3)(1 - x) < x - 1$
   
   f  $(4x + 3)(3x - 1) < 2x(x + 3)$

   g  $(x + 4)^2 \geqslant 25$
   
   h  $(x - 2)^2 > 14 - x$
   
   i  $6x(x + 1) < 5(7 - x)$

4  Find the range of values of $x$ for which $\dfrac{5}{2x^2 + x - 15} < 0$.

5  Find the set of values of $x$ for which:

   a  $x^2 - 3x \geqslant 10$ $\quad$ and $\quad (x - 5)^2 < 4$

   b  $x^2 + 4x - 21 \leqslant 0$ $\quad$ and $\quad x^2 - 9x + 8 > 0$

   c  $x^2 + x - 2 > 0$ $\quad$ and $\quad x^2 - 2x - 3 \geqslant 0$

6  Find the range of values of $x$ for which $2^{x^2 - 3x - 40} > 1$.

**E**

7 Solve:

a $\dfrac{x}{x-1} \geqslant 3$

b $\dfrac{x(x-1)}{x+1} > x$

c $\dfrac{x^2-9}{x-1} \geqslant 4$

d $\dfrac{x^2-2x-15}{x-2} \geqslant 0$

e $\dfrac{x^2+4x-5}{x^2-4} \leqslant 0$

f $\dfrac{x-3}{x+4} \geqslant \dfrac{x+2}{x-5}$

## 1.8 The number of roots of a quadratic equation

If f$(x)$ is a function, then we call the solutions to the equation f$(x) = 0$ the **roots** of f$(x)$.

Consider solving the following three quadratic equations of the form $ax^2 + bx + c = 0$ using the formula $x = \dfrac{-b \pm \sqrt{b^2-4ac}}{2a}$.

| | | |
|---|---|---|
| $x^2+2x-8=0$ | $x^2+6x+9=0$ | $x^2+2x+6=0$ |
| $x=\dfrac{-2\pm\sqrt{2^2-4\times1\times(-8)}}{2\times1}$ | $x=\dfrac{-6\pm\sqrt{6^2-4\times1\times9}}{2\times1}$ | $x=\dfrac{-2\pm\sqrt{2^2-4\times1\times6}}{2\times1}$ |
| $x=\dfrac{-2\pm\sqrt{36}}{2}$ | $x=\dfrac{-6\pm\sqrt{0}}{2}$ | $x=\dfrac{-2\pm\sqrt{-20}}{2}$ |
| $x=2$ or $x=-4$ | $x=-3$ or $x=-3$ | no real solution |
| **two distinct real roots** | **two equal real roots** | **no real roots** |

The part of the quadratic formula underneath the square root sign is called the **discriminant**.

**🔍 KEY POINT 1.5**

The discriminant of $ax^2 + bx + c = 0$ is $b^2 - 4ac$.

The sign (positive, zero or negative) of the discriminant tells us how many roots there are for a particular quadratic equation.

| $b^2-4ac$ | Nature of roots |
|---|---|
| $>0$ | two distinct real roots |
| $=0$ | two equal real roots (or 1 repeated real root) |
| $<0$ | no real roots |

There is a connection between the roots of the quadratic equation $ax^2 + bx + c = 0$ and the corresponding curve $y = ax^2 + bx + c$.

| $b^2-4ac$ | Nature of roots of $ax^2+bx+c=0$ | Shape of curve $y=ax^2+bx+c$ |
|---|---|---|
| $>0$ | two distinct real roots | The curve cuts the $x$-axis at two distinct points. |
| $=0$ | two equal real roots (or 1 repeated real root) | The curve touches the $x$-axis at one point. |
| $<0$ | no real roots | The curve is entirely above or entirely below the $x$-axis. |

24

**WORKED EXAMPLE 1.17**

Find the values of $k$ for which the equation $4x^2 + kx + 1 = 0$ has two equal roots.

**Answer**

For two equal roots: $\qquad b^2 - 4ac = 0$

$$k^2 - 4 \times 4 \times 1 = 0$$

$$k^2 = 16$$

$$k = -4 \quad \text{or} \quad k = 4$$

**WORKED EXAMPLE 1.18**

Find the values of $k$ for which $x^2 - 5x + 9 = k(5 - x)$ has two equal roots.

**Answer**

$$x^2 - 5x + 9 = k(5 - x) \quad \cdots\cdots\cdots\cdots \quad \boxed{\text{Rearrange the equation into the form } ax^2 + bx + c = 0.}$$

$$x^2 - 5x + 9 - 5k + kx = 0$$

$$x^2 + (k - 5)x + 9 - 5k = 0$$

For two equal roots: $b^2 - 4ac = 0$

$$(k - 5)^2 - 4 \times 1 \times (9 - 5k) = 0$$

$$k^2 - 10k + 25 - 36 + 20k = 0$$

$$k^2 + 10k - 11 = 0$$

$$(k + 11)(k - 1) = 0$$

$$k = -11 \quad \text{or} \quad k = 1$$

25

**WORKED EXAMPLE 1.19**

Find the values of $k$ for which $kx^2 - 2kx + 8 = 0$ has two distinct roots.

**Answer**

$$kx^2 - 2kx + 8 = 0$$

For two distinct roots:

$$b^2 - 4ac > 0$$

$$(-2k)^2 - 4 \times k \times 8 > 0$$

$$4k^2 - 32k > 0$$

$$4k(k - 8) > 0$$

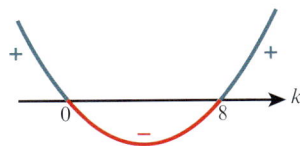

Critical values are 0 and 8.

Note that the critical values are where $4k(k - 8) = 0$.

Hence, $k < 0$ or $k > 8$.

1 Find the discriminant for each equation and, hence, decide if the equation has two distinct roots, two equal roots or no real roots.

a $x^2 - 12x + 36 = 0$
b $x^2 + 5x - 36 = 0$
c $x^2 + 9x + 2 = 0$

d $4x^2 - 4x + 1 = 0$
e $2x^2 - 7x + 8 = 0$
f $3x^2 + 10x - 2 = 0$

2 Use the discriminant to determine the nature of the roots of $2 - 5x = \dfrac{4}{x}$.

3 The equation $x^2 + bx + c = 0$ has roots −5 and 7.

Find the value of $b$ and the value of $c$.

4 Find the values of $k$ for which the following equations have two equal roots.

a $x^2 + kx + 4 = 0$
b $4x^2 + 4(k - 2)x + k = 0$

c $(k + 2)x^2 + 4k = (4k + 2)x$
d $x^2 - 2x + 1 = 2k(k - 2)$

e $(k + 1)x^2 + kx - 2k = 0$
f $4x^2 - (k - 2)x + 9 = 0$

5 Find the values of $k$ for which the following equations have two distinct roots.

a $x^2 + 8x + 3 = k$
b $2x^2 - 5x = 4 - k$

c $kx^2 - 4x + 2 = 0$
d $kx^2 + 2(k - 1)x + k = 0$

e $2x^2 = 2(x - 1) + k$
f $kx^2 + (2k - 5)x = 1 - k$

6 Find the values of $k$ for which the following equations have no real roots.

a $kx^2 - 4x + 8 = 0$
b $3x^2 + 5x + k + 1 = 0$

c $2x^2 + 8x - 5 = kx^2$
d $2x^2 + k = 3(x - 2)$

e $kx^2 + 2kx = 4x - 6$
f $kx^2 + kx = 3x - 2$

7 The equation $kx^2 + px + 5 = 0$ has repeated real roots.

Find $k$ in terms of $p$.

8 Find the range of values of $k$ for which the equation $kx^2 - 5x + 2 = 0$ has real roots.

P 9 Prove that the roots of the equation $2kx^2 + 5x - k = 0$ are real and distinct for all real values of $k$.

P 10 Prove that the roots of the equation $x^2 + (k - 2)x - 2k = 0$ are real and distinct for all real values of $k$.

P 11 Prove that $x^2 + kx + 2 = 0$ has real roots if $k \geqslant 2\sqrt{2}$.

For which other values of $k$ does the equation have real roots?

WEB LINK

Try the *Discriminating* resource on the Underground Mathematics website.

## 1.9 Intersection of a line and a quadratic curve

When considering the intersection of a straight line and a parabola, there are three possible situations.

| Situation 1 | Situation 2 | Situation 3 |
|---|---|---|
| 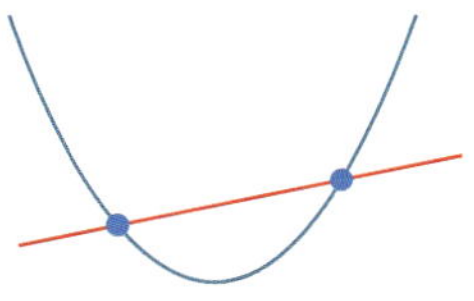 | 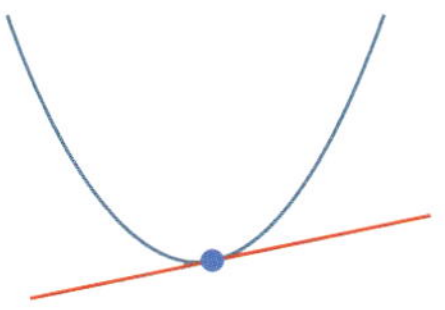 | 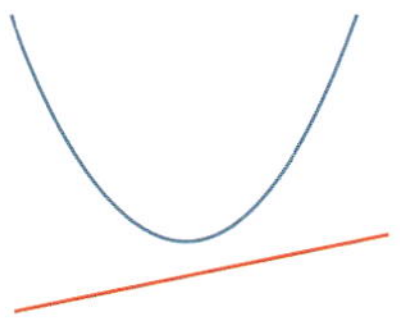 |
| two points of intersection | one point of intersection | no points of intersection |
| The line cuts the curve at two distinct points. | The line touches the curve at one point. This means that the line is a **tangent** to the curve. | The line does not intersect the curve. |

We have already learnt that to find the points of intersection of a straight line and a quadratic curve, we solve their equations simultaneously.

The discriminant of the resulting equation then enables us to say how many points of intersection there are. The three possible situations are shown in the following table.

| $b^2 - 4ac$ | Nature of roots | Line and curve |
|---|---|---|
| $> 0$ | two distinct real roots | two distinct points of intersection |
| $= 0$ | two equal real roots (repeated roots) | one point of intersection (line is a tangent) |
| $< 0$ | no real roots | no points of intersection |

**WORKED EXAMPLE 1.20**

Find the value of $k$ for which $y = x + k$ is a tangent to the curve $y = x^2 + 5x + 2$.

**Answer**

$$x^2 + 5x + 2 = x + k$$
$$x^2 + 4x + (2 - k) = 0$$

Since the line is a tangent to the curve, the discriminant of the quadratic must be zero, so:

$$b^2 - 4ac = 0$$
$$4^2 - 4 \times 1 \times (2 - k) = 0$$
$$16 - 8 + 4k = 0$$
$$4k = -8$$
$$k = -2$$

**WORKED EXAMPLE 1.21**

Find the set of values of $k$ for which $y = kx - 1$ intersects the curve $y = x^2 - 2x$ at two distinct points.

**Answer**

$$x^2 - 2x = kx - 1$$
$$x^2 - (k + 2)x + 1 = 0$$

Since the line intersects the curve at two distinct points, we must have discriminant $> 0$.

$$b^2 - 4ac > 0$$
$$(k + 2)^2 - 4 \times 1 \times 1 > 0$$
$$k^2 + 4k > 0$$
$$k(k + 4) > 0$$

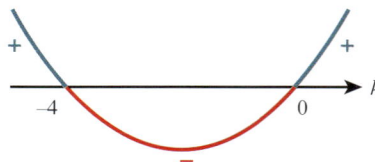

Critical values are $-4$ and $0$.

Hence, $k < -4$ or $k > 0$.

This next example involves a more general quadratic equation. Our techniques for finding the conditions for intersection of a straight line and a quadratic equation will work for this more general quadratic equation too.

**WORKED EXAMPLE 1.22**

Find the set of values of $k$ for which the line $2x + y = k$ does not intersect the curve $xy = 8$.

**Answer**

Substituting $y = k - 2x$ into $xy = 8$ gives:

$$x(k - 2x) = 8$$
$$2x^2 - kx + 8 = 0$$

Since the line and curve do not intersect, we must have discriminant $< 0$.

$$b^2 - 4ac < 0$$
$$(-k)^2 - 4 \times 2 \times 8 < 0$$
$$k^2 - 64 < 0$$
$$(k + 8)(k - 8) < 0$$

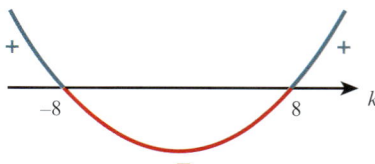

Critical values are $-8$ and $8$.

Hence, $-8 < k < 8$.

1   Find the values of $k$ for which the line $y = kx + 1$ is a tangent to the curve $y = x^2 - 7x + 2$.

2   Find the values of $k$ for which the $x$-axis is a tangent to the curve $y = x^2 - (k + 3)x + (3k + 4)$.

3   Find the value of $k$ for which the line $x + ky = 12$ is a tangent to the curve $y = \dfrac{5}{x - 2}$.

   Can you explain graphically why there is only one such value of $k$? (You may want to use graph-drawing software to help with this.)

4   The line $y = k - 3x$ is a tangent to the curve $x^2 + 2xy - 20 = 0$.

   a   Find the possible values of $k$.

   b   For each of these values of $k$, find the coordinates of the point of contact of the tangent with the curve.

5   Find the values of $m$ for which the line $y = mx + 6$ is a tangent to the curve $y = x^2 - 4x + 7$.

   For each of these values of $m$, find the coordinates of the point where the line touches the curve.

6   Find the set of values of $k$ for which the line $y = 2x - 1$ intersects the curve $y = x^2 + kx + 3$ at two distinct points.

7   Find the set of values of $k$ for which the line $x + 2y = k$ intersects the curve $xy = 6$ at two distinct points.

8   Find the set of values of $k$ for which the line $y = k - x$ cuts the curve $y = 5 - 3x - x^2$ at two distinct points.

9   Find the set of values of $m$ for which the line $y = mx + 5$ does not meet the curve $y = x^2 - x + 6$.

10  Find the set of values of $k$ for which the line $y = 2x - 10$ does not meet the curve $y = x^2 - 6x + k$.

11  Find the value of $k$ for which the line $y = kx + 6$ is a tangent to the curve $x^2 + y^2 - 10x + 8y = 84$.

**P**  12  The line $y = mx + c$ is a tangent to the curve $y = x^2 - 4x + 4$.

   Prove that $m^2 + 8m + 4c = 0$.

**P**  13  The line $y = mx + c$ is a tangent to the curve $ax^2 + by^2 = c$, where $a, b, c$ and $m$ are constants.

   Prove that $m^2 = \dfrac{abc - a}{b}$.

29

# Checklist of learning and understanding

**Quadratic equations can be solved by:**

- factorisation
- completing the square
- using the quadratic formula $x = \dfrac{-b \pm \sqrt{b^2 - 4ac}}{2a}$.

**Solving simultaneous equations where one is linear and one is quadratic**

- Rearrange the linear equation to make either $x$ or $y$ the subject.
- Substitute this for $x$ or $y$ in the quadratic equation and then solve.

**Maximum and minimum points and lines of symmetry**

For a quadratic function $f(x) = ax^2 + bx + c$ that is written in the form $f(x) = a(x - h)^2 + k$:

- the line of symmetry is $x = h = -\dfrac{b}{2a}$
- if $a > 0$, there is a minimum point at $(h, k)$
- if $a < 0$, there is a maximum point at $(h, k)$.

**Quadratic equation $ax^2 + bx + c = 0$ and corresponding curve $y = ax^2 + bx + c$**

- Discriminant $= b^2 - 4ac$.
- If $b^2 - 4ac > 0$, then the equation $ax^2 + bx + c = 0$ has two distinct real roots.
- If $b^2 - 4ac = 0$, then the equation $ax^2 + bx + c = 0$ has two equal real roots.
- If $b^2 - 4ac < 0$, then the equation $ax^2 + bx + c = 0$ has no real roots.
- The condition for a quadratic equation to have real roots is $b^2 - 4ac \geqslant 0$.

**Intersection of a line and a general quadratic curve**

- If a line and a general quadratic curve intersect at one point, then the line is a tangent to the curve at that point.
- Solving simultaneously the equations for the line and the curve gives an equation of the form $ax^2 + bx + c = 0$.
- $b^2 - 4ac$ gives information about the intersection of the line and the curve.

| $b^2 - 4ac$ | Nature of roots | Line and parabola |
|---|---|---|
| $> 0$ | two distinct real roots | two distinct points of intersection |
| $= 0$ | two equal real roots | one point of intersection (line is a tangent) |
| $< 0$ | no real roots | no points of intersection |

**END-OF-CHAPTER REVIEW EXERCISE 1**

1   A curve has equation $y = 2xy + 5$ and a line has equation $2x + 5y = 1$.

   The curve and the line intersect at the points $A$ and $B$. Find the coordinates of the midpoint of the line $AB$.   [4]

2   a   Express $9x^2 - 15x$ in the form $(3x - a)^2 - b$.   [2]

   b   Find the set of values of $x$ that satisfy the inequality $9x^2 - 15x < 6$.   [2]

3   Find the real roots of the equation $\dfrac{36}{x^4} + 4 = \dfrac{25}{x^2}$.   [4]

4   Find the set of values of $k$ for which the line $y = kx - 3$ intersects the curve $y = x^2 - 9x$ at two distinct points.   [4]

5   Find the set of values of the constant $k$ for which the line $y = 2x + k$ meets the curve $y = 1 + 2kx - x^2$ at two distinct points.   [5]

6   a   Find the coordinates of the vertex of the parabola $y = 4x^2 - 12x + 7$.   [4]

   b   Find the values of the constant $k$ for which the line $y = kx + 3$ is a tangent to the curve $y = 4x^2 - 12x + 7$.   [3]

7   A curve has equation $y = 5 - 2x + x^2$ and a line has equation $y = 2x + k$, where $k$ is a constant.

   a   Show that the $x$-coordinates of the points of intersection of the curve and the line are given by the equation $x^2 - 4x + (5 - k) = 0$.   [1]

   b   For one value of $k$, the line intersects the curve at two distinct points, $A$ and $B$, where the coordinates of $A$ are $(-2, 13)$. Find the coordinates of $B$.   [3]

   c   For the case where the line is a tangent to the curve at a point $C$, find the value of $k$ and the coordinates of $C$.   [4]

8   A curve has equation $y = x^2 - 5x + 7$ and a line has equation $y = 2x - 3$.

   a   Show that the curve lies above the $x$-axis.   [3]

   b   Find the coordinates of the points of intersection of the line and the curve.   [3]

   c   Write down the set of values of $x$ that satisfy the inequality $x^2 - 5x + 7 < 2x - 3$.   [1]

9   A curve has equation $y = 10x - x^2$.

   a   Express $10x - x^2$ in the form $a - (x + b)^2$.   [3]

   b   Write down the coordinates of the vertex of the curve.   [2]

   c   Find the set of values of $x$ for which $y \le 9$.   [3]

10   A line has equation $y = kx + 6$ and a curve has equation $y = x^2 + 3x + 2k$, where $k$ is a constant.

   i   For the case where $k = 2$, the line and the curve intersect at points $A$ and $B$.

      Find the distance $AB$ and the coordinates of the mid-point of $AB$.   [5]

   ii   Find the two values of $k$ for which the line is a tangent to the curve.   [4]

   *Cambridge International AS & A Level Mathematics 9709 Paper 11 Q9 November 2011*

**11** A curve has equation $y = x^2 - 4x + 4$ and a line has the equation $y = mx$, where $m$ is a constant.

   **i**   For the case where $m = 1$, the curve and the line intersect at the points $A$ and $B$.

       Find the coordinates of the mid-point of $AB$.    **[4]**

   **ii**  Find the non-zero value of $m$ for which the line is a tangent to the curve, and find the coordinates of the point where the tangent touches the curve.    **[5]**

*Cambridge International AS & A Level Mathematics 9709 Paper 11 Q7 June 2013*

**12 i**   Express $2x^2 - 4x + 1$ in the form $a(x + b)^2 + c$ and hence state the coordinates of the minimum point, $A$, on the curve $y = 2x^2 - 4x + 1$.    **[4]**

   The line $x - y + 4 = 0$ intersects the curve $y = 2x^2 - 4x + 1$ at the points $P$ and $Q$.

   It is given that the coordinates of $P$ are $(3, 7)$.

   **ii**  Find the coordinates of $Q$.    **[3]**

   **iii** Find the equation of the line joining $Q$ to the mid-point of $AP$.    **[3]**

*Cambridge International AS & A Level Mathematics 9709 Paper 11 Q10 June 2011*

# Chapter 2
# Functions

**In this chapter you will learn how to:**

- understand the terms function, domain, range, one-one function, inverse function and composition of functions
- identify the range of a given function in simple cases, and find the composition of two given functions
- determine whether or not a given function is one-one, and find the inverse of a one-one function in simple cases
- illustrate in graphical terms the relation between a one-one function and its inverse
- understand and use the transformations of the graph $y = f(x)$ given by $y = f(x) + a$, $y = f(x + a)$, $y = af(x)$, $y = f(ax)$ and simple combinations of these.

## PREREQUISITE KNOWLEDGE

| Where it comes from | What you should be able to do | Check your skills |
|---|---|---|
| IGCSE / O Level Mathematics | Find an output for a given function. | 1  If $f(x) = 3x - 2$, find $f(4)$. |
| IGCSE / O Level Mathematics | Find a composite function. | 2  If $f(x) = 2x + 1$ and $g(x) = 1 - x$, find $fg(x)$. |
| IGCSE / O Level Mathematics | Find the inverse of a simple function. | 3  If $f(x) = 5x + 4$, find $f^{-1}(x)$. |
| Chapter 1 | Complete the square. | 4  Express $2x^2 - 12x + 5$ in the form $a(x + b)^2 + c$. |

## Why do we study functions?

At IGCSE / O Level, you learnt how to interpret expressions as functions with inputs and outputs and find simple composite functions and simple inverse functions.

There are many situations in the real world that can be modelled as functions. Some examples are:

- the temperature of a hot drink as it cools over time
- the height of a valve on a bicycle tyre as the bicycle travels along a horizontal road
- the depth of water in a conical container as it is filled from a tap
- the number of bacteria present after the start of an experiment.

Modelling these situations using appropriate functions enables us to make predictions about real-life situations, such as: How long will it take for the number of bacteria to exceed 5 billion?

In this chapter we will develop a deeper understanding of functions and their special properties.

> **WEB LINK**
>
> Try the *Thinking about functions* and *Combining functions* resources on the Underground Mathematics website.

## 2.1 Definition of a function

A **function** is a relation that uniquely associates members of one set with members of another set.

An alternative name for a function is a **mapping**.

A function can be either a **one-one** function or a **many-one** function.

The function $x \mapsto x + 2$, where $x \in \mathbb{R}$ is an example of a one-one function.

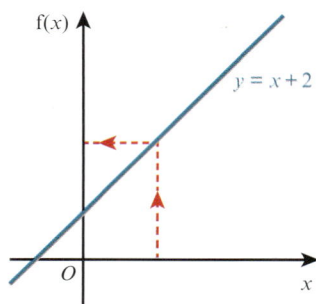

> **TIP**
>
> $x \in \mathbb{R}$ means that $x$ belongs to the set of real numbers.

A one-one function has one output value for each input value. Equally important is the fact that for each output value appearing there is only one input value resulting in this output value.

We can write this function as $f : x \mapsto x + 2$ for $x \in \mathbb{R}$ or $f(x) = x + 2$ for $x \in \mathbb{R}$.

$f : x \mapsto x + 2$ is read as 'the function f is such that $x$ is mapped to $x + 2$' or 'f maps $x$ to $x + 2$'.

$f(x)$ is the output value of the function f when the input value is $x$. For example, when $f(x) = x + 2$, $f(5) = 5 + 2 = 7$.

The function $x \mapsto x^2$, where $x \in \mathbb{R}$ is a many-one function.

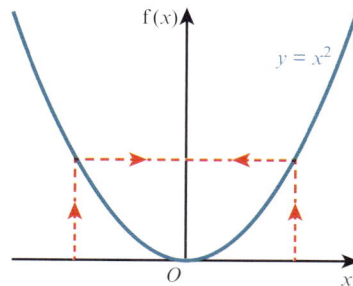

A many-one function has one output value for each input value but each output value can have more than one input value.

We can write this function as $f : x \mapsto x^2$ for $x \in \mathbb{R}$ or $f(x) = x^2$ for $x \in \mathbb{R}$.

$f : x \mapsto x^2$ is read as 'the function f is such that $x$ is mapped to $x^2$' or 'f maps $x$ to $x^2$'.

If we now consider the graph of $y^2 = x$:

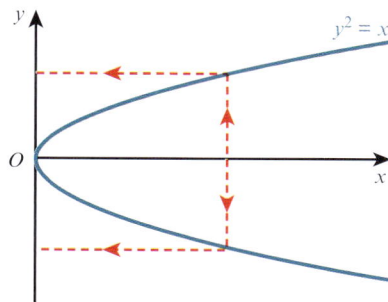

We can see that the input value shown has two output values. This means that this relation is not a function.

The set of input values for a function is called the **domain** of the function.

When defining a function, it is important to also specify its domain.

The set of output values for a function is called the **range** (or codomain) of the function.

**WORKED EXAMPLE 2.1**

$f(x) = 5 - 2x$ for $x \in \mathbb{R}$, $-4 \leqslant x \leqslant 5$.

   **a** Write down the domain of the function $f$.

   **b** Sketch the graph of the function $f$.

   **c** Write down the range of the function $f$.

**Answer**

   **a** The domain is $-4 \leqslant x \leqslant 5$.

   **b** The graph of $y = 5 - 2x$ is a straight line with gradient $-2$ and $y$-intercept $5$.

      When $x = -4$, $y = 5 - 2(-4) = 13$

      When $x = 5$, $y = 5 - 2(5) = -5$

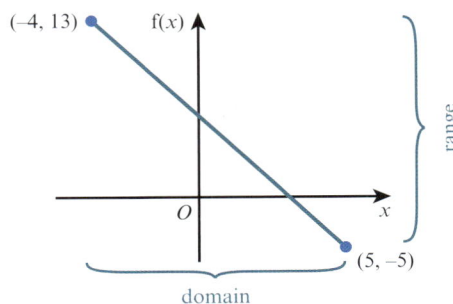

   **c** The range is $-5 \leqslant f(x) \leqslant 13$.

**WORKED EXAMPLE 2.2**

The function $f$ is defined by $f(x) = (x - 3)^2 + 8$ for $-1 \leqslant x \leqslant 9$.

Sketch the graph of the function.

Find the range of $f$.

**Answer**

$f(x) = (x - 3)^2 + 8$ is a positive quadratic function so the graph

will be of the form $\bigcup$.

$\boxed{(x - 3)^2} + 8$  · · · · · · · · · · · · · · · · · · · · · · · · · · ·

 

The minimum value of the expression is $0 + 8 = 8$ and this minimum occurs when $x = 3$.

So the function $f(x) = (x - 3)^2 + 8$ will have a minimum point at the point $(3, 8)$.

When $x = -1$, $y = (-1 - 3)^2 + 8 = 24$

When $x = 9$, $y = (9 - 3)^2 + 8 = 44$

> The circled part of the expression is a square so it will always be $\geqslant 0$.
> The smallest value it can be is $0$.
> This occurs when $x = 3$.

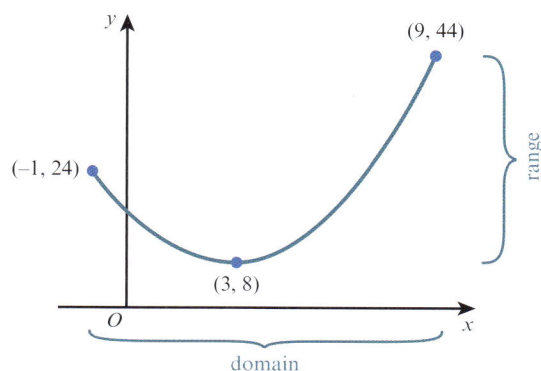

The range is $8 \leq f(x) \leq 44$.

### EXERCISE 2A

1  Which of these graphs represent functions? If the graph represents a function, state whether it is a one-one function or a many-one function.

a  $y = 2x - 3$  for  $x \in \mathbb{R}$

b  $y = x^2 - 3$  for  $x \in \mathbb{R}$

c  $y = 2x^3 - 1$ for  $x \in \mathbb{R}$

d  $y = 2^x$   for  $x \in \mathbb{R}$

e  $y = \dfrac{10}{x}$   for  $x \in \mathbb{R}$, $x > 0$

f  $y = 3x^2 + 4$ for  $x \in \mathbb{R}$, $x \geq 0$

g  $y = \sqrt{x}$   for  $x \in \mathbb{R}$, $x \geq 0$

h  $y^2 = 4x$   for  $x \in \mathbb{R}$

> **TIP**
>
> $x \in \mathbb{R}$, $x \geq 0$ is sometimes shortened to just $x \geq 0$.

2  a  Represent on a graph the function:

$$x \mapsto \begin{cases} 9 - x^2 & \text{for } x \in \mathbb{R}, -3 \leq x \leq 2 \\ 2x + 1 & \text{for } x \in \mathbb{R}, \ 2 \leq x \leq 4 \end{cases}$$

b  State the nature of the function.

3  a  Represent on a graph the relation:

$$y = \begin{cases} x^2 + 1 & \text{for } 0 \leq x \leq 2 \\ 2x - 3 & \text{for } 2 \leq x \leq 4 \end{cases}$$

b  Explain why this relation is not a function.

4  State the domain and range for the functions represented by these two graphs.

a

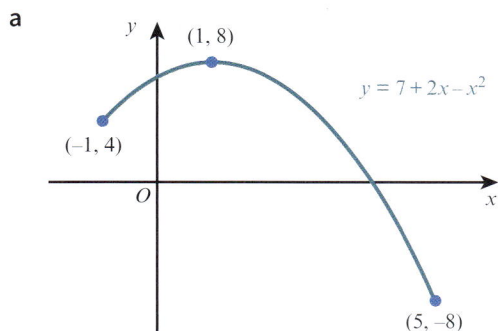

b

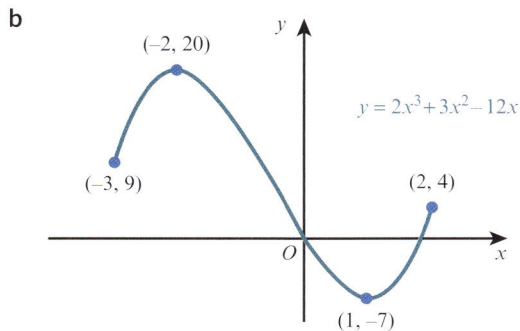

37

**5** Find the range for each of these functions.

**a** $f(x) = x + 4$ for $x > 8$

**b** $f(x) = 2x - 7$ for $-3 \leqslant x \leqslant 2$

**c** $f(x) = 7 - 2x$ for $-1 \leqslant x \leqslant 4$

**d** $f : x \mapsto 2x^2$ for $1 \leqslant x \leqslant 4$

**e** $f(x) = 2^x$ for $-5 \leqslant x \leqslant 4$

**f** $f(x) = \dfrac{12}{x}$ for $1 \leqslant x \leqslant 8$

**6** Find the range for each of these functions.

**a** $f(x) = x^2 - 2$ for $x \in \mathbb{R}$

**b** $f : x \mapsto x^2 + 3$ for $-2 \leqslant x \leqslant 5$

**c** $f(x) = 3 - 2x^2$ for $x \leqslant 2$

**d** $f(x) = 7 - 3x^2$ for $-1 \leqslant x \leqslant 2$

**7** Find the range for each of these functions.

**a** $f(x) = (x - 2)^2 + 5$ for $x \geqslant 2$

**b** $f(x) = (2x - 1)^2 - 7$ for $x \geqslant \dfrac{1}{2}$

**c** $f : x \mapsto 8 - (x - 5)^2$ for $4 \leqslant x \leqslant 10$

**d** $f(x) = 1 + \sqrt{x - 4}$ for $x \geqslant 4$

**8** Express each function in the form $a(x + b)^2 + c$, where $a$, $b$ and $c$ are constants and, hence, state the range of each function.

**a** $f(x) = x^2 + 6x - 11$ for $x \in \mathbb{R}$

**b** $f(x) = 3x^2 - 10x + 2$ for $x \in \mathbb{R}$

**9** Express each function in the form $a - b(x + c)^2$, where $a$, $b$ and $c$ are constants and, hence, state the range of each function.

**a** $f(x) = 7 - 8x - x^2$ for $x \in \mathbb{R}$

**b** $f(x) = 2 - 6x - 3x^2$ for $x \in \mathbb{R}$

**10 a** Represent, on a graph, the function:

$$f(x) = \begin{cases} 3 - x^2 & \text{for} \quad 0 \leqslant x \leqslant 2 \\ 3x - 7 & \text{for} \quad 2 \leqslant x \leqslant 4 \end{cases}$$

**b** Find the range of the function.

**11** The function $f : x \mapsto x^2 + 6x + k$, where $k$ is a constant, is defined for $x \in \mathbb{R}$.

Find the range of f in terms of $k$.

**12** The function $g : x \mapsto 5 - ax - 2x^2$, where $a$ is a constant, is defined for $x \in \mathbb{R}$.

Find the range of g in terms of $a$.

**13** $f(x) = x^2 - 2x - 3$ for $x \in \mathbb{R}$, $-a \leqslant x \leqslant a$

If the range of the function f is $-4 \leqslant f(x) \leqslant 5$, find the value of $a$.

**14** $f(x) = x^2 + x - 4$ for $x \in \mathbb{R}$, $a \leqslant x \leqslant a + 3$

If the range of the function f is $-2 \leqslant f(x) \leqslant 16$, find the possible values of $a$.

**15** $f(x) = 2x^2 - 8x + 5$ for $x \in \mathbb{R}$, $0 \leqslant x \leqslant k$

**a** Express $f(x)$ in the form $a(x + b)^2 + c$.

**b** State the value of $k$ for which the graph of $y = f(x)$ has a line of symmetry.

**c** For your value of $k$ from part **b**, find the range of f.

**16** Find the largest possible domain for each function and state the corresponding range.

**a** $f(x) = 3x - 1$

**b** $f(x) = x^2 + 2$

**c** $f(x) = 2^x$

**d** $f(x) = \dfrac{1}{x}$

**e** $f(x) = \dfrac{1}{x - 2}$

**f** $f(x) = \sqrt{x - 3} - 2$

## 2.2 Composite functions

Most functions that we meet can be described as combinations of two or more functions.

For example, the function $x \mapsto 3x - 7$ is the function 'multiply by 3 and then subtract 7'.
It is a combination of the two functions g and f, where:

$g : x \mapsto 3x$          (the function 'multiply by 3')

$f : x \mapsto x - 7$       (the function 'subtract 7')

So, $x \mapsto 3x - 7$ can be described as the function 'first do g, then do f'.

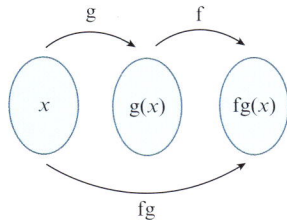

When one function is followed by another function, the resulting function is called a **composite function**.

> 🔍 **KEY POINT 2.1**
>
> $fg(x)$ means the function g acts on $x$ first, then f acts on the result.

There are three important points to remember about composite functions:

> 🔍 **KEY POINT 2.2**
>
> fg only exists if the range of g is contained within the domain of f.
>
> In general, $fg(x) \neq gf(x)$.
>
> $ff(x)$ means you apply the function f twice.

**EXPLORE 2.1**

$f(x) = 2x - 5$ for $x \in \mathbb{R}$          $g(x) = 3x - 1$ for $x \in \mathbb{R}$

Three students are asked to find the composite function $gf(x)$.

Here are their solutions.

| Student A | Student B | Student C |
|---|---|---|
| $gf(x) = (3x - 1)(2x - 5)$ | $gf(x) = 2(3x - 1) - 5$ | $gf(x) = 3(2x - 5) - 1$ |
| $= 6x^2 - 17x + 5$ | $= 6x - 7$ | $= 6x - 16$ |

Discuss these solutions with your classmates.

Which student is correct? What error has each of the other students made?

**WORKED EXAMPLE 2.3**

$f(x) = (x - 4)^2 - 1$ for $x \in \mathbb{R}$          $g(x) = \dfrac{2x + 3}{x - 2}$ for $x \in \mathbb{R}, x > 2$

Find $fg(4)$.

**Answer**

$fg(4) = f\left(\dfrac{11}{2}\right)$ ........................................... g acts on 4 first and $g(4) = \dfrac{2(4) + 3}{4 - 2} = \dfrac{11}{2}$.

$\quad = \left(\dfrac{11}{2} - 4\right)^2 - 1$

$\quad = 1\frac{1}{4}$ ........................................... f is the function 'subtract 4, square and then subtract 1'.

**WORKED EXAMPLE 2.4**

$f(x) = 2x + 3$ for $x \in \mathbb{R}$          $g(x) = x^2 - 1$ for $x \in \mathbb{R}$

Find:   **a**   $fg(x)$          **b**   $gf(x)$          **c**   $ff(x)$

**Answer**

**a**   $fg(x) = f(x^2 - 1)$ ........................................... g acts on $x$ first and $g(x) = x^2 - 1$.

$\quad = 2(x^2 - 1) + 3$ ...........................

$\quad = 2x^2 + 1$ ........................................... f is the function 'double and add 3'.

**b**   $gf(x) = g(2x + 3)$ ........................................... f acts on $x$ first and $f(x) = 2x + 3$.

$\quad = (2x + 3)^2 - 1$ ...........................

$\quad = 4x^2 + 12x + 9 - 1$ ........................................... g is the function 'square and subtract 1'.

$\quad = 4x^2 + 12x + 8$

**c**   $ff(x) = f(2x + 3)$ ........................................... f is the function 'double and add 3'.

$\quad = 2(2x + 3) + 3$

$\quad = 4x + 9$

**WORKED EXAMPLE 2.5**

$f : x \mapsto \dfrac{5}{x - 2}$ for $x \in \mathbb{R}, x \neq 2$          $g(x) = 3 - x^2$ for $x \in \mathbb{R}$

Find:   **a**   $fg(x)$          **b**   $ff(x)$

**Answer**

**a**   $fg(x) = f(3 - x^2)$ ........................................... g acts on $x$ first and $g(x) = 3 - x^2$.

$\quad = \dfrac{5}{(3 - x^2) - 2}$ ........................................... f is the function 'subtract 2 and then divide into 5'.

$\quad = \dfrac{5}{1 - x^2}$

**b** $\quad \mathrm{ff}(x) = \mathrm{f}\left(\dfrac{5}{x-2}\right)$

$\qquad = \dfrac{5}{\dfrac{5}{x-2} - 2}$ .......................... Multiply numerator and denominator by $(x-2)$.

$\qquad = \dfrac{5(x-2)}{5 - 2(x-2)}$

$\qquad = \dfrac{5x - 10}{9 - 2x}$

## WORKED EXAMPLE 2.6

$\mathrm{f}(x) = x^2 + 4x \;$ for $\; x \in \mathbb{R}$ $\qquad\qquad\qquad$ $\mathrm{g}(x) = 3x - 1 \;$ for $\; x \in \mathbb{R}$

Find the values of $k$ for which the equation $\mathrm{fg}(x) = k$ has real solutions.

**Answer**

$\mathrm{fg}(x) = (3x - 1)^2 + 4(3x - 1)$ .......................... Expand brackets and simplify.

$\qquad = 9x^2 + 6x - 3$

When $\mathrm{fg}(x) = k$,

$\qquad 9x^2 + 6x - 3 = k$ .......................... Rearrange and simplify.

$9x^2 + 6x + (-3 - k) = 0$

For real solutions: $\qquad\qquad b^2 - 4ac \geqslant 0$

$\qquad\qquad\qquad 6^2 - 4 \times 9 \times (-3 - k) \geqslant 0$

$\qquad\qquad\qquad\qquad 144 + 36k \geqslant 0$

$\qquad\qquad\qquad\qquad\qquad k \geqslant -4$

## EXERCISE 2B

**1** $\quad \mathrm{f}(x) = x^2 + 6 \;$ for $\; x \in \mathbb{R}$ $\qquad\qquad\qquad$ $\mathrm{g}(x) = \sqrt{x+3} - 2 \;$ for $\; x \in \mathbb{R}, x \geqslant -3$

$\qquad$ Find: $\quad$ **a** $\quad \mathrm{fg}(6)$ $\qquad\qquad$ **b** $\quad \mathrm{gf}(4)$ $\qquad\qquad$ **c** $\quad \mathrm{ff}(-3)$

**2** $\quad \mathrm{h}: x \mapsto x + 5 \;$ for $\; x \in \mathbb{R}, x > 0$ $\qquad\qquad$ $\mathrm{k}: x \mapsto \sqrt{x} \;$ for $\; x \in \mathbb{R}, x > 0$

$\qquad$ Express each of the following in terms of h and/or k.

$\qquad$ **a** $\quad x \mapsto \sqrt{x} + 5$ $\qquad\qquad$ **b** $\quad x \mapsto \sqrt{x+5}$ $\qquad\qquad$ **c** $\quad x \mapsto x + 10$

**3** $\quad \mathrm{f}(x) = ax + b \;$ for $\; x \in \mathbb{R}$

$\qquad$ Given that $\mathrm{f}(5) = 3$ and $\mathrm{f}(3) = -3$:

$\qquad$ **a** $\quad$ find the value of $a$ and the value of $b$

$\qquad$ **b** $\quad$ solve the equation $\mathrm{ff}(x) = 4$.

**4** $\quad \mathrm{f}: x \mapsto 2x + 3 \;$ for $\; x \in \mathbb{R}$ $\qquad\qquad$ $\mathrm{g}: x \mapsto \dfrac{12}{1-x} \;$ for $\; x \in \mathbb{R}, x \neq 1$

$\qquad$ **a** $\quad$ Find $\mathrm{gf}(x)$.

$\qquad$ **b** $\quad$ Solve the equation $\mathrm{gf}(x) = 2$.

5  $g(x) = x^2 - 2$ for $x \in \mathbb{R}$          $h(x) = 2x + 5$ for $x \in \mathbb{R}$

   **a**  Find $gh(x)$.

   **b**  Solve the equation $gh(x) = 14$.

6  $f(x) = x^2 + 1$ for $x \in \mathbb{R}$          $g(x) = \dfrac{3}{x-2}$ for $x \in \mathbb{R},\ x \neq 2$

   Solve the equation $fg(x) = 5$.

7  $g(x) = \dfrac{2}{x+1}$ for $x \in \mathbb{R},\ x \neq -1$          $h(x) = (x+2)^2 - 5$ for $x \in \mathbb{R}$

   Solve the equation $hg(x) = 11$.

8  $f: x \mapsto \dfrac{x+1}{2}$ for $x \in \mathbb{R}$          $g: x \mapsto \dfrac{2x+3}{x-1}$ for $x \in \mathbb{R},\ x \neq 1$

   Solve the equation $gf(x) = 1$.

9  $f(x) = \dfrac{x+1}{2x+5}$ for $x \in \mathbb{R},\ x > 0$

   Find an expression for $ff(x)$, giving your answer as a single fraction in its simplest form.

10  $f: x \mapsto x^2$ for $x \in \mathbb{R}$          $g: x \mapsto x + 1$ for $x \in \mathbb{R}$

   Express each of the following as a composite function, using only f and/or g.

   **a**  $x \mapsto (x+1)^2$          **b**  $x \mapsto x^2 + 1$          **c**  $x \mapsto x + 2$

   **d**  $x \mapsto x^4$          **e**  $x \mapsto x^2 + 2x + 2$          **f**  $x \mapsto x^4 + 2x^2 + 1$

11  $f(x) = x^2 - 3x$ for $x \in \mathbb{R}$          $g(x) = 2x + 5$ for $x \in \mathbb{R}$

   Show that the equation $gf(x) = 0$ has no real solutions.

12  $f(x) = k - 2x$ for $x \in \mathbb{R}$          $g(x) = \dfrac{2}{x}$ for $x \in \mathbb{R},\ x \neq 0$

   Find the values of $k$ for which the equation $fg(x) = x$ has two equal roots.

13  $f(x) = x^2 - 3x$ for $x \in \mathbb{R}$          $g(x) = 2x - 5$ for $x \in \mathbb{R}$

   Find the values of $k$ for which the equation $gf(x) = k$ has real solutions.

14  $f(x) = \dfrac{x+5}{2x-1}$ for $x \in \mathbb{R},\ x \neq \dfrac{1}{2}$

   Show that $ff(x) = x$.

15  $f(x) = 2x^2 + 4x - 8$ for $x \in \mathbb{R},\ x \geqslant k$

   **a**  Express $2x^2 + 4x - 8$ in the form $a(x+b)^2 + c$.

   **b**  Find the least value of $k$ for which the function is one-one.

16  $f(x) = x^2 - 2x + 4$ for $x \in \mathbb{R}$

   **a**  Find the set of values of $x$ for which $f(x) \geqslant 7$.

   **b**  Express $x^2 - 2x + 4$ in the form $(x-a)^2 + b$.

   **c**  Write down the range of $f$.

17  $f(x) = x^2 - 5x$ for $x \in \mathbb{R}$          $g(x) = 2x + 3$ for $x \in \mathbb{R}$

   **a**  Find $fg(x)$.

   **b**  Find the range of the function $fg(x)$.

42

**18** $f(x) = \dfrac{2}{x+1}$ for $x \in \mathbb{R}$, $x \neq -1$

   **a** Find $ff(x)$ and state the domain of this function.

   **b** Show that if $f(x) = ff(x)$ then $x^2 + x - 2 = 0$.

   **c** Find the values of $x$ for which $f(x) = ff(x)$.

**PS** **19**

| $P(x) = x^2 - 1$ for $x \in \mathbb{R}$ | $Q(x) = x + 2$ for $x \in \mathbb{R}$ |

| $R(x) = \dfrac{1}{x}$ for $x \in \mathbb{R}$, $x \neq 0$ | $S(x) = \sqrt{x+1} - 1$ for $x \in \mathbb{R}$, $x \geqslant -1$ |

Functions P, Q, R and S are composed in some way to make a new function, $f(x)$.

For each of the following, write $f(x)$ in terms of the functions P, Q, R and/or S, and state the domain and range for each composite function.

   **a** $f(x) = x^2 + 4x + 3$    **b** $f(x) = x^2 + 1$    **c** $f(x) = x$    **d** $f(x) = \dfrac{1}{x^2} + 1$

   **e** $f(x) = \dfrac{1}{x+4}$    **f** $f(x) = x - 2\sqrt{x+1} + 1$    **g** $f(x) = x - 1$

## 2.3 Inverse functions

The **inverse of a function** $f(x)$ is the function that undoes what $f(x)$ has done.

We write the inverse of the function $f(x)$ as $f^{-1}(x)$.

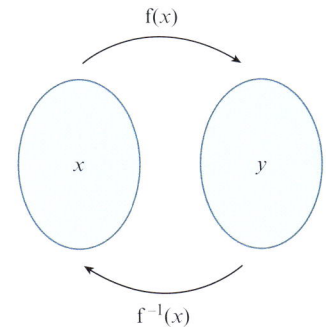

> **KEY POINT 2.3**
>
> $ff^{-1}(x) = f^{-1}f(x) = x$
>
> The domain of $f^{-1}(x)$ is the range of $f(x)$.
>
> The range of $f^{-1}(x)$ is the domain of $f(x)$.

It is important to remember that not every function has an inverse.

> **KEY POINT 2.4**
>
> An inverse function $f^{-1}(x)$ exists if, and only if, the function $f(x)$ is a one-one mapping.

You should already know how to find the inverse function of some simple one-one mappings.

We want to find the function $f^{-1}(x)$, so if we write $y = f^{-1}(x)$, then $f(y) = f(f^{-1}(x)) = x$, because $f$ and $f^{-1}$ are inverse functions. So if we write $x = f(y)$ and then rearrange it to get $y = \ldots$, then the right-hand side will be $f^{-1}(x)$.

We find the inverse of the function $f(x) = 3x - 1$ by following these steps:

**Step 1:** Write the function as $y =$ $\longrightarrow$ $y = 3x - 1$

**Step 2:** Interchange the $x$ and $y$ variables. $\longrightarrow$ $x = 3y - 1$

**Step 3:** Rearrange to make $y$ the subject. $\longrightarrow$ $y = \dfrac{x+1}{3}$

Hence, if $f(x) = 3x - 1$, then $f^{-1}(x) = \dfrac{x+1}{3}$.

If f and $f^{-1}$ are the same function, then f is called a **self-inverse function**.

For example, if $f(x) = \dfrac{1}{x}$ for $x \neq 0$, then $f^{-1}(x) = \dfrac{1}{x}$ for $x \neq 0$.

So $f(x) = \dfrac{1}{x}$ for $x \neq 0$ is a self-inverse function.

## EXPLORE 2.2

The diagram shows the function $f(x) = (x-2)^2 + 1$ for $x \in \mathbb{R}$.
Discuss the following questions with your classmates.

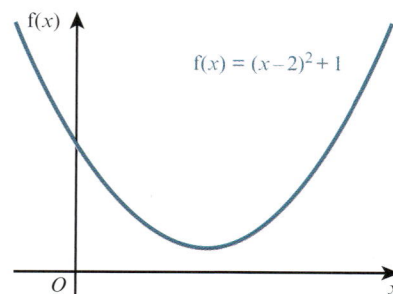

1  What type of mapping is this function?

2  What are the coordinates of the vertex of the parabola?

3  What is the domain of the function?

4  What is the range of the function?

5  Does this function have an inverse?

6  If f has an inverse, what is it? If not, then how could you change the domain of f so that the function does have an inverse?

## WORKED EXAMPLE 2.7

$f(x) = \sqrt{x+2} - 7$ for $x \in \mathbb{R}, x \geqslant -2$

**a** Find an expression for $f^{-1}(x)$.

**b** Solve the equation $f^{-1}(x) = f(62)$.

**Answer**

**a**  $f(x) = \sqrt{x+2} - 7$

**Step 1:** Write the function as $y =$ $\longrightarrow$ $y = \sqrt{x+2} - 7$

**Step 2:** Interchange the $x$ and $y$ variables. $\longrightarrow$ $x = \sqrt{y+2} - 7$

**Step 3:** Rearrange to make $y$ the subject. $\longrightarrow$ $x + 7 = \sqrt{y+2}$

$(x+7)^2 = y + 2$

$y = (x+7)^2 - 2$

$f^{-1}(x) = (x+7)^2 - 2$

**b** $f(62) = \sqrt{62 + 2} - 7 = 1$

$(x + 7)^2 - 2 = 1$

$(x + 7)^2 = 3$

$x + 7 = \pm\sqrt{3}$

$x = -7 \pm \sqrt{3}$

$x = -7 - \sqrt{3}$ or $x = -7 + \sqrt{3}$

The range of f is $f(x) \geq -7$ so the domain of $f^{-1}$ is $x \geq -7$.

Hence, the only solution of $f^{-1}(x) = f(62)$ is $x = -7 + \sqrt{3}$.

**WORKED EXAMPLE 2.8**

$f(x) = 5 - (x - 2)^2$ for $x \in \mathbb{R}, k \leq x \leq 6$

**a** State the smallest value of $k$ for which f has an inverse.

**b** For this value of $k$ find an expression for $f^{-1}(x)$, and state the domain and range of $f^{-1}$.

**Answer**

**a** The vertex of the graph of $y = 5 - (x - 2)^2$ is at the point (2, 5).

When $x = 6$, $y = 5 - 4^2 = -11$

For the function f to have an inverse it must be a one-one function.

Hence, the smallest value of $k$ is 2.

**b** $f(x) = 5 - (x - 2)^2$

**Step 1:** Write the function as $y =$ → $y = 5 - (x - 2)^2$

**Step 2:** Interchange the $x$ and $y$ variables. → $x = 5 - (y - 2)^2$

**Step 3:** Rearrange to make $y$ the subject. → $(y - 2)^2 = 5 - x$

$y - 2 = \sqrt{5 - x}$

$y = 2 + \sqrt{5 - x}$

Hence, $f^{-1}(x) = 2 + \sqrt{5 - x}$.

The domain of $f^{-1}$ is the same as the range of f.

Hence, the domain of $f^{-1}$ is $-11 \leq x \leq 5$.

The range of $f^{-1}$ is the same as the domain of f.

Hence, the range of $f^{-1}$ is $2 \leq f^{-1}(x) \leq 6$.

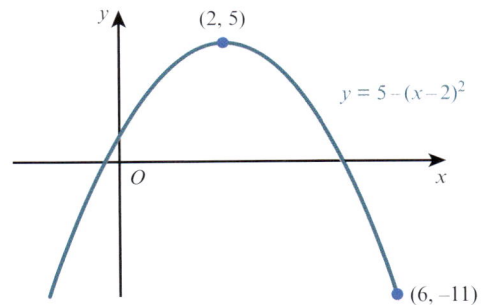

**EXERCISE 2C**

1   Find an expression for $f^{-1}(x)$ for each of the following functions.

   **a**   $f(x) = 5x - 8$ for $x \in \mathbb{R}$

   **b**   $f(x) = x^2 + 3$ for $x \in \mathbb{R}$, $x \geq 0$

   **c**   $f(x) = (x - 5)^2 + 3$ for $x \in \mathbb{R}$, $x \geq 5$

   **d**   $f(x) = \dfrac{8}{x - 3}$ for $x \in \mathbb{R}$, $x \neq 3$

   **e**   $f(x) = \dfrac{x + 7}{x + 2}$ for $x \in \mathbb{R}$, $x \neq -2$

   **f**   $f(x) = (x - 2)^3 - 1$ for $x \in \mathbb{R}$, $x \geq 2$

2   $f : x \mapsto x^2 + 4x$ for $x \in \mathbb{R}, x \geq -2$

   **a**   State the domain and range of $f^{-1}$.

   **b**   Find an expression for $f^{-1}(x)$.

3   $f : x \mapsto \dfrac{5}{2x + 1}$ for $x \in \mathbb{R}, x \geq 2$

   **a**   Find an expression for $f^{-1}(x)$.

   **b**   Find the domain of $f^{-1}$.

4   $f : x \mapsto (x + 1)^3 - 4$ for $x \in \mathbb{R}, x \geq 0$

   **a**   Find an expression for $f^{-1}(x)$.

   **b**   Find the domain of $f^{-1}$.

5   $g : x \mapsto 2x^2 - 8x + 10$ for $x \in \mathbb{R}, x \geq 3$

   **a**   Explain why g has an inverse.

   **b**   Find an expression for $g^{-1}(x)$.

6   $f : x \mapsto 2x^2 + 12x - 14$ for $x \in \mathbb{R}, x \geq k$

   **a**   Find the least value of $k$ for which f is one-one.

   **b**   Find an expression for $f^{-1}(x)$.

7   $f : x \mapsto x^2 - 6x$ for $x \in \mathbb{R}$

   **a**   Find the range of f.

   **b**   State, with a reason, whether f has an inverse.

8   $f(x) = 9 - (x - 3)^2$ for $x \in \mathbb{R}, k \leq x \leq 7$

   **a**   State the smallest value of $k$ for which f has an inverse.

   **b**   For this value of $k$:

      **i**   find an expression for $f^{-1}(x)$

      **ii**   state the domain and range of $f^{-1}$.

**9**

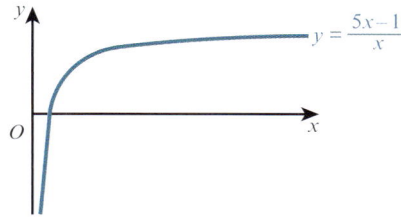

The diagram shows the graph of $y = f^{-1}(x)$, where $f^{-1}(x) = \dfrac{5x - 1}{x}$ for $x \in \mathbb{R}, 0 < x \leqslant 3$.

**a** Find an expression for $f(x)$.

**b** State the domain of $f$.

**10** $f(x) = 3x + a$ for $x \in \mathbb{R}$ $\qquad\qquad\qquad$ $g(x) = b - 5x$ for $x \in \mathbb{R}$

Given that $gf(-1) = 2$ and $g^{-1}(7) = 1$, find the value of $a$ and the value of $b$.

**11** $f(x) = 3x - 1$ for $x \in \mathbb{R}$ $\qquad\qquad$ $g(x) = \dfrac{3}{2x - 4}$ for $x \in \mathbb{R}, x \neq 2$

**a** Find expressions for $f^{-1}(x)$ and $g^{-1}(x)$.

**b** Show that the equation $f^{-1}(x) = g^{-1}(x)$ has two real roots.

**12** $f : x \mapsto (2x - 1)^3 - 3$ for $x \in \mathbb{R}, 1 \leqslant x \leqslant 3$

**a** Find an expression for $f^{-1}(x)$.

**b** Find the domain of $f^{-1}$.

**13** $f : x \mapsto x^2 - 10x$ for $x \in \mathbb{R}, x \geqslant 5$

**a** Express $f(x)$ in the form $(x - a)^2 - b$.

**b** Find an expression for $f^{-1}(x)$ and state the domain of $f^{-1}$.

**14** $f(x) = \dfrac{1}{x - 1}$ for $x \in \mathbb{R}, \ x \neq 1$

**a** Find an expression for $f^{-1}(x)$.

**b** Show that if $f(x) = f^{-1}(x)$, then $x^2 - x - 1 = 0$.

**c** Find the values of $x$ for which $f(x) = f^{-1}(x)$.

Give your answer in surd form.

**15** Determine which of the following functions are self-inverse functions.

**a** $f(x) = \dfrac{1}{3 - x}$ for $x \in \mathbb{R}, \ x \neq 3$ $\qquad\qquad$ **b** $f(x) = \dfrac{2x + 1}{x - 2}$ for $x \in \mathbb{R}, \ x \neq 2$

**c** $f(x) = \dfrac{3x + 5}{4x - 3}$ for $x \in \mathbb{R}, \ x \neq \dfrac{3}{4}$

**16** $f : x \mapsto 3x - 5$ for $x \in \mathbb{R}$ $\qquad\qquad\qquad$ $g : x \mapsto 4 - 2x$ for $x \in \mathbb{R}$

**a** Find an expression for $(fg)^{-1}(x)$.

**b** Find expressions for:

$\quad$ **i** $f^{-1} g^{-1}(x)$ $\qquad\qquad$ **ii** $g^{-1} f^{-1}(x)$.

**c** Comment on your results in part **b**.

Investigate if this is true for other functions.

47

## 2.4 The graph of a function and its inverse

Consider the function defined by $f(x) = 2x + 1$ for $x \in \mathbb{R}$, $-4 \leqslant x \leqslant 2$.

$f(-4) = -7$ and $f(2) = 5$.

The domain of f is $-4 \leqslant x \leqslant 2$ and the range is $-7 \leqslant f(x) \leqslant 5$.

The inverse of this function is $f^{-1}(x) = \dfrac{x-1}{2}$.

The domain of $f^{-1}$ is the same as the range of f.

Hence, the domain of $f^{-1}$ is $-7 \leqslant x \leqslant 5$.

The range of $f^{-1}$ is the same as the domain of f.

Hence, the range of $f^{-1}$ is $-4 \leqslant f^{-1}(x) \leqslant 2$.

The representation of f and $f^{-1}$ on the same graph can be seen in the diagram opposite.

It is important to note that the graphs of f and $f^{-1}$ are reflections of each other in the line $y = x$. This is true for each one-one function and its inverse functions.

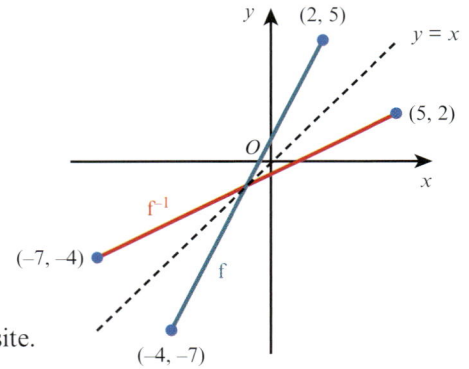

> **KEY POINT 2.5**
>
> The graphs of f and $f^{-1}$ are reflections of each other in the line $y = x$.
>
> This is because $ff^{-1}(x) = x = f^{-1}f(x)$
>
> When a function f is self-inverse, the graph of f will be symmetrical about the line $y = x$.

**48**

**WORKED EXAMPLE 2.9**

$f(x) = (x-1)^2 - 2$ for $x \in \mathbb{R}, 1 \leqslant x \leqslant 4$

On the same axes, draw the graph of f and the graph of $f^{-1}$.

**Answer**

$y = (x-1)^2 - 2$

When $x = 4$, $y = 7$.

The function is one-one, so the inverse function exists.

The circled part of the expression is a square so it will always be $\geqslant 0$. The smallest value it can be is 0. This occurs when $x = 1$. The vertex is at the point $(1, -2)$.

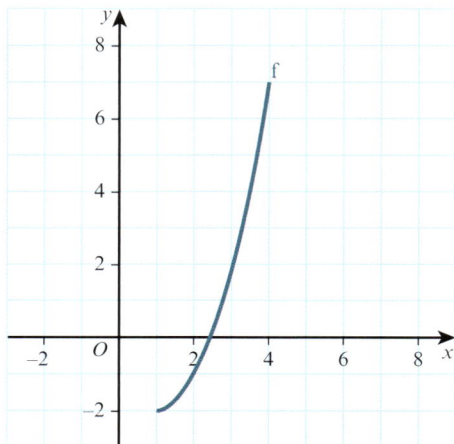

Reflect f in $y = x$

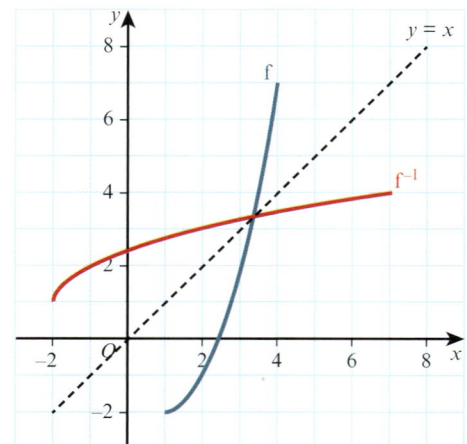

**WORKED EXAMPLE 2.10**

$f : x \mapsto \dfrac{2x + 7}{x - 2}$ for $x \in \mathbb{R}, x \neq 2$

**a** Find an expression for $f^{-1}(x)$.

**b** State what your answer to part **a** tells you about the symmetry of the graph of $y = f(x)$.

**Answer**

**a** $f : x \mapsto \dfrac{2x + 7}{x - 2}$

**Step 1:** Write the function as $y = \quad\quad\longrightarrow\quad y = \dfrac{2x + 7}{x - 2}$

**Step 2:** Interchange the $x$ and $y$ variables. $\longrightarrow\quad x = \dfrac{2y + 7}{y - 2}$

**Step 3:** Rearrange to make $y$ the subject. $\longrightarrow\quad xy - 2x = 2y + 7$

$$y(x - 2) = 2x + 7$$

$$y = \dfrac{2x + 7}{x - 2}$$

Hence $f^{-1}(x) = \dfrac{2x + 7}{x - 2}$.

**b** $f^{-1}(x) = f(x)$, so the function f is self-inverse.

The graph of $y = f(x)$ is symmetrical about the line $y = x$.

**EXPLORE 2.3**

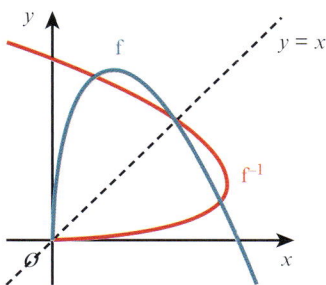

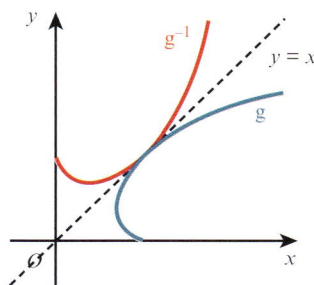

Ali states that:

The diagrams show the functions f and g, together with their inverse functions $f^{-1}$ and $g^{-1}$.

Is Ali correct?

Explain your answer.

49

**1** On a copy of each grid, draw the graph of $f^{-1}(x)$ if it exists.

a

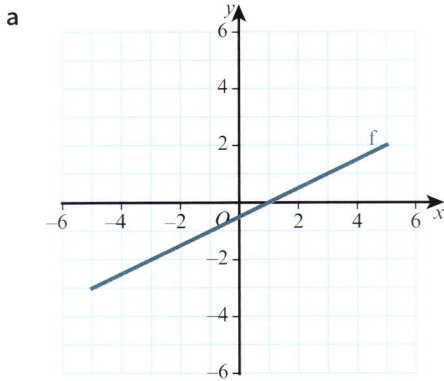

b

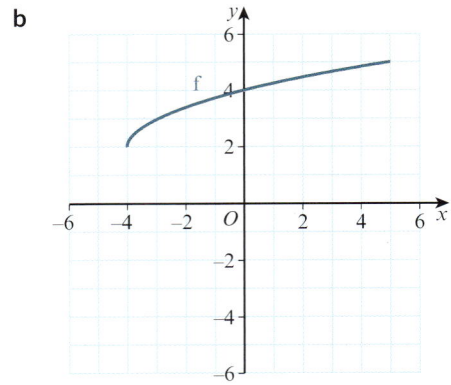

c

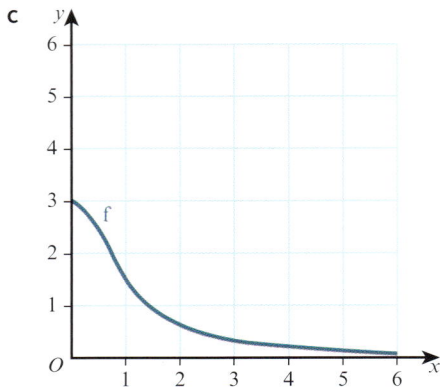

d

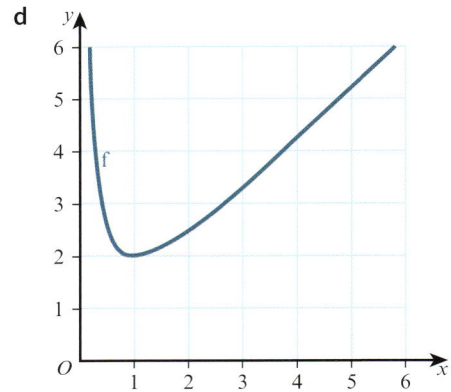

**2** $f : x \mapsto 2x - 1$ for $x \in \mathbb{R}, -1 \leqslant x \leqslant 3$

  **a** Find an expression for $f^{-1}(x)$.

  **b** State the domain and range of $f^{-1}$.

  **c** Sketch, on the same diagram, the graphs of $y = f(x)$ and $y = f^{-1}(x)$, making clear the relationship between the graphs.

**3** The diagram shows the graph of $y = f(x)$, where $f(x) = \dfrac{4}{x + 2}$ for $x \in \mathbb{R}, x \geqslant 0$.

  **a** State the range of $f$.

  **b** Find an expression for $f^{-1}(x)$.

  **c** State the domain and range of $f^{-1}$.

  **d** On a copy of the diagram, sketch the graph of $y = f^{-1}(x)$, making clear the relationship between the graphs.

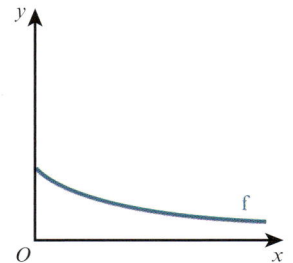

**4** For each of the following functions, find an expression for $f^{-1}(x)$ and, hence, decide if the graph of $y = f(x)$ is symmetrical about the line $y = x$.

  **a** $f(x) = \dfrac{x + 5}{2x - 1}$ for $x \in \mathbb{R}, x \neq \dfrac{1}{2}$

  **b** $f(x) = \dfrac{2x - 3}{x - 5}$ for $x \in \mathbb{R}, x \neq 5$

  **c** $f(x) = \dfrac{3x - 1}{2x - 3}$ for $x \in \mathbb{R}, x \neq \dfrac{3}{2}$

  **d** $f(x) = \dfrac{4x + 5}{3x - 4}$ for $x \in \mathbb{R}, x \neq \dfrac{4}{3}$

50

**P** 5 a $f(x) = \dfrac{x + a}{bx - 1}$ for $x \in \mathbb{R}$, $x \neq \dfrac{1}{b}$, where $a$ and $b$ are constants.

Prove that this function is self-inverse.

b $g(x) = \dfrac{ax + b}{cx + d}$ for $x \in \mathbb{R}$, $x \neq -\dfrac{d}{c}$, where $a$, $b$, $c$ and $d$ are constants.

Find the condition for this function to be self-inverse.

## 2.5 Transformations of functions

At IGCSE / O Level you met various transformations that can be applied to two-dimensional shapes. These included translations, reflections, rotations and enlargements. In this section you will learn how translations, reflections and stretches (and combinations of these) can be used to transform the graph of a function.

### EXPLORE 2.4

1 a Use graphing software to draw the graphs of $y = x^2$, $y = x^2 + 2$ and $y = x^2 - 3$.

Discuss your observations with your classmates and explain how the second and third graphs could be obtained from the first graph.

b Repeat part **a** using the graphs $y = \sqrt{x}$, $y = \sqrt{x} + 1$ and $y = \sqrt{x} - 2$.

c Repeat part **a** using the graphs $y = \dfrac{12}{x}$, $y = \dfrac{12}{x} + 5$ and $y = \dfrac{12}{x} - 4$.

d Can you generalise your results?

2 a Use graphing software to draw the graphs of $y = x^2$, $y = (x + 2)^2$ and $y = (x - 5)^2$.

Discuss your observations with your classmates and explain how the second and third graphs could be obtained from the first graph.

b Repeat part **a** using the graphs $y = x^3$, $y = (x + 1)^3$ and $y = (x - 4)^3$.

c Can you generalise your results?

3 a Use graphing software to draw the graphs of $y = x^2$ and $y = -x^2$.

Discuss your observations with your classmates and explain how the second graph could be obtained from the first graph.

b Repeat part **a** using the graphs $y = x^3$ and $y = -x^3$.

c Repeat part **a** using the graphs $y = 2^x$ and $y = -2^x$.

d Can you generalise your results?

4 a Use graphing software to draw the graphs of $y = 5 + x$ and $y = 5 - x$.

Discuss your observations with your classmates and explain how the second graph could be obtained from the first graph.

b Repeat part **a** using the graphs $y = \sqrt{2 + x}$ and $y = \sqrt{2 - x}$.

c Can you generalise your results?

5 a Use graphing software to draw the graphs of $y = x^2$ and $y = 2x^2$ and $y = (2x)^2$.

Discuss your observations with your classmates and explain how the second graph could be obtained from the first graph.

b Repeat part **a** using the graphs $y = \sqrt{x}$, $y = 2\sqrt{x}$ and $y = \sqrt{2x}$.

c Repeat part **a** using the graphs $y = 3^x$, $y = 2 \times 3^x$ and $y = 3^{2x}$.

d Can you generalise your results?

## Translations

The diagram shows the graphs of two functions that differ only by a constant.

$$y = x^2 - 2x + 1$$

$$y = x^2 - 2x + 4$$

When the $x$-coordinates on the two graphs are the same ($x = x$) the $y$-coordinates differ by 3 ($y = y + 3$).

This means that the two curves have exactly the same shape but that they are separated by 3 units in the positive $y$ direction.

Hence, the graph of $y = x^2 - 2x + 4$ is a translation of the graph of $y = x^2 - 2x + 1$ by the vector $\begin{pmatrix} 0 \\ 3 \end{pmatrix}$.

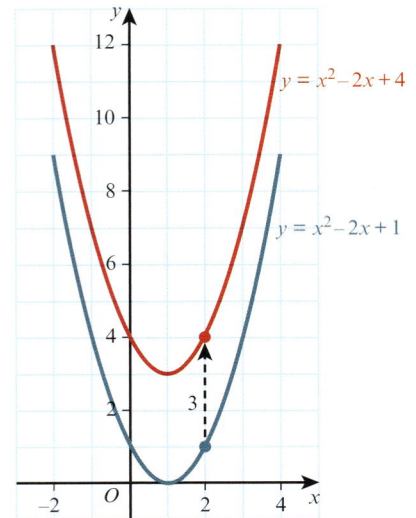

**KEY POINT 2.6**

The graph of $y = f(x) + a$ is a translation of the graph $y = f(x)$ by the vector $\begin{pmatrix} 0 \\ a \end{pmatrix}$.

Now consider the two functions:

$$y = x^2 - 2x + 1$$

$$y = (x - 3)^2 - 2(x - 3) + 1$$

We obtain the second function by replacing $x$ by $x - 3$ in the first function.

The graphs of these two functions are:

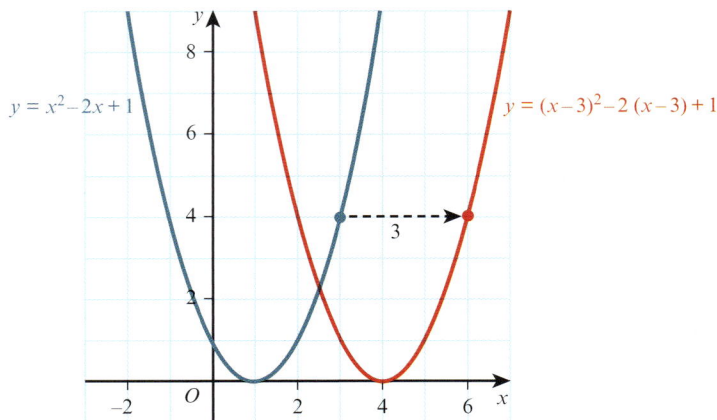

The curves have exactly the same shape but this time they are separated by 3 units in the positive $x$-direction.

You may be surprised that the curve has moved in the positive $x$-direction. Note, however, that a way of obtaining $y = y$ is to have $x = x - 3$ or equivalently $x = x + 3$. This means that the two curves are at the same height when the red curve is 3 units to the right of the blue curve.

Hence, the graph of $y = (x - 3)^2 - 2(x - 3) + 1$ is a translation of the graph of $y = x^2 - 2x + 1$ by the vector $\begin{pmatrix} 3 \\ 0 \end{pmatrix}$.

### KEY POINT 2.7

The graph of $y = f(x - a)$ is a translation of the graph $y = f(x)$ by the vector $\begin{pmatrix} a \\ 0 \end{pmatrix}$.

Combining these two results gives:

### KEY POINT 2.8

The graph of $y = f(x - a) + b$ is a translation of the graph $y = f(x)$ by the vector $\begin{pmatrix} a \\ b \end{pmatrix}$.

### WORKED EXAMPLE 2.11

The graph of $y = x^2 + 5x$ is translated 2 units to the right. Find the equation of the resulting graph. Give your answer in the form $y = ax^2 + bx + c$.

**Answer**

$y = x^2 + 5x$         Replace all occurrences of $x$ by $x - 2$.

$y = (x - 2)^2 + 5(x - 2)$         Expand and simplify.

$y = x^2 + x - 6$

### WORKED EXAMPLE 2.12

The graph of $y = \sqrt{2x}$ is translated by the vector $\begin{pmatrix} -5 \\ 3 \end{pmatrix}$. Find the equation of the resulting graph.

**Answer**

$y = \sqrt{2x}$         Replace $x$ by $x + 5$, and add 3 to the resulting function.

$y = \sqrt{2(x + 5)} + 3$

$y = \sqrt{2x + 10} + 3$

1 Find the equation of each graph after the given transformation.

   a   $y = 2x^2$          after translation by $\begin{pmatrix} 0 \\ 4 \end{pmatrix}$

   b   $y = 5\sqrt{x}$         after translation by $\begin{pmatrix} 0 \\ -2 \end{pmatrix}$

   c   $y = 7x^2 - 2x$   after translation by $\begin{pmatrix} 0 \\ 1 \end{pmatrix}$

   d   $y = x^2 - 1$      after translation by $\begin{pmatrix} 0 \\ 2 \end{pmatrix}$

   e   $y = \dfrac{2}{x}$         after translation by $\begin{pmatrix} -5 \\ 0 \end{pmatrix}$

   f   $y = \dfrac{x}{x+1}$    after translation by $\begin{pmatrix} 3 \\ 0 \end{pmatrix}$

   g   $y = x^2 + x$     after translation by $\begin{pmatrix} -1 \\ 0 \end{pmatrix}$

   h   $y = 3x^2 - 2$   after translation by $\begin{pmatrix} 2 \\ 3 \end{pmatrix}$

2 Find the translation that transforms the graph.

   a   $y = x^2 + 5x - 2$    to the graph  $y = x^2 + 5x + 2$

   b   $y = x^3 + 2x^2 + 1$  to the graph  $y = x^3 + 2x^2 - 4$

   c   $y = x^2 - 3x$        to the graph  $y = (x+1)^2 - 3(x+1)$

   d   $y = x + \dfrac{6}{x}$     to the graph  $y = x - 2 + \dfrac{6}{x-2}$

   e   $y = \sqrt{2x+5}$    to the graph  $y = \sqrt{2x+3}$

   f   $y = \dfrac{5}{x^2} - 3x$    to the graph  $y = \dfrac{5}{(x-2)^2} - 3x + 10$

3 The diagram shows the graph of $y = f(x)$.

   Sketch the graphs of each of the following functions.

   a   $y = f(x) - 4$

   b   $y = f(x - 2)$

   c   $y = f(x + 1) - 5$

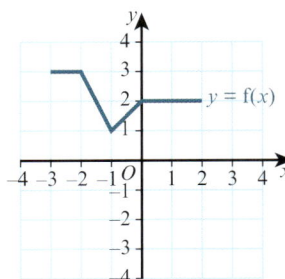

4   a   On the same diagram, sketch the graphs of $y = 2x$ and $y = 2x + 2$.

     b   $y = 2x$ can be transformed to $y = 2x + 2$ by a translation of $\begin{pmatrix} 0 \\ a \end{pmatrix}$.

        Find the value of $a$.

     c   $y = 2x$ can be transformed to $y = 2x + 2$ by a translation of $\begin{pmatrix} b \\ 0 \end{pmatrix}$.

        Find the value of $b$.

5 A cubic graph has equation $y = (x + 3)(x - 2)(x - 5)$.

Write, in a similar form, the equation of the graph after a translation of $\begin{pmatrix} 2 \\ 0 \end{pmatrix}$.

6 The graph of $y = x^2 - 4x + 1$ is translated by the vector $\begin{pmatrix} 1 \\ 2 \end{pmatrix}$.

Find, in the form $y = ax^2 + bx + c$, the equation of the resulting graph.

**PS** 7 The graph of $y = ax^2 + bx + c$ is translated by the vector $\begin{pmatrix} 2 \\ -5 \end{pmatrix}$.

The resulting graph is $y = 2x^2 - 11x + 10$. Find the value of $a$, the value of $b$ and the value of $c$.

> **WEB LINK**
>
> Try the *Between the lines* resource on the Underground Mathematics website.

## 2.6 Reflections

The diagram shows the graphs of the two functions:

$y = x^2 - 2x + 1$

$y = -(x^2 - 2x + 1)$

When the $x$-coordinates on the two graphs are the same ($x = x$), the $y$-coordinates are negative of each other ($y = -y$).

This means that, when the $x$-coordinates are the same, the red curve is the same vertical distance from the $x$-axis as the blue curve but it is on the opposite side of the $x$-axis.

Hence, the graph of $y = -(x^2 - 2x + 1)$ is a reflection of the graph of $y = x^2 - 2x + 1$ in the $x$-axis.

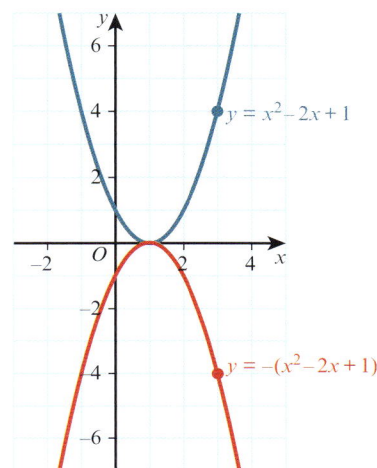

> 🔍 **KEY POINT 2.9**
>
> The graph of $y = -f(x)$ is a reflection of the graph $y = f(x)$ in the $x$-axis.

Now consider the two functions:

$y = x^2 - 2x + 1$

$y = (-x)^2 - 2(-x) + 1$

We obtain the second function by replacing $x$ by $-x$ in the first function.

The graphs of these two functions are demonstrated in the diagram.

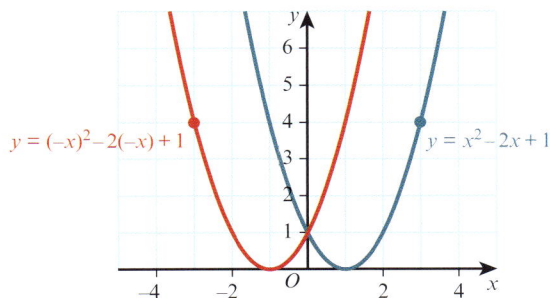

The curves are at the same height $(y = y)$ when $x = -x$ or equivalently $x = -x$.

This means that the heights of the two graphs are the same when the red graph has the same horizontal displacement from the $y$-axis as the blue graph but is on the opposite side of the $y$-axis.

Hence, the graph of $y = (-x)^2 - 2(-x) + 1$ is a reflection of the graph of $y = x^2 - 2x + 1$ in the $y$-axis.

> ### 🔍 KEY POINT 2.10
>
> The graph of $y = f(-x)$ is a reflection of the graph $y = f(x)$ in the $y$-axis.

### WORKED EXAMPLE 2.13

The quadratic graph $y = f(x)$ has a minimum at the point $(5, -7)$. Find the coordinates of the vertex and state whether it is a maximum or minimum of the graph for each of the following graphs.

**a**   $y = -f(x)$                               **b**   $y = f(-x)$

**Answer**

**a**   $y = -f(x)$ is a reflection of $y = f(x)$ in the $x$-axis.

    The turning point is $(5, 7)$. It is a maximum point.

**b**   $y = f(-x)$ is a reflection of $y = f(x)$ in the $y$-axis.

    The turning point is $(-5, -7)$. It is a minimum point.

### EXERCISE 2F

**1**   The diagram shows the graph of $y = g(x)$.

    Sketch the graphs of each of the following functions.

    **a**   $y = -g(x)$                **b**   $y = g(-x)$

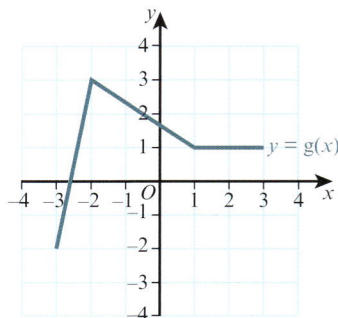

**2**   Find the equation of each graph after the given transformation.

    **a**   $y = 5x^2$ after reflection in the $x$-axis.

    **b**   $y = 2x^4$ after reflection in the $y$-axis.

    **c**   $y = 2x^2 - 3x + 1$ after reflection in the $y$-axis.

    **d**   $y = 5 + 2x - 3x^2$ after reflection in the $x$-axis.

3   Describe the transformation that maps the graph:

a   $y = x^2 + 7x - 3$ onto the graph $y = -x^2 - 7x + 3$

b   $y = x^2 - 3x + 4$ onto the graph $y = x^2 + 3x + 4$

c   $y = 2x - 5x^2$ onto the graph $y = 5x^2 - 2x$

d   $y = x^3 + 2x^2 - 3x + 1$ onto the graph $y = -x^3 - 2x^2 + 3x - 1$.

## 2.7 Stretches

The diagram shows the graphs of the two functions:

$$y = x^2 - 2x - 3$$

$$y = 2(x^2 - 2x - 3)$$

When the $x$-coordinates on the two graphs are the same ($x = x$), the $y$-coordinate on the red graph is double the $y$-coordinate on the blue graph ($y = 2y$).

This means that, when the $x$-coordinates are the same, the red curve is twice the distance of the blue graph from the $x$-axis.

Hence, the graph of $y = 2(x^2 - 2x - 3)$ is a stretch of the graph of $y = x^2 - 2x - 3$ from the $x$-axis. We say that it has been stretched with stretch factor 2 parallel to the $y$-axis.

Note: there are alternative ways of expressing this transformation:

- a stretch with scale factor 2 with the line $y = 0$ invariant
- a stretch with stretch factor 2 with the $x$-axis invariant
- a stretch with stretch factor 2 relative to the $x$-axis
- a vertical stretch with stretch factor 2.

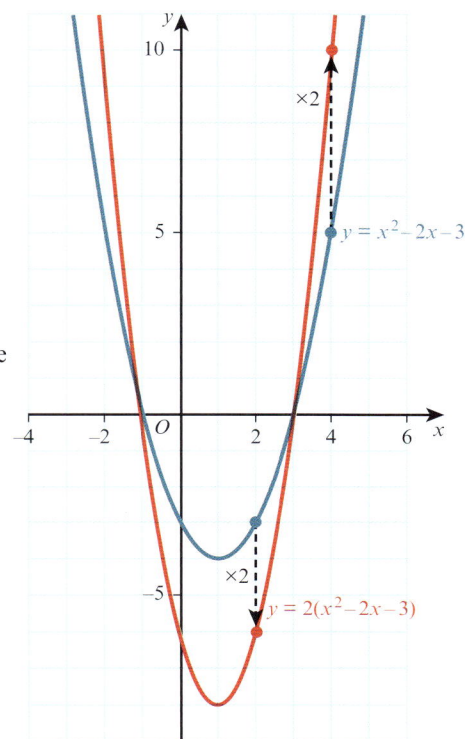

**KEY POINT 2.11**

The graph of $y = af(x)$ is a stretch of the graph $y = f(x)$ with stretch factor $a$ parallel to the $y$-axis.

Note: if $a < 0$, then $y = af(x)$ can be considered to be a stretch of $y = f(x)$ with a negative scale factor or as a stretch with positive scale factor followed by a reflection in the $x$-axis.

57

Now consider the two functions:

$$y = x^2 - 2x - 3$$

$$y = (2x)^2 - 2(2x) - 3$$

We obtain the second function by replacing $x$ by $2x$ in the first function.

The two curves are at the same height ($y = y$) when $x = 2x$ or equivalently $x = \dfrac{1}{2}x$.

This means that the heights of the two graphs are the same when the red graph has half the horizontal displacement from the $y$-axis as the blue graph.

Hence, the graph of $y = (2x)^2 - 2(2x) - 3$ is a stretch of the graph of $y = x^2 - 2x - 3$ from the $y$-axis. We say that it has been stretched with stretch factor $\dfrac{1}{2}$ parallel to the $x$-axis.

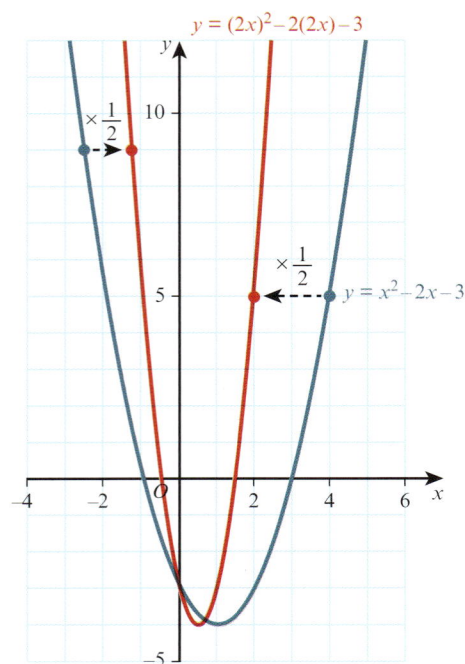

58

---

**KEY POINT 2.12**

The graph of $y = f(ax)$ is a stretch of the graph $y = f(x)$ with stretch factor $\dfrac{1}{a}$ parallel to the $x$-axis.

---

**WORKED EXAMPLE 2.14**

The graph of $y = 5 - \dfrac{1}{2}x^2$ is stretched with stretch factor 4 parallel to the $y$-axis.

Find the equation of the resulting graph.

**Answer**

Let $f(x) = 5 - \dfrac{1}{2}x^2$ . . . . . . . . . . . . . . . . . . . . . . . . . A stretch parallel to the $y$-axis, factor 4, gives the function $4f(x)$.

$$4f(x) = 20 - 2x^2$$

The equation of the resulting graph is $y = 20 - 2x^2$.

---

**WORKED EXAMPLE 2.15**

Describe the single transformation that maps the graph of $y = x^2 - 3x - 5$ to the graph of $y = 4x^2 - 6x - 5$.

**Answer**

Let $f(x) = x^2 - 3x - 5$ . . . . . . . . . . . . . . . . . . . . . . . . . . Express $4x^2 - 6x - 5$ in terms of $f(x)$.

$$4x^2 - 6x - 5 = (2x)^2 - 3(2x) - 5$$

$$= f(2x)$$

The transformation is a stretch parallel to the $x$-axis with stretch factor $\dfrac{1}{2}$.

**EXERCISE 2G**

1   The diagram shows the graph of $y = f(x)$.

Sketch the graphs of each of the following functions.

a   $y = 3f(x)$

b   $y = f(2x)$

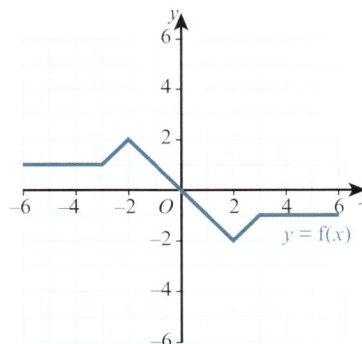

2   Find the equation of each graph after the given transformation.

a   $y = 3x^2$ after a stretch parallel to the $y$-axis with stretch factor 2.

b   $y = x^3 - 1$ after a stretch parallel to the $y$-axis with stretch factor 3.

c   $y = 2^x + 4$ after a stretch parallel to the $y$-axis with stretch factor $\frac{1}{2}$.

d   $y = 2x^2 - 8x + 10$ after a stretch parallel to the $x$-axis with stretch factor 2.

e   $y = 6x^3 - 36x$ after a stretch parallel to the $x$-axis with stretch factor $\frac{1}{3}$.

3   Describe the single transformation that maps the graph:

a   $y = x^2 + 2x - 5$ onto the graph $y = 4x^2 + 4x - 5$

b   $y = x^2 - 3x + 2$ onto the graph $y = 3x^2 - 9x + 6$

c   $y = 2^x + 1$ onto the graph $y = 2^{x+1} + 2$

d   $y = \sqrt{x - 6}$ onto the graph $y = \sqrt{3x - 6}$

## 2.8 Combined transformations

In this section you will learn how to apply simple combinations of transformations.

The transformations of the graph of $y = f(x)$ that you have studied so far can each be categorised as either vertical or horizontal transformations.

| Vertical transformations | |
| --- | --- |
| $y = f(x) + a$ | translation $\begin{pmatrix} 0 \\ a \end{pmatrix}$ |
| $y = -f(x)$ | reflection in the $x$-axis |
| $y = af(x)$ | vertical stretch, factor $a$ |

| Horizontal transformations | |
| --- | --- |
| $y = f(x + a)$ | translation $\begin{pmatrix} -a \\ 0 \end{pmatrix}$ |
| $y = f(-x)$ | reflection in the $y$-axis |
| $y = f(ax)$ | horizontal stretch, factor $\frac{1}{a}$ |

When combining transformations care must be taken with the order in which the transformations are applied.

59

### EXPLORE 2.5

Apply the transformations in the given order to triangle $T$ and for each question comment on whether the final images are the same or different.

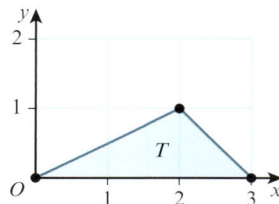

1 **Combining two vertical transformations**

  **a** **i** Translate $\begin{pmatrix} 0 \\ 3 \end{pmatrix}$, then stretch vertically with factor $\frac{1}{2}$.

    **ii** Stretch vertically with factor $\frac{1}{2}$, then translate $\begin{pmatrix} 0 \\ 3 \end{pmatrix}$.

  **b** Investigate for other pairs of vertical transformations.

2 **Combining one vertical and one horizontal transformation**

  **a** **i** Reflect in the $x$-axis, then translate $\begin{pmatrix} -2 \\ 0 \end{pmatrix}$.

    **ii** Translate $\begin{pmatrix} -2 \\ 0 \end{pmatrix}$, then reflect in the $x$-axis.

  **b** Investigate for other pairs of transformations where one is vertical and the other is horizontal.

3 **Combining two horizontal transformations**

  **a** **i** Stretch horizontally with factor 2, then translate $\begin{pmatrix} 2 \\ 0 \end{pmatrix}$.

    **ii** Translate $\begin{pmatrix} 2 \\ 0 \end{pmatrix}$, then stretch horizontally with factor 2.

  **b** Investigate for other pairs of horizontal transformations.

From the Explore activity, you should have found that:

### KEY POINT 2.13

- When two vertical transformations or two horizontal transformations are combined, the order in which they are applied may affect the outcome.

- When one horizontal and one vertical transformation are combined, the order in which they are applied does **not** affect the outcome.

## Combining two vertical transformations

We will now consider how the graph of $y = f(x)$ is transformed to the graph $y = af(x) + k$.

This can be shown in a flow diagram as:

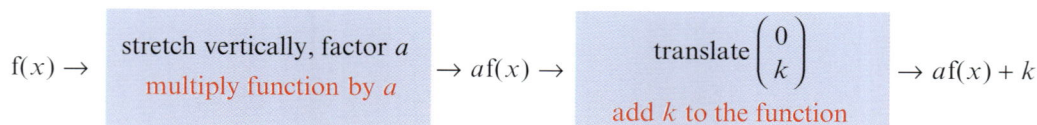

$$f(x) \to \boxed{\begin{array}{c} \text{stretch vertically, factor } a \\ \text{multiply function by } a \end{array}} \to af(x) \to \boxed{\begin{array}{c} \text{translate } \begin{pmatrix} 0 \\ k \end{pmatrix} \\ \text{add } k \text{ to the function} \end{array}} \to af(x) + k$$

This leads to the important result:

Vertical transformations follow the 'normal' order of operations, as used in arithmetic.

## Combining two horizontal transformations

Now consider how the graph of $y = f(x)$ is transformed to the graph $y = f(bx + c)$.

$$f(x) \rightarrow \boxed{\begin{array}{c} \text{translate} \begin{pmatrix} -c \\ 0 \end{pmatrix} \\ \text{replace } x \text{ with } x + c \end{array}} \rightarrow f(x + c) \rightarrow \boxed{\begin{array}{c} \text{stretch horizontally, factor } \dfrac{1}{b} \\ \text{replace } x \text{ with } bx \end{array}} \rightarrow f(bx + c)$$

This leads to the important result:

Horizontal transformations follow the **opposite** order to the 'normal' order of operations, as used in arithmetic.

**WORKED EXAMPLE 2.16**

The diagram shows the graph of $y = f(x)$.

Sketch the graph of $y = 2f(x) - 3$.

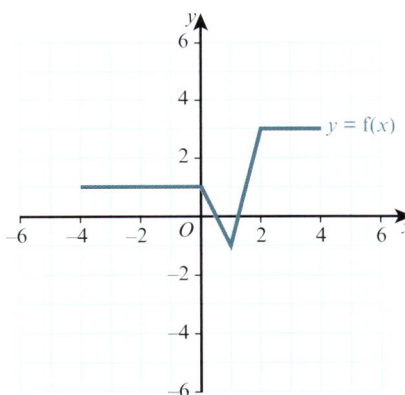

**Answer**

$y = 2f(x) - 3$ is a combination of two vertical transformations of $y = f(x)$, hence the transformations follow the 'normal' order of operations.

**Step 1:** Sketch the graph $y = 2f(x)$:

Stretch $y = f(x)$ vertically with stretch factor 2.

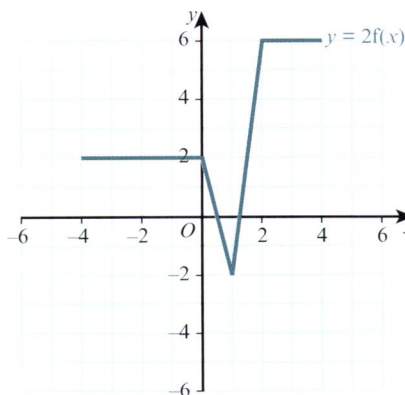

61

**Step 2:** Sketch the graph $y = 2f(x) - 3$:

Translate $y = 2f(x)$ by the vector $\begin{pmatrix} 0 \\ -3 \end{pmatrix}$.

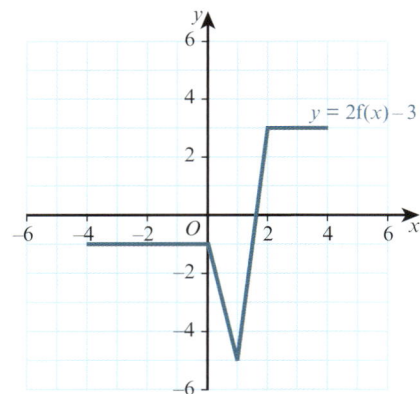

**WORKED EXAMPLE 2.17**

The diagram shows the graph of $y = x^2$ and its image, $y = g(x)$, after a combination of transformations.

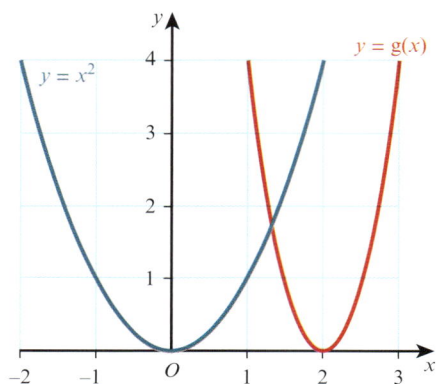

a   Find two different ways of describing the combination of transformations.

b   Write down the equation of the graph $y = g(x)$.

**Answer**

a   Translation of $\begin{pmatrix} 4 \\ 0 \end{pmatrix}$ followed by a horizontal stretch, stretch factor $\dfrac{1}{2}$.

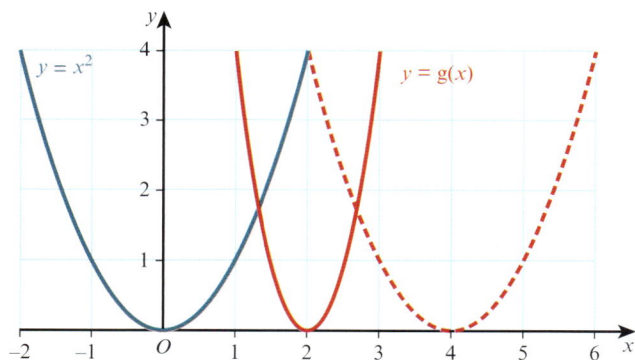

**OR**

Horizontal stretch, factor $\dfrac{1}{2}$, followed by a translation of $\begin{pmatrix} 2 \\ 0 \end{pmatrix}$.

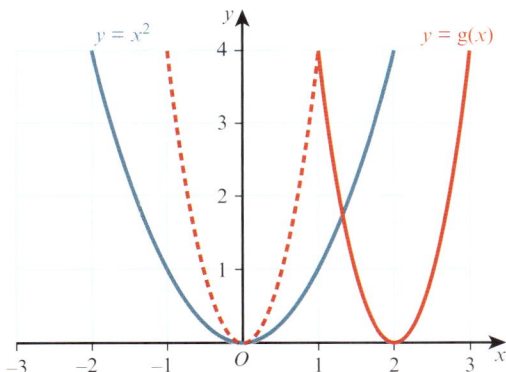

**b** Using the first combination of transformations:

Translation of $\begin{pmatrix} 4 \\ 0 \end{pmatrix}$ means 'replace $x$ by $x - 4$'.

$y = x^2$      becomes      $y = (x - 4)^2$

Horizontal stretch, factor $\dfrac{1}{2}$ means 'replace $x$ by $2x$'.

$y = (x - 4)^2$   becomes   $y = (2x - 4)^2$

Hence, $g(x) = (2x - 4)^2$.

**TIP**

The same answer will be obtained when using the second combination of transformations. You may wish to check this yourself.

63

## EXERCISE 2H

**1** The diagram shows the graph of $y = g(x)$.

Sketch the graph of each of the following.

**a**   $y = g(x + 2) + 3$      **b**   $y = 2g(x) + 1$

**c**   $y = 2 - g(x)$      **d**   $y = 2g(-x) + 1$

**e**   $y = -2g(x) - 1$      **f**   $y = g(2x) + 3$

**g**   $y = g(2x - 6)$      **h**   $y = g(-x + 1)$

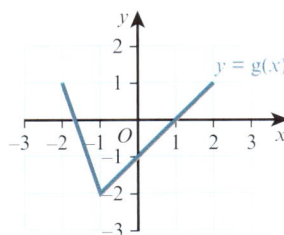

2   The diagram shows the graph of $y = f(x)$.

Write down, in terms of $f(x)$, the equation of the graph of each of the following diagrams.

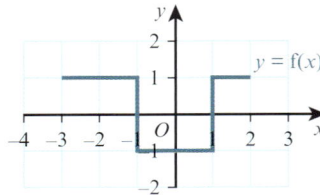

a

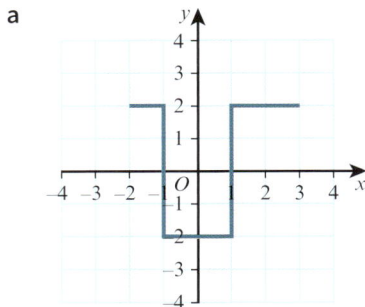

b

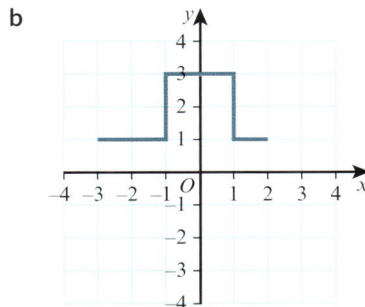

c

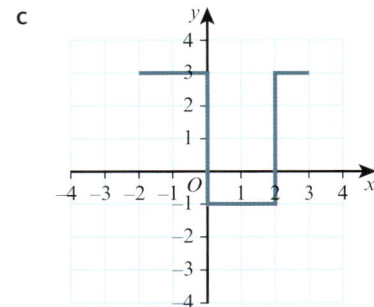

3   Given that $y = x^2$, find the image of the curve $y = x^2$ after each of the following combinations of transformations.

a   a stretch in the $y$-direction with factor 3 followed by a translation by the vector $\begin{pmatrix} 1 \\ 0 \end{pmatrix}$

b   a translation by the vector $\begin{pmatrix} 1 \\ 0 \end{pmatrix}$ followed by a stretch in the $y$-direction with factor 3

4   Find the equation of the image of the curve $y = x^2$ after each of the following combinations of transformations and, in each case, sketch the graph of the resulting curve.

a   a stretch in the $x$-direction with factor 2 followed by a translation by the vector $\begin{pmatrix} 5 \\ 0 \end{pmatrix}$

b   a translation by the vector $\begin{pmatrix} 5 \\ 0 \end{pmatrix}$ followed by a stretch in the $x$-direction with factor 2

c   On a graph show the curve $y = x^2$ and each of your answers to **parts a** and **b**.

5   Given that $f(x) = x^2 + 1$, find the image of $y = f(x)$ after each of the following combinations of transformations.

a   translation $\begin{pmatrix} 0 \\ -5 \end{pmatrix}$ followed by a stretch parallel to the $y$-axis with stretch factor 2

b   translation $\begin{pmatrix} 2 \\ 0 \end{pmatrix}$ followed by a reflection in the $x$-axis

6 a The graph of $y = g(x)$ is reflected in the $y$-axis and then stretched with stretch factor 2 parallel to the $y$-axis. Write down the equation of the resulting graph.

   b The graph of $y = f(x)$ is translated by the vector $\begin{pmatrix} 2 \\ -3 \end{pmatrix}$ and then reflected in the $x$-axis. Write down the equation of the resulting graph.

7 Determine the sequence of transformations that maps $y = f(x)$ to each of the following functions.

   a $y = \dfrac{1}{2}f(x) + 3$   b $y = -f(x) + 2$   c $y = f(2x - 6)$   d $y = 2f(x) - 8$

8 Determine the sequence of transformations that maps:

   a the curve $y = x^3$ onto the curve $y = \dfrac{1}{2}(x + 5)^3$

   b the curve $y = x^3$ onto the curve $y = -\dfrac{1}{2}(x + 1)^3 - 2$

   c the curve $y = \sqrt[3]{x}$ onto the curve $y = -2\sqrt[3]{x - 3} + 4$

9 Given that $f(x) = \sqrt{x}$, write down the equation of the image of $f(x)$ after:

   a reflection in the $x$-axis, followed by translation $\begin{pmatrix} 0 \\ 3 \end{pmatrix}$, followed by translation $\begin{pmatrix} 1 \\ 0 \end{pmatrix}$, followed by a stretch parallel to the $x$-axis with stretch factor 2

   b translation $\begin{pmatrix} 0 \\ 3 \end{pmatrix}$, followed by a stretch parallel to the $x$-axis with stretch factor 2, followed by a reflection in the $x$-axis, followed by translation $\begin{pmatrix} 1 \\ 0 \end{pmatrix}$.

10 Given that $g(x) = x^2$, write down the equation of the image of $g(x)$ after:

   a translation $\begin{pmatrix} -4 \\ 0 \end{pmatrix}$, followed by a reflection in the $y$-axis, followed by translation $\begin{pmatrix} 0 \\ 2 \end{pmatrix}$, followed by a stretch parallel to the $y$-axis with stretch factor 3

   b a stretch parallel to the $y$-axis with stretch factor 3, followed by translation $\begin{pmatrix} 0 \\ 2 \end{pmatrix}$, followed by reflection in the $y$-axis, followed by translation $\begin{pmatrix} -4 \\ 0 \end{pmatrix}$.

PS 11 Find two different ways of describing the combination of transformations that maps the graph of $f(x) = \sqrt{x}$ onto the graph $g(x) = \sqrt{-x - 2}$ and sketch the graphs of $y = f(x)$ and $y = g(x)$.

PS 12 Find two different ways of describing the sequence of transformations that maps the graph of $y = f(x)$ onto the graph of $y = f(2x + 10)$.

⊕ WEB LINK

Try the *Transformers* resource on the Underground Mathematics website.

# Checklist of learning and understanding

**Functions**

- A function is a rule that maps each $x$ value to just one $y$ value for a defined set of input values.
- A function can be either one-one or many-one.
- The set of input values for a function is called the domain of the function.
- The set of output values for a function is called the range (or image set) of the function.

**Composite functions**

- $\text{fg}(x)$ means the function g acts on $x$ first, then f acts on the result.
- fg only exists if the range of g is contained within the domain of f.
- In general, $\text{fg}(x) \neq \text{gf}(x)$.

**Inverse functions**

- The inverse of a function $\text{f}(x)$ is the function that undoes what $\text{f}(x)$ has done.
  $\text{f}\,\text{f}^{-1}(x) = \text{f}^{-1}\,\text{f}(x) = x$ or if $y = \text{f}(x)$ then $x = \text{f}^{-1}(y)$
- The inverse of the function $\text{f}(x)$ is written as $\text{f}^{-1}(x)$.
- The steps for finding the inverse function are:

  **Step 1:** Write the function as $y =$
  **Step 2:** Interchange the $x$ and $y$ variables.
  **Step 3:** Rearrange to make $y$ the subject.

- The domain of $\text{f}^{-1}(x)$ is the range of $\text{f}(x)$.
- The range of $\text{f}^{-1}(x)$ is the domain of $\text{f}(x)$.
- An inverse function $\text{f}^{-1}(x)$ can exist if, and only if, the function $\text{f}(x)$ is one-one.
- The graphs of f and $\text{f}^{-1}$ are reflections of each other in the line $y = x$.
- If $\text{f}(x) = \text{f}^{-1}(x)$, then the function f is called a self-inverse function.
- If f is self-inverse then $\text{ff}(x) = x$.
- The graph of a self-inverse function has $y = x$ as a line of symmetry.

**Transformations of functions**

- The graph of $y = \text{f}(x) + a$ is a translation of $y = \text{f}(x)$ by the vector $\begin{pmatrix} 0 \\ a \end{pmatrix}$.

- The graph of $y = \text{f}(x + a)$ is a translation of $y = \text{f}(x)$ by the vector $\begin{pmatrix} -a \\ 0 \end{pmatrix}$.

- The graph of $y = -\text{f}(x)$ is a reflection of the graph $y = \text{f}(x)$ in the $x$-axis.
- The graph of $y = \text{f}(-x)$ is a reflection of the graph $y = \text{f}(x)$ in the $y$-axis.
- The graph of $y = a\text{f}(x)$ is a stretch of $y = \text{f}(x)$, stretch factor $a$, parallel to the $y$-axis.
- The graph of $y = \text{f}(ax)$ is a stretch of $y = \text{f}(x)$, stretch factor $\dfrac{1}{a}$, parallel to the $x$-axis.

**Combining transformations**

- When two vertical transformations or two horizontal transformations are combined, the order in which they are applied may affect the outcome.
- When one horizontal and one vertical transformation are combined, the order in which they are applied does not affect the outcome.
- Vertical transformations follow the 'normal' order of operations, as used in arithmetic
- Horizontal transformations follow the **opposite** order to the 'normal' order of operations, as used in arithmetic.

## END-OF-CHAPTER REVIEW EXERCISE 2

1  Functions f and g are defined for $x \in \mathbb{R}$ by:

$$f : x \mapsto 3x - 1$$
$$g : x \mapsto 5x - x^2$$

Express $gf(x)$ in the form $a - b(x - c)^2$, where $a$, $b$ and $c$ are constants.   [5]

2

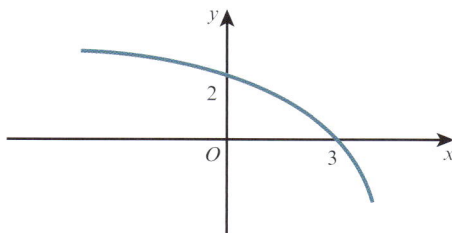

The diagram shows a sketch of the curve with equation $y = f(x)$.

   a  Sketch the graph of $y = -f\left(\dfrac{1}{2}x\right)$.   [3]

   b  Describe fully a sequence of two transformations that maps the graph of $y = f(x)$ onto the graph of $y = f(3 - x)$.   [2]

3  A curve has equation $y = x^2 + 6x + 8$.

   a  Sketch the curve, showing the coordinates of any axes crossing points.   [2]

   b  The curve is translated by the vector $\begin{pmatrix} 2 \\ 0 \end{pmatrix}$, then stretched vertically with stretch factor 3.

   Find the equation of the resulting curve, giving your answer in the form $y = ax^2 + bx$.   [4]

4  The function $f : x \mapsto x^2 - 2$ is defined for the domain $x \geqslant 0$.

   a  Find $f^{-1}(x)$ and state the domain of $f^{-1}$.   [3]

   b  On the same diagram, sketch the graphs of $f$ and $f^{-1}$.   [3]

5  i  Express $-x^2 + 6x - 5$ in the form $a(x + b)^2 + c$, where $a$, $b$ and $c$ are constants.   [3]

   The function $f : x \mapsto -x^2 + 6x - 5$ is defined for $x \geqslant m$, where $m$ is a constant.

   ii  State the smallest possible value of $m$ for which $f$ is one-one.   [1]

   iii  For the case where $m = 5$, find an expression for $f^{-1}(x)$ and state the domain of $f^{-1}$.   [4]

   *Cambridge International AS & A Level Mathematics 9709 Paper 11 Q9 November 2015*

6  The function $f : x \mapsto x^2 - 4x + k$ is defined for the domain $x \geqslant p$, where $k$ and $p$ are constants.

   i  Express $f(x)$ in the form $(x + a)^2 + b + k$, where $a$ and $b$ are constants.   [2]

   ii  State the range of $f$ in terms of $k$.   [1]

   iii  State the smallest value of $p$ for which $f$ is one-one.   [1]

   iv  For the value of $p$ found in part **iii**, find an expression for $f^{-1}(x)$ and state the domain of $f^{-1}$, giving your answer in terms of $k$.   [4]

   *Cambridge International AS & A Level Mathematics 9709 Paper 11 Q8 June 2012*

**7**

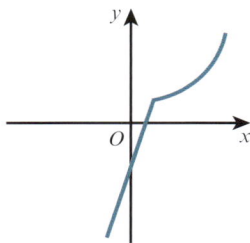

The diagram shows the function f defined for $-1 \leqslant x \leqslant 4$, where

$$f(x) = \begin{cases} 3x - 2 & \text{for } -1 \leqslant x \leqslant 1, \\ \dfrac{4}{5 - x} & \text{for } 1 < x \leqslant 4. \end{cases}$$

i   State the range of f. [1]

ii  Copy the diagram and on your copy sketch the graph of $y = f^{-1}(x)$. [2]

iii Obtain expressions to define the function $f^{-1}$, giving also the set of values for which each expression is valid. [6]

*Cambridge International AS & A Level Mathematics 9709 Paper 11 Q10 June 2014*

**8**  The function f is defined by $f(x) = 4x^2 - 24x + 11$, for $x \in \mathbb{R}$.

i   Express $f(x)$ in the form $a(x - b)^2 + c$ and hence state the coordinates of the vertex of the graph of $y = f(x)$. [4]

The function g is defined by $g(x) = 4x^2 - 24x + 11$, for $x \leqslant 1$.

ii  State the range of g. [2]

iii Find an expression for $g^{-1}(x)$ and state the domain of $g^{-1}$. [4]

*Cambridge International AS & A Level Mathematics 9709 Paper 11 Q10 November 2012*

**9**  i   Express $2x^2 - 12x + 13$ in the form $a(x + b)^2 + c$, where $a$, $b$ and $c$ are constants. [3]

ii  The function f is defined by $f(x) = 2x^2 - 12x + 13$, for $x \geqslant k$, where $k$ is a constant. It is given that f is a one-one function. State the smallest possible value of $k$. [1]

The value of $k$ is now given to be 7.

iii Find the range of f. [1]

iv  Find the expression for $f^{-1}(x)$ and state the domain of $f^{-1}$. [5]

*Cambridge International AS & A Level Mathematics 9709 Paper 11 Q8 June 2013*

**10** i   Express $x^2 - 2x - 15$ in the form $(x + a)^2 + b$. [2]

The function f is defined for $p \leqslant x \leqslant q$, where $p$ and $q$ are positive constants, by

$$f : x \mapsto x^2 - 2x - 15.$$

The range of f is given by $c \leqslant f(x) \leqslant d$, where $c$ and $d$ are constants.

ii  State the smallest possible value of $c$. [1]

For the case where $c = 9$ and $d = 65$,

   **iii** find $p$ and $q$,                                         **[4]**

   **iv** find an expression for $f^{-1}(x)$.                       **[3]**

*Cambridge International AS & A Level Mathematics 9709 Paper 11 Q10 November 2014*

**11** The function f is defined by $f : x \mapsto 2x^2 - 12x + 7$ for $x \in \mathbb{R}$.

   **i** Express $f(x)$ in the form $a(x - b)^2 - c$.               **[3]**

   **ii** State the range of f.                          **[1]**

   **iii** Find the set of values of $x$ for which $f(x) < 21$.       **[3]**

   The function g is defined by $g : x \mapsto 2x + k$ for $x \in \mathbb{R}$.

   **iv** Find the value of the constant $k$ for which the equation $gf(x) = 0$ has two equal roots.    **[4]**

*Cambridge International AS & A Level Mathematics 9709 Paper 11 Q9 June 2010*

**12** Functions f and g are defined for $x \in \mathbb{R}$ by

$$f : x \mapsto 2x + 1,$$
$$g : x \mapsto x^2 - 2.$$

   **i** Find and simplify expressions for $fg(x)$ and $gf(x)$.       **[2]**

   **ii** Hence find the value of $a$ for which $fg(a) = gf(a)$.       **[3]**

   **iii** Find the value of $b$ ($b \neq a$) for which $g(b) = b$.       **[2]**

   **iv** Find and simplify an expression for $f^{-1}g(x)$.       **[2]**

   The function h is defined by

$$h : x \mapsto x^2 - 2, \text{ for } x \leq 0.$$

   **v** Find an expression for $h^{-1}(x)$.                     **[2]**

*Cambridge International AS & A Level Mathematics 9709 Paper 11 Q11 June 2011*

**13** Functions f and g are defined by

$$f : x \mapsto 2x^2 - 8x + 10 \text{ for } 0 \leq x \leq 2,$$
$$g : x \mapsto x \qquad\qquad \text{ for } 0 \leq x \leq 10.$$

   **i** Express $f(x)$ in the form $a(x + b)^2 + c$, where $a$, $b$ and $c$ are constants.    **[3]**

   **ii** State the range of f.                          **[1]**

   **iii** State the domain of $f^{-1}$.                      **[1]**

   **iv** Sketch on the same diagram the graphs of $y = f(x)$, $y = g(x)$ and $y = f^{-1}(x)$, making clear the relationship between the graphs.    **[4]**

   **v** Find an expression for $f^{-1}(x)$.                 **[3]**

*Cambridge International AS & A Level Mathematics 9709 Paper 11 Q11 November 2011*

# Chapter 3
# Coordinate geometry

**In this chapter you will learn how to:**

- find the equation of a straight line when given sufficient information
- interpret and use any of the forms $y = mx + c$, $y - y_1 = m(x - x_1)$, $ax + by + c = 0$ in solving problems
- understand that the equation $(x - a)^2 + (y - b)^2 = r^2$ represents the circle with centre $(a, b)$ and radius $r$
- use algebraic methods to solve problems involving lines and circles
- understand the relationship between a graph and its associated algebraic equation, and use the relationship between points of intersection of graphs and solutions of equations.

## PREREQUISITE KNOWLEDGE

| Where it comes from | What you should be able to do | Check your skills |
|---|---|---|
| IGCSE / O Level Mathematics | Find the midpoint and length of a line segment. | 1  Find the midpoint and length of the line segment joining $(-7, 4)$ and $(-2, -8)$. |
| IGCSE / O Level Mathematics | Find the gradient of a line and state the gradient of a line that is perpendicular to the line. | 2  a  Find the gradient of the line joining $A(-1, 3)$ and $B(5, 2)$.<br>b  State the gradient of the line that is perpendicular to the line $AB$. |
| IGCSE / O Level Mathematics | Interpret and use equations of lines of the form $y = mx + c$. | 3  The equation of a line is $y = \frac{2}{3}x - 5$. Write down:<br>a  the gradient of the line<br>b  the $y$-intercept<br>c  the $x$-intercept. |
| Chapter 1 | Complete the square and solve quadratic equations. | 4  a  Complete the square for $x^2 - 8x - 5$.<br>b  Solve $x^2 - 8x - 5 = 0$. |

## Why do we study coordinate geometry?

This chapter builds on the coordinate geometry work that you learnt at IGCSE / O Level. You shall also learn about the Cartesian equation of a circle. Circles are one of a collection of mathematical shapes called conics or conic sections.

A conic section is a curve obtained from the intersection of a plane with a cone. The three types of conic section are the ellipse, the parabola and the hyperbola. The circle is a special case of the ellipse. Conic sections provide a rich source of fascinating and beautiful results that mathematicians have been studying for thousands of years.

Conic sections are very important in the study of astronomy. We also use their reflective properties in the design of satellite dishes, searchlights, and optical and radio telescopes.

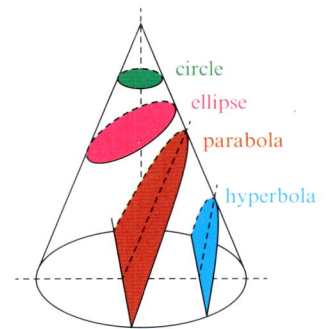

### WEB LINK

The *Geometry of equations* and *Circles* stations on the Underground Mathematics website have many useful resources for studying this topic.

## 3.1 Length of a line segment and midpoint

At IGCSE / O Level you learnt how to find the midpoint, $M$, of a line segment joining the points $P(x_1, y_1)$ and $Q(x_2, y_2)$ and the length of the line segment, $PQ$, using the two formulae in Key point 3.1. You need to know how to apply these formulae to solve problems.

> **TIP**
>
> It is important to remember to show appropriate calculations in coordinate geometry questions. Answers from scale drawings are not accepted.

> **KEY POINT 3.1**
>
> To find the midpoint, $M$, of the line segment $PQ$: $M = \left( \dfrac{x_1 + x_2}{2}, \dfrac{y_1 + y_2}{2} \right)$
>
> To find the length of $PQ$: $PQ = \sqrt{(x_2 - x_1)^2 + (y_2 - y_1)^2}$
>
> $Q(x_2, y_2)$
> $M$
> $P(x_1, y_1)$

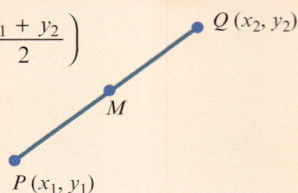

**WORKED EXAMPLE 3.1**

The point $M\left( \dfrac{3}{2}, -11 \right)$ is the midpoint of the line segment joining the points $P(-7, 4)$ and $Q(a, b)$.

Find the value of $a$ and the value of $b$.

**Answer**

**Method 1:** Using algebra

$(-7, 4) \qquad (a, \quad b)$
$\uparrow \ \uparrow \qquad\quad \uparrow \ \uparrow$
$(x_1, y_1) \qquad (x_2, y_2)$ ............... Decide which values to use for $x_1$, $y_1$, $x_2$, $y_2$.

Using $\left( \dfrac{x_1 + x_2}{2}, \dfrac{y_1 + y_2}{2} \right)$ and midpoint $= \left( \dfrac{3}{2}, -11 \right)$

$$\left( \frac{-7 + a}{2}, \frac{4 + b}{2} \right) = \left( \frac{3}{2}, -11 \right)$$

Equating the $x$-coordinates: $\dfrac{-7 + a}{2} = \dfrac{3}{2}$

$$-7 + a = 3$$
$$a = 10$$

Equating the $y$-coordinates: $\dfrac{4 + b}{2} = -11$

$$4 + b = -22$$
$$b = -26$$

Hence, $a = 10$ and $b = -26$.

**Method 2:** Using vectors

$$\overrightarrow{PM} = \begin{pmatrix} 8\frac{1}{2} \\ -15 \end{pmatrix}$$

$$\therefore \overrightarrow{MQ} = \begin{pmatrix} 8\frac{1}{2} \\ -15 \end{pmatrix}$$

$\therefore a = \frac{3}{2} + 8\frac{1}{2}$ and $b = -11 + (-15)$

$\therefore a = 10$ and $b = -26$.

*P* (−7, 4)

$M\left(\frac{3}{2}, -11\right)$

*Q* (*a*, *b*)

---

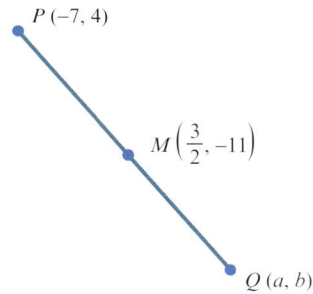

**WORKED EXAMPLE 3.2**

Three of the vertices of a parallelogram, $ABCD$, are $A(-5, -1)$, $B(-1, -4)$ and $C(6, -2)$.

    **a** Find the midpoint of $AC$.

    **b** Find the coordinates of $D$.

**Answer**

    **a** Midpoint of $AC = \left( \dfrac{-5 + 6}{2}, \dfrac{-1 + -2}{2} \right) = \left( \dfrac{1}{2}, -\dfrac{3}{2} \right)$

    **b** Let the coordinates of $D$ be $(m, n)$.

       Since $ABCD$ is a parallelogram, the midpoint of $BD$ is the same as the midpoint of $AC$.

       Midpoint of $BD = \left( \dfrac{-1 + m}{2}, \dfrac{-4 + n}{2} \right) = \left( \dfrac{1}{2}, -\dfrac{3}{2} \right)$

       Equating the $x$-coordinates: $\dfrac{-1 + m}{2} = \dfrac{1}{2}$

$$-1 + m = 1$$
$$m = 2$$

       Equating the $y$-coordinates: $\dfrac{-4 + n}{2} = -\dfrac{3}{2}$

$$-4 + n = -3$$
$$n = 1$$

    $D$ is the point (2, 1).

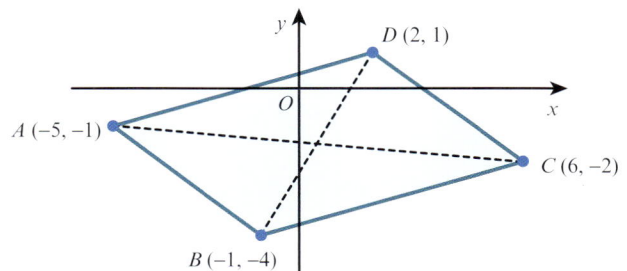

*D* (2, 1)

*A* (−5, −1)

*O*

*C* (6, −2)

*B* (−1, −4)

---

**WORKED EXAMPLE 3.3**

The distance between two points $P(-2, a)$ and $Q(a - 2, -7)$ is 17.

Find the two possible values of $a$.

**Answer**

$\begin{array}{cc} (-2, a) & (a - 2, -7) \\ \uparrow \ \uparrow & \uparrow \quad \uparrow \\ (x_1, y_1) & (x_2, y_2) \end{array}$

Decide which values to use for $x_1, y_1, x_2, y_2$.

73

Using $PQ = \sqrt{(x_2 - x_1)^2 + (y_2 - y_1)^2} = 17$

$$\sqrt{(a - 2 + 2)^2 + (-7 - a)^2} = 17 \qquad \text{Square both sides.}$$

$$a^2 + (-7 - a)^2 = 289 \qquad \text{Expand brackets.}$$

$$a^2 + 49 + 14a + a^2 = 289 \qquad \text{Collect terms on one side.}$$

$$2a^2 + 14a - 240 = 0 \qquad \text{Divide both sides by 2.}$$

$$a^2 + 7a - 120 = 0 \qquad \text{Factorise.}$$

$$(a - 8)(a + 15) = 0 \qquad \text{Solve.}$$

$$a - 8 = 0 \quad \text{or} \quad a + 15 = 0$$

$$a = 8 \quad \text{or} \quad a = -15$$

**EXPLORE 3.1**

The triangle has sides of length $2\sqrt{7}$ cm, $4\sqrt{3}$ cm and $5\sqrt{3}$ cm.

Tamar says that this triangle is right angled.

Discuss whether he is correct.

Explain your reasoning.

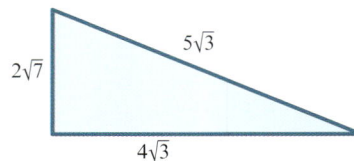

**EXERCISE 3A**

1 Calculate the lengths of the sides of the triangle $PQR$.

Use your answers to determine whether or not the triangle is right angled.

a $P(-4, 6)$, $Q(6, 1)$, $R(2, 9)$

b $P(-5, 2)$, $Q(9, 3)$, $R(-2, 8)$

2 $P(1, 6)$, $Q(-2, 1)$ and $R(3, -2)$.

Show that triangle $PQR$ is a right-angled isosceles triangle and calculate the area of the triangle.

3 The distance between two points, $P(a, -1)$ and $Q(-5, a)$, is $4\sqrt{5}$.

Find the two possible values of $a$.

4 The distance between two points, $P(-3, -2)$ and $Q(b, 2b)$, is 10.

Find the two possible values of $b$.

5 The point $(-2, -3)$ is the midpoint of the line segment joining $P(-6, -5)$ and $Q(a, b)$.

Find the value of $a$ and the value of $b$.

6 Three of the vertices of a parallelogram, $ABCD$, are $A(-7, 3)$, $B(-3, -11)$ and $C(3, -5)$.

a Find the midpoint of $AC$.

b Find the coordinates of $D$.

c Find the length of the diagonals $AC$ and $BD$.

**7**  The point $P(k, 2k)$ is equidistant from $A(8, 11)$ and $B(1, 12)$.
Find the value of $k$.

**8**  Triangle $ABC$ has vertices at $A(-6, 3)$, $B(3, 5)$ and $C(1, -4)$.
Show that triangle $ABC$ is isosceles and find the area of this triangle.

**9**  Triangle $ABC$ has vertices at $A(-7, 8)$, $B(3, k)$ and $C(8, 5)$.
Given that $AB = 2BC$, find the value of $k$.

**10**  The line $x + y = 4$ meets the curve $y = 8 - \dfrac{5}{x}$ at the points $A$ and $B$.
Find the coordinates of the midpoint of $AB$.

**11**  The line $y = x - 3$ meets the curve $y^2 = 4x$ at the points $A$ and $B$.

    **a**  Find the coordinates of the midpoint of $AB$.

    **b**  Find the length of the line segment $AB$.

**PS**  **12**  In triangle $ABC$, the midpoints of the sides $AB$, $BC$ and $AC$ are $(1, 4)$, $(2, 0)$ and $(-4, 1)$, respectively.
Find the coordinates of points $A$, $B$ and $C$.

## 3.2 Parallel and perpendicular lines

At IGCSE / O Level you learnt how to find the gradient of the line joining the points
$P(x_1, y_1)$ and $Q(x_2, y_2)$ using the formula in Key point 3.2.

> **KEY POINT 3.2**
>
> Gradient of $PQ = \dfrac{y_2 - y_1}{x_2 - x_1}$
>
> 

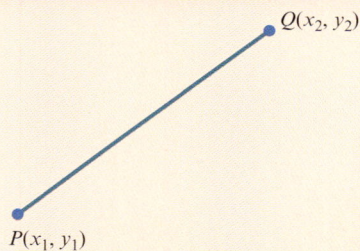

You also learnt the following rules about parallel and perpendicular lines.

| Parallel lines | Perpendicular lines |
|---|---|
|  |  gradient $= m$ <br> gradient $= -\dfrac{1}{m}$ |
| If two lines are parallel, then their gradients are equal. | If a line has gradient $m$, then every line perpendicular to it has gradient $-\dfrac{1}{m}$. |

We can also write the rule for perpendicular lines as:

> ### 🔍 KEY POINT 3.3
>
> If the gradients of two perpendicular lines are $m_1$ and $m_2$, then $m_1 \times m_2 = -1$.

You need to know how to apply the rules for gradients to solve problems involving parallel and perpendicular lines.

### WORKED EXAMPLE 3.4

The coordinates of three points are $A(k - 5, -15)$, $B(10, k)$ and $C(6, -k)$.

Find the two possible values of $k$ if $A$, $B$ and $C$ are collinear.

**Answer**

If $A$, $B$ and $C$ are collinear, then they lie on the same line.

gradient of $AB$ = gradient of $BC$

$$\frac{k - (-15)}{10 - (k - 5)} = \frac{-k - k}{6 - 10}$$     Simplify.

$$\frac{k + 15}{15 - k} = \frac{k}{2}$$     Cross-multiply.

$$2(k + 15) = k(15 - k)$$     Expand brackets.

$$2k + 30 = 15k - k^2$$     Collect terms on one side.

$$k^2 - 13k + 30 = 0$$     Factorise.

$$(k - 3)(k - 10) = 0$$     Solve.

$$k - 3 = 0 \quad \text{or} \quad k - 10 = 0$$

$$\therefore k = 3 \quad \text{or} \quad k = 10$$

### WORKED EXAMPLE 3.5

The vertices of triangle $ABC$ are $A(11, 3)$, $B(2k, k)$ and $C(-1, -11)$.

   **a**   Find the two possible values of $k$ if angle $ABC$ is $90°$.

   **b**   Draw diagrams to show the two possible triangles.

**Answer**

   **a**   Since angle $ABC$ is $90°$, gradient of $AB \times$ gradient of $BC = -1$.

$$\frac{k - 3}{2k - 11} \times \frac{-11 - k}{-1 - 2k} = -1$$     Simplify the second fraction.

$$\frac{k-3}{2k-11} \times \frac{k+11}{2k+1} = -1$$ ⟷ Multiply both sides by $(2k-11)(2k+1)$.

$$(k-3)(k+11) = -(2k-11)(2k+1)$$ ⟷ Expand brackets.

$$k^2 + 8k - 33 = -4k^2 + 20k + 11$$ ⟷ Collect terms on one side.

$$5k^2 - 12k - 44 = 0$$ ⟷ Factorise.

$$(5k - 22)(k + 2) = 0$$

$$5k - 22 = 0 \quad \text{or} \quad k + 2 = 0$$

$$\therefore k = 4.4 \quad \text{or} \quad k = -2$$

**b**  If $k = 4.4$, then $B$ is the point $(8.8, 4.4)$.

If $k = -2$, then $B$ is the point $(-4, 2)$.

The two possible triangles are:

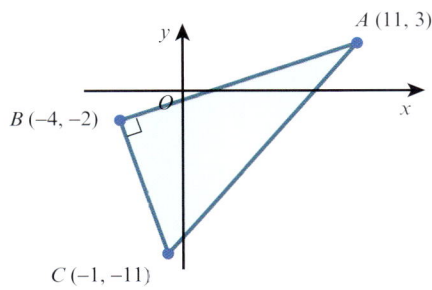

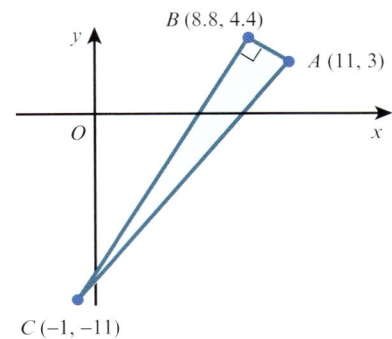

---

### EXERCISE 3B

1   The coordinates of three points are $A(-6, 4)$, $B(4, 6)$ and $C(10, 7)$.

   **a**   Find the gradient of $AB$ and the gradient of $BC$.

   **b**   Use your answer to part **a** to decide whether or not the points $A$, $B$ and $C$ are collinear.

2   The midpoint of the line segment joining $P(-4, 5)$ and $Q(6, 1)$ is $M$.

   The point $R$ has coordinates $(-3, -7)$.

   Show that $RM$ is perpendicular to $PQ$.

3   Two vertices of a rectangle, $ABCD$, are $A(-6, -4)$ and $B(4, -8)$.

   Find the gradient of $CD$ and the gradient of $BC$.

4   The coordinates of three of the vertices of a trapezium, $ABCD$, are $A(3, 5)$, $B(-5, 4)$ and $C(1, -5)$.

   $AD$ is parallel to $BC$ and angle $ADC$ is $90°$.

   Find the coordinates of $D$.

5   The coordinates of three points are $A(5, 8)$, $B(k, 5)$ and $C(-k, 4)$.

   Find the value of $k$ if $A$, $B$ and $C$ are collinear.

6   The vertices of triangle $ABC$ are $A(-9, 2k - 8)$, $B(6, k)$ and $C(k, 12)$.

   Find the two possible values of $k$ if angle $ABC$ is $90°$.

7   $A$ is the point $(0, 8)$ and $B$ is the point $(8, 6)$.

    Find the point $C$ on the $y$-axis such that angle $ABC$ is $90°$.

8   Three points have coordinates $A(7, 4)$, $B(19, 8)$ and $C(k, 2k)$.

    Find the value of the constant $k$ for which:

    **a**   $C$ lies on the line that passes through the points $A$ and $B$

    **b**   angle $CAB$ is $90°$.

9   The line $\dfrac{x}{a} - \dfrac{y}{b} = 1$, where $a$ and $b$ are positive constants, meets the $x$-axis at $P$ and the $y$-axis at $Q$.

    The gradient of the line $PQ$ is $\dfrac{2}{5}$ and the length of the line $PQ$ is $2\sqrt{29}$.

    Find the value of $a$ and the value of $b$.

10   $P$ is the point $(a, a - 2)$ and $Q$ is the point $(4 - 3a, -a)$.

    **a**   Find the gradient of the line $PQ$.

    **b**   Find the gradient of a line perpendicular to $PQ$.

    **c**   Given that the distance $PQ$ is $10\sqrt{5}$, find the two possible values of $a$.

11   The diagram shows a rhombus $ABCD$.

    $M$ is the midpoint of $BD$.

    **a**   Find the coordinates of $M$.

    **b**   Find the value of $a$, the value of $b$ and the value of $c$.

    **c**   Find the perimeter of the rhombus.

    **d**   Find the area of the rhombus.

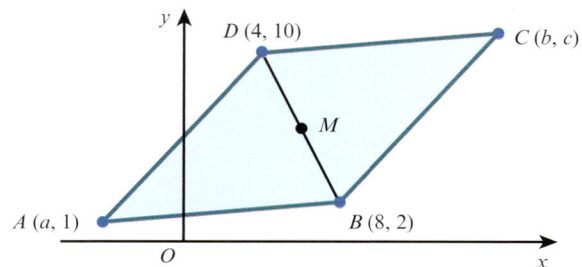

## 3.3 Equations of straight lines

At IGCSE / O Level you learnt the equation of a straight line is:

> ### 🔍 KEY POINT 3.4
>
> $y = mx + c$, where $m$ is the gradient and $c$ is the $y$-intercept, when the line is non-vertical.
> $x = b$ when the line is vertical, where $b$ is the $x$-intercept.

There is an alternative formula that we can use when we know the gradient of a straight line and a point on the line.

Consider a line, with gradient $m$, that passes through the known point $A(x_1, y_1)$ and whose general point is $P(x, y)$.

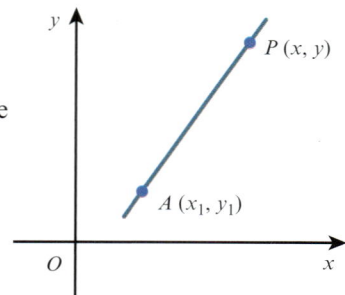

Gradient of $AP = m$, hence $\dfrac{y - y_1}{x - x_1} = m$        Multiply both sides by $(x - x_1)$.

$$y - y_1 = m(x - x_1)$$

---

### 🔍 KEY POINT 3.5

The equation of a straight line, with gradient $m$, that passes through the point $(x_1, y_1)$ is:

$$y - y_1 = m(x - x_1)$$

---

### WORKED EXAMPLE 3.6

Find the equation of the straight line with gradient $-2$ that passes through the point $(4, 1)$.

**Answer**

Using $y - y_1 = m(x - x_1)$ with $m = -2$, $x_1 = 4$ and $y_1 = 1$:

$$y - 1 = -2(x - 4)$$
$$y - 1 = -2x + 8$$
$$2x + y = 9$$

---

### WORKED EXAMPLE 3.7

Find the equation of the straight line passing through the points $(-4, 3)$ and $(6, -2)$.

**Answer**

$(-4, 3)$      $(6, -2)$

↑ ↑       ↑ ↑                Decide which values to use for $x_1$, $y_1$, $x_2$, $y_2$.

$(x_1, y_1)$    $(x_2, y_2)$

Gradient $= m = \dfrac{y_2 - y_1}{x_2 - x_1} = \dfrac{(-2) - 3}{6 - (-4)} = -\dfrac{1}{2}$

Using $y - y_1 = m(x - x_1)$ with $m = -\dfrac{1}{2}$, $x_1 = -4$ and $y_1 = 3$:

$$y - 3 = -\dfrac{1}{2}(x + 4)$$
$$2y - 6 = -x - 4$$
$$x + 2y = 2$$

**WORKED EXAMPLE 3.8**

Find the equation of the perpendicular bisector of the line segment joining $A(-5, 1)$ and $B(7, -2)$.

**Answer**

Gradient of $AB = \dfrac{-2-1}{7-(-5)} = \dfrac{-3}{12} = -\dfrac{1}{4}$ 〜 Use gradient $= \dfrac{y_2 - y_1}{x_2 - x_1}$.

Gradient of the perpendicular $= 4$ 〜 Use $m_1 \times m_2 = -1$.

Midpoint of $AB = \left(\dfrac{-5+7}{2}, \dfrac{1+(-2)}{2}\right) = \left(1, -\dfrac{1}{2}\right)$ 〜 Use midpoint $= \left(\dfrac{x_1 + x_2}{2}, \dfrac{y_1 + y_2}{2}\right)$.

$\therefore$ The perpendicular bisector is the line with gradient 4 passing through the point $\left(1, -\dfrac{1}{2}\right)$.

Using $y - y_1 = m(x - x_1)$ with $x_1 = 1$, $y_1 = -\dfrac{1}{2}$ and $m = 4$:

$y + \dfrac{1}{2} = 4(x - 1)$ 〜 Expand brackets and simplify.

$y = 4x - 4\frac{1}{2}$ 〜 Multiply both sides by 2.

$2y = 8x - 9$

**EXERCISE 3C**

80

1 Find the equation of the line with:

a gradient 2 passing through the point (4, 9)

b gradient −3 passing through the point (1, −4)

c gradient $-\dfrac{2}{3}$ passing through the point (−4, 3).

2 Find the equation of the line passing through each pair of points.

a $(1, 0)$ and $(5, 6)$

b $(3, -5)$ and $(-2, 4)$

c $(3, -1)$ and $(-3, -5)$

3 Find the equation of the line:

a parallel to the line $y = 3x - 5$, passing through the point (1, 7)

b parallel to the line $x + 2y = 6$, passing through the point (4, −6)

c perpendicular to the line $y = 2x - 3$, passing through the point (6, 1)

d perpendicular to the line $2x - 3y = 12$, passing through the point (8, −3).

4 Find the equation of the perpendicular bisector of the line segment joining the points:

a $(5, 2)$ and $(-3, 6)$

b $(-2, -5)$ and $(8, 1)$

c $(-2, -7)$ and $(5, -4)$.

5   The line $l_1$ passes through the points $P(-10, 1)$ and $Q(2, 10)$. The line $l_2$ is parallel to $l_1$ and passes through the point $(4, -1)$. The point $R$ lies on $l_2$, such that $QR$ is perpendicular to $l_2$. Find the coordinates of $R$.

6   $P$ is the point $(-4, 2)$ and $Q$ is the point $(5, -4)$.

   A line, $l$, is drawn through $P$ and perpendicular to $PQ$ to meet the $y$-axis at the point $R$.

   **a**   Find the equation of the line $l$.

   **b**   Find the coordinates of the point $R$.

   **c**   Find the area of triangle $PQR$.

7   The line $l_1$ has equation $3x - 2y = 12$ and the line $l_2$ has equation $y = 15 - 2x$. The lines $l_1$ and $l_2$ intersect at the point $A$.

   **a**   Find the coordinates of $A$.

   **b**   Find the equation of the line through $A$ that is perpendicular to the line $l_1$.

8   The perpendicular bisector of the line joining $A(-10, 5)$ and $B(-2, -1)$ intersects the $x$-axis at $P$ and the $y$-axis at $Q$.

   **a**   Find the equation of the line $PQ$.

   **b**   Find the coordinates of $P$ and $Q$.

   **c**   Find the length of $PQ$.

9   The line $l_1$ has equation $2x + 5y = 10$.

   The line $l_2$ passes through the point $A(-9, -6)$ and is perpendicular to the line $l_1$.

   **a**   Find the equation of the line $l_2$.

   **b**   Given that the lines $l_1$ and $l_2$ intersect at the point $B$, find the area of triangle $ABO$, where $O$ is the origin.

10  The diagram shows the points $E$, $F$ and $G$ lying on the line $x + 2y = 16$. The point $G$ lies on the $x$-axis and $EF = FG$. The line $FH$ is perpendicular to $EG$. Find the coordinates of $E$ and $F$.

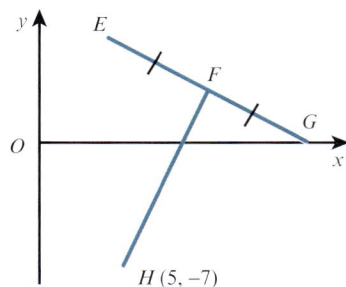

11  The coordinates of three points are $A(-4, -1)$, $B(8, -9)$ and $C(k, 7)$. $M$ is the midpoint of $AB$ and $MC$ is perpendicular to $AB$. Find the value of $k$.

12  The point $P$ is the reflection of the point $(-2, 10)$ in the line $4x - 3y = 12$.

   Find the coordinates of $P$.

13 The coordinates of triangle $ABC$ are $A(-7, 3)$, $B(3, -7)$ and $C(8, 8)$.
   $P$ is the foot of the perpendicular from $B$ to $AC$.

   a   Find the equation of the line $BP$.

   b   Find the coordinates of $P$.

   c   Find the lengths of $AC$ and $BP$.

   d   Use your answers to part **c** to find the area of triangle $ABC$.

14 The coordinates of triangle $PQR$ are $P(1, 1)$, $Q(1, 8)$ and $R(6, 6)$.

   a   Find the equation of the perpendicular bisectors of:

   i   $PQ$              ii   $PR$

   b   Find the coordinates of the point that is equidistant from $P$, $Q$ and $R$.

**PS**   15 The equations of two of the sides of triangle $ABC$ are $x + 2y = 8$ and $2x + y = 1$.
   Given that $A$ is the point $(2, -3)$ and that angle $ABC = 90°$, find:

   a   the equation of the third side

   b   the coordinates of the point $B$.

**PS**   16 Find two straight lines whose $x$-intercepts differ by 7, whose $y$-intercepts differ by 5 and whose gradients differ by 2.

   Is your solution unique? Investigate further.

   [This question is based upon *Straight line pairs* on the Underground Mathematics website.]

WEB LINK

Try the following resources on the Underground Mathematics website:
• *Lots of lines!*
• *Straight lines*
• *Simultaneous squares*
• *Straight line pairs*.

## 3.4 The equation of a circle

In this section you will learn about the equation of a circle. A circle is defined as the locus of all the points in a plane that are a fixed distance (the radius) from a given point (the centre).

### EXPLORE 3.2

1   Use graphing software to draw each of the following circles. From your graphs find the coordinates of the centre and the radius of each circle, and copy and complete the following table.

|   | Equation of circle | Centre | Radius |
|---|---|---|---|
| a | $x^2 + y^2 = 25$ | | |
| b | $(x-2)^2 + (y-1)^2 = 9$ | | |
| c | $(x+3)^2 + (y+5)^2 = 16$ | | |
| d | $(x-8)^2 + (y+6)^2 = 49$ | | |
| e | $x^2 + (y+4)^2 = 4$ | | |
| f | $(x+6)^2 + y^2 = 64$ | | |

2   Discuss your results with your classmates and explain how you can find the coordinates of the centre of a circle and the radius of a circle just by looking at the equation of the circle.

To find the equation of a circle, we let $P(x, y)$ be any point on the circumference of a circle with centre $C(a, b)$ and radius $r$.

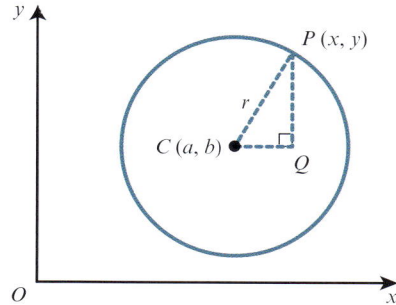

Using Pythagoras' theorem on triangle $CQP$ gives $CQ^2 + PQ^2 = r^2$.

Substituting $CQ = x - a$ and $PQ = y - b$ into $CQ^2 + PQ^2 = r^2$ gives:

$(x - a)^2 + (y - b)^2 = r^2$

---

**KEY POINT 3.6**

The equation of a circle with centre $(a, b)$ and radius $r$ can be written in **completed square form** as:

$(x - a)^2 + (y - b)^2 = r^2$

---

83

**EXPLORE 3.3**

The completed square form for the equation of a circle with centre $(a, b)$ and radius $r$ is $(x - a)^2 + (y - b)^2 = r^2$. Use graphing software to investigate the effects of:

**a** increasing the value of $a$          **b** decreasing the value of $a$

**c** increasing the value of $b$          **d** decreasing the value of $b$.

---

**WORKED EXAMPLE 3.9**

Write down the coordinates of the centre and the radius of each of these circles.

   **a** $x^2 + y^2 = 4$

   **b** $(x - 2)^2 + (y - 4)^2 = 100$

   **c** $(x + 1)^2 + (y - 8)^2 = 12$

**Answer**

   **a**  Centre $= (0, 0)$,  radius $= \sqrt{4} = 2$

   **b**  Centre $= (2, 4)$,  radius $= \sqrt{100} = 10$

   **c**  Centre $= (-1, 8)$,  radius $= \sqrt{12} = 2\sqrt{3}$

**DID YOU KNOW?**

In the 17th century, the French philosopher and mathematician René Descartes developed the idea of using equations to represent geometrical shapes. The Cartesian coordinate system is named after this famous mathematician.

Find the equation of the circle with centre $(-4, 3)$ and radius 6.

**Answer**

Equation of circle is $(x - a)^2 + (y - b)^2 = r^2$, where $a = -4$, $b = 3$ and $r = 6$.

$$(x - (-4))^2 + (y - 3)^2 = 6^2$$
$$(x + 4)^2 + (y - 3)^2 = 36$$

**WORKED EXAMPLE 3.11**

$A$ is the point $(3, 0)$ and $B$ is the point $(7, -4)$.

Find the equation of the circle that has $AB$ as a diameter.

**Answer**

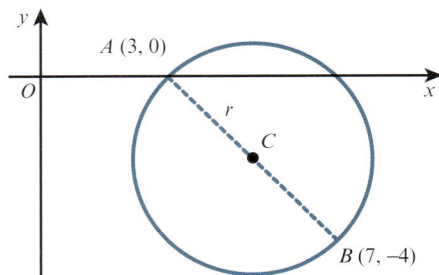

The centre of the circle, $C$, is the midpoint of $AB$.

$$C = \left( \frac{3 + 7}{2}, \frac{0 + (-4)}{2} \right) = (5, -2)$$

Radius of circle, $r$, is equal to $AC$.

$$r = \sqrt{(5 - 3)^2 + (-2 - 0)^2} = \sqrt{8}$$

Equation of circle is $(x - a)^2 + (y - b)^2 = r^2$, where $a = 5$, $b = -2$ and $r = \sqrt{8}$.

$$(x - 5)^2 + (y + 2)^2 = \left( \sqrt{8} \right)^2$$
$$(x - 5)^2 + (y + 2)^2 = 8$$

Expanding the equation $(x - a)^2 + (y - b)^2 = r^2$ gives:

$$x^2 - 2ax + a^2 + y^2 - 2by + b^2 = r^2$$

Rearranging gives:

$$x^2 + y^2 - 2ax - 2by + (a^2 + b^2 - r^2) = 0$$

When we write the equation of a circle in this form, we can note some important characteristics of the equation of a circle. For example:

- the coefficients of $x^2$ and $y^2$ are equal
- there is no $xy$ term.

We often write the expanded form of a circle as:

> ### KEY POINT 3.7
>
> $$x^2 + y^2 + 2gx + 2fy + c = 0$$
>
> where $(-g, -f)$ is the centre and $\sqrt{g^2 + f^2 - c}$ is the radius.
>
> This is the equation of a circle in expanded **general form**.

> **TIP**
>
> You should not try to memorise the formulae for the centre and radius of a circle in this form, but rather work them out if needed, as shown in Worked example 3.12.

### WORKED EXAMPLE 3.12

Find the centre and the radius of the circle $x^2 + y^2 + 10x - 8y - 40 = 0$.

**Answer**

We answer this question by first completing the square.

$$x^2 + 10x + y^2 - 8y - 40 = 0 \qquad \text{Complete the square.}$$

$$(x + 5)^2 - 5^2 + (y - 4)^2 - 4^2 - 40 = 0 \qquad \text{Collect constant terms together.}$$

$$(x + 5)^2 + (y - 4)^2 = 81 \qquad \text{Compare with } (x - a)^2 + (y - b)^2 = r^2.$$

$$a = -5 \qquad b = 4 \qquad r^2 = 81$$

Centre $= (-5, 4)$ and radius $= 9$.

It is useful to remember the three following right angle facts for circles.

| | | |
|---|---|---|
| The angle in a semicircle is a right angle. | The perpendicular from the centre of a circle to a chord bisects the chord. | The tangent to a circle at a point is perpendicular to the radius at that point. |

From these statements we can conclude that:

- If triangle $ABC$ is right angled at $B$, then the points $A$, $B$ and $C$ lie on the circumference of a circle with $AC$ as diameter.

- The perpendicular bisector of a chord passes through the centre of the circle.

- If a radius and a line at a point, $P$, on the circumference are at right angles, then the line must be a tangent to the curve.

85

**WORKED EXAMPLE 3.13**

A circle passes through the points $P(-1, 4)$, $Q(1, 6)$ and $R(5, 4)$.

Find the equation of the circle.

**Answer**

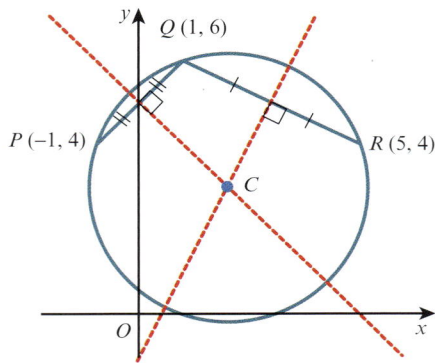

The centre of the circle lies on the perpendicular bisector of $PQ$ and on the perpendicular bisector of $QR$.

Midpoint of $PQ = \left( \dfrac{-1+1}{2}, \dfrac{4+6}{2} \right) = (0, 5)$

Gradient of $PQ = \dfrac{6-4}{1-(-1)} = 1$

Gradient of perpendicular bisector of $PQ = -1$

Equation of perpendicular bisector of $PQ$ is:

$$(y - 5) = -1(x - 0)$$
$$y = -x + 5 \quad\text{------------ (1)}$$

Midpoint of $QR = \left( \dfrac{1+5}{2}, \dfrac{6+4}{2} \right) = (3, 5)$

Gradient of $QR = \dfrac{4-6}{5-1} = -\dfrac{1}{2}$

Gradient of perpendicular bisector of $QR = 2$

Equation of perpendicular bisector of $QR$ is:

$$(y - 5) = 2(x - 3)$$
$$y = 2x - 1 \quad\text{----------- (2)}$$

Solving equations (1) and (2) gives:

$$x = 2, \; y = 3$$

Centre of circle $= (2, 3)$

Radius $= CR = \sqrt{(5-2)^2 + (4-3)^2} = \sqrt{10}$

Hence, the equation of the circle is $(x - 2)^2 + (y - 3)^2 = 10$.

**Alternative method:**

The equation of the circle is $(x - a)^2 + (y - b)^2 = r^2$.

The points $(-1, 4)$, $(1, 6)$ and $(5, 4)$ lie on the circle, so substituting gives:

$(-1 - a)^2 + (4 - b)^2 = r^2$

$a^2 + 2a + b^2 - 8b + 17 = r^2$ ------(1)

and similar for the other two points, giving equations (2) and (3).

Then subtracting (1)−(3) and (2)−(3) gives two simultaneous equations for $a$ and $b$, which can then be solved.

Finally, substituting into (1) gives $r^2$.

## EXERCISE 3D

1   Find the centre and the radius of each of the following circles.

   a   $x^2 + y^2 = 16$

   b   $2x^2 + 2y^2 = 9$

   c   $x^2 + (y - 2)^2 = 25$

   d   $(x - 5)^2 + (y + 3)^2 = 4$

   e   $(x + 7)^2 + y^2 = 18$

   f   $2(x - 3)^2 + 2(y + 4)^2 = 45$

   g   $x^2 + y^2 - 8x + 20y + 110 = 0$

   h   $2x^2 + 2y^2 - 14x - 10y - 163 = 0$

2   Find the equation of each of the following circles.

   a   centre $(0, 0)$, radius 8

   b   centre $(5, -2)$, radius 4

   c   centre $(-1, 3)$, radius $\sqrt{7}$

   d   centre $\left( \dfrac{1}{2}, -\dfrac{3}{2} \right)$, radius $\dfrac{5}{2}$

3   Find the equation of the circle with centre $(2, 5)$ passing through the point $(6, 8)$.

4   A diameter of a circle has its end points at $A(-6, 8)$ and $B(2, -4)$.

   Find the equation of the circle.

5   Sketch the circle $(x - 3)^2 + (y + 2)^2 = 9$.

6   Find the equation of the circle that touches the $x$-axis and whose centre is $(6, -5)$.

7   The points $P(1, -2)$ and $Q(7, 1)$ lie on the circumference of a circle.

   Show that the centre of the circle lies on the line $4x + 2y = 15$.

8   A circle passes through the points $(3, 2)$ and $(7, 2)$ and has radius $2\sqrt{2}$.

   Find the two possible equations for this circle.

9   A circle passes through the points $O(0, 0)$, $A(8, 4)$ and $B(6, 6)$.

   Show that $OA$ is a diameter of the circle and find the equation of this circle.

10  Show that $x^2 + y^2 - 6x + 2y = 6$ can be written in the form $(x - a)^2 + (y - b)^2 = r^2$, where $a$, $b$ and $r$ are constants to be found. Hence, write down the coordinates of the centre of the circle and also the radius of the circle.

87

11 The equation of a circle is $(x - 3)^2 + (y + 2)^2 = 25$. Show that the point $A(6, -6)$ lies on the circle and find the equation of the tangent to the circle at the point $A$.

12 The line $2x + 5y = 20$ cuts the $x$-axis at $A$ and the $y$-axis at $B$. The point $C$ is the midpoint of the line $AB$. Find the equation of the circle that has centre $C$ and that passes through the points $A$ and $B$. Show that this circle also passes through the point $O(0, 0)$.

13 The points $P(-5, 6)$, $Q(-3, 8)$ and $R(3, 2)$ are joined to form a triangle.

    a  Show that angle $PQR$ is a right angle.

    b  Find the equation of the circle that passes through the points $P$, $Q$ and $R$.

14 Find the equation of the circle that passes through the points $(7, 3)$ and $(11, -1)$ and has its centre lying on the line $2x + y = 7$.

15 A circle passes through the points $O(0, 0)$, $P(3, 9)$ and $Q(11, 11)$.

    Find the equation of the circle.

**WEB LINK**

Try the following resources on the Underground Mathematics website:
• *Olympic rings*
• *Teddy bear.*

16 A circle has radius 10 units and passes through the point $(5, -16)$. The $x$-axis is a tangent to the circle. Find the possible equations of the circle.

**(PS)** 17 a  The design shown is made from four green circles and one orange circle.

       i  The radius of each green circle is 1 unit. Find the radius of the orange circle.

      ii  Use graphing software to draw the design.

    b  The design in part **a** is extended, as shown.

       i  The radius of each green circle is 1 unit. Find the radius of the blue circle.

      ii  Use graphing software to draw this extended design.

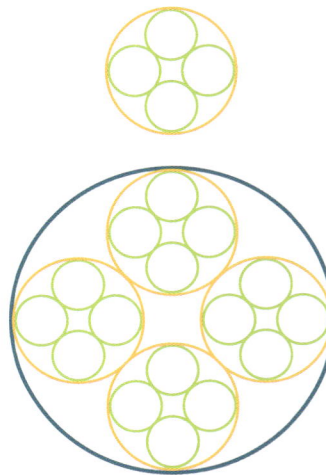

## 3.5 Problems involving intersections of lines and circles

In Chapter 1 you learnt that the points of intersection of a line and a curve can be found by solving their equations simultaneously. You also learnt that if the resulting equation is of the form $ax^2 + bx + c = 0$, then $b^2 - 4ac$ gives information about the line and the curve.

| $b^2 - 4ac$ | Nature of roots | Line and parabola |
|---|---|---|
| $> 0$ | two distinct real roots | two distinct points of intersection |
| $= 0$ | two equal real roots | one point of intersection (line is a tangent) |
| $< 0$ | no real roots | no points of intersection |

**TIP**

We can also describe an equation that has 'two equal real roots' as having 'one repeated (real) root'.

In this section you will solve problems involving the intersection of lines and circles.

**WORKED EXAMPLE 3.14**

The line $x = 3y + 10$ intersects the circle $x^2 + y^2 = 20$ at the points $A$ and $B$.

a   Find the coordinates of the points $A$ and $B$.

b   Find the equation of the perpendicular bisector of $AB$ and show that it passes through the centre of the circle.

c   The perpendicular bisector of $AB$ intersects the circle at the points $P$ and $Q$.

Find the exact coordinates of $P$ and $Q$.

**Answer**

a   $\qquad x^2 + y^2 = 20$ ............................ Substitute $3y + 10$ for $x$.

$\qquad (3y + 10)^2 + y^2 = 20$ ............................ Expand and simplify.

$\qquad y^2 + 6y + 8 = 0$ ............................ Factorise.

$\qquad (y + 2)(y + 4) = 0$

$\qquad y = -2 \quad$ or $\quad y = -4$

When $y = -2$, $x = 4$ and when $y = -4$, $x = -2$.

$A$ and $B$ are the points $(-2, -4)$ and $(4, -2)$.

b   Gradient of $AB = \dfrac{-2 - (-4)}{4 - (-2)} = \dfrac{1}{3}$

So the gradient of the perpendicular bisector $= -3$.

Midpoint of $AB = \left( \dfrac{-2 + 4}{2}, \dfrac{-4 + (-2)}{2} \right) = (1, -3)$

$\qquad y - y_1 = m(x - x_1)$ ............................ Use $m = -3$, $x_1 = 1$ and $y_1 = -3$.

$\qquad y - (-3) = -3(x - 1)$

Perpendicular bisector is $y = -3x$.

When $x = 0$, $y = -3(0) = 0$.

Hence, the perpendicular bisector of $AB$ passes through the point $(0, 0)$, the centre of the circle $x^2 + y^2 = 20$.

c   $x^2 + y^2 = 20$ ............................ Substitute $-3x$ for $y$.

$\qquad 10x^2 = 20$

$\qquad x = \pm\sqrt{2}$

When $x = -\sqrt{2}$, $y = 3\sqrt{2}$ and when $x = \sqrt{2}$, $y = -3\sqrt{2}$.

$P$ and $Q$ are the points $\left(-\sqrt{2}, 3\sqrt{2}\right)$ and $\left(\sqrt{2}, -3\sqrt{2}\right)$, respectively.

89

**WORKED EXAMPLE 3.15**

Show that the line $y = x - 13$ is a tangent to the circle $x^2 + y^2 - 8x + 6y + 7 = 0$.

**Answer**

$$x^2 + y^2 - 8x + 6y + 7 = 0 \qquad \text{Substitute } x - 13 \text{ for } y.$$

$$x^2 + (x - 13)^2 - 8x + 6(x - 13) + 7 = 0 \qquad \text{Expand and simplify.}$$

$$x^2 - 14x + 49 = 0 \qquad \text{Factorise.}$$

$$(x - 7)(x - 7) = 0$$

$$x = 7 \text{ or } x = 7$$

The equation has one repeated root, hence $y = x - 13$ is a tangent.

**EXERCISE 3E**

1  Find the points of intersection of the line $y = x - 3$ and the circle $(x - 3)^2 + (y + 2)^2 = 20$.

2  The line $2x - y + 3 = 0$ intersects the circle $x^2 + y^2 - 4x + 6y - 12 = 0$ at two points, $D$ and $E$.

   Find the length of $DE$.

3  Show that the line $3x + y = 6$ is a tangent to the circle $x^2 + y^2 + 4x + 16y + 28 = 0$.

4  Find the set of values of $m$ for which the line $y = mx + 1$ intersects the circle $(x - 7)^2 + (y - 5)^2 = 20$ at two distinct points.

5  The line $2y - x = 12$ intersects the circle $x^2 + y^2 - 10x - 12y + 36 = 0$ at the points $A$ and $B$.

   a  Find the coordinates of the points $A$ and $B$.

   b  Find the equation of the perpendicular bisector of $AB$.

   c  The perpendicular bisector of $AB$ intersects the circle at the points $P$ and $Q$.

      Find the exact coordinates of $P$ and $Q$.

   d  Find the exact area of quadrilateral $APBQ$.

**PS**  6  Show that the circles $x^2 + y^2 = 25$ and $x^2 + y^2 - 24x - 18y + 125 = 0$ touch each other.

   Find the coordinates of the point where they touch.

   [This question is taken from *Can we show that these two circles touch?* on the Underground Mathematics website.]

**PS**  7  Two circles have the following properties:

   ● the $x$-axis is a common tangent to the circles

   ● the point $(8, 2)$ lies on both circles

   ● the centre of each circle lies on the line $x + 2y = 22$.

   a  Find the equation of each circle.

   b  Prove that the line $4x + 3y = 88$ is a common tangent to these circles.

   [Inspired by *Can we find the two circles that satisfy these three conditions?* on the Underground Mathematics website.]

# Checklist of learning and understanding

**Midpoint, gradient and length of line segment**

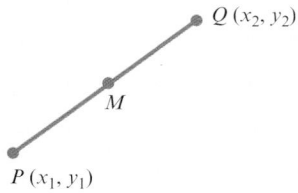

- Midpoint, $M$, of $PQ$ is $\left( \dfrac{x_1 + x_2}{2}, \dfrac{y_1 + y_2}{2} \right)$.

- Gradient of $PQ$ is $\dfrac{y_2 - y_1}{x_2 - x_1}$.

- Length of segment $PQ$ is $\sqrt{(x_2 - x_1)^2 + (y_2 - y_1)^2}$

**Parallel and perpendicular lines**

- If the gradients of two parallel lines are $m_1$ and $m_2$, then $m_1 = m_2$.
- If the gradients of two perpendicular lines are $m_1$ and $m_2$, then $m_1 \times m_2 = -1$.

**The equation of a straight line is:**

- $y - y_1 = m(x - x_1)$, where $m$ is the gradient and $(x_1, y_1)$ is a point on the line.

**The equation of a circle is:**

- $(x - a)^2 + (y - b)^2 = r^2$, where $(a, b)$ is the centre and $r$ is the radius.
- $x^2 + y^2 + 2gx + 2fy + c = 0$, where $(-g, -f)$ is the centre and $\sqrt{g^2 + f^2 - c}$ is the radius.

1   A line has equation $2x + y = 20$ and a curve has equation $y = a + \dfrac{18}{x-3}$, where $a$ is a constant.

Find the set of values of $a$ for which the line does not intersect the curve. [4]

2
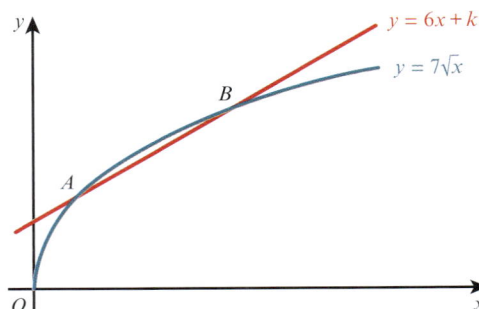

The diagram shows the curve $y = 7\sqrt{x}$ and the line $y = 6x + k$, where $k$ is a constant.

The curve and the line intersect at the points $A$ and $B$.

i   For the case where $k = 2$, find the $x$-coordinates of $A$ and $B$. [4]

ii   Find the value of $k$ for which $y = 6x + k$ is a tangent to the curve $y = 7\sqrt{x}$. [2]

*Cambridge International AS & A Level Mathematics 9709 Paper 11 Q5 June 2012*

3   $A$ is the point $(a, 3)$ and $B$ is the point $(4, b)$.

The length of the line segment $AB$ is $4\sqrt{5}$ units and the gradient is $-\dfrac{1}{2}$.

Find the possible values of $a$ and $b$. [6]

4   The curve $y = 3\sqrt{x - 2}$ and the line $3x - 4y + 3 = 0$ intersect at the points $P$ and $Q$.

Find the length of $PQ$. [6]

5   The line $ax - 2y = 30$ passes through the points $A(10, 10)$ and $B(b, 10b)$, where $a$ and $b$ are constants.

a   Find the values of $a$ and $b$. [3]

b   Find the coordinates of the midpoint of $AB$. [1]

c   Find the equation of the perpendicular bisector of the line $AB$. [3]

6   The line with gradient $-2$ passing through the point $P(3t, 2t)$ intersects the $x$-axis at $A$ and the $y$-axis at $B$.

i   Find the area of triangle $AOB$ in terms of $t$. [3]

The line through $P$ perpendicular to $AB$ intersects the $x$-axis at $C$.

ii   Show that the mid-point of $PC$ lies on the line $y = x$. [4]

*Cambridge International AS & A Level Mathematics 9709 Paper 11 Q6 June 2015*

7   The point $P$ is the reflection of the point $(-7, 5)$ in the line $5x - 3y = 18$.

Find the coordinates of $P$. You must show all your working. [7]

8   The curve $y = x + 2 - \dfrac{4}{x}$ and the line $x - 2y + 6 = 0$ intersect at the points $A$ and $B$.

    **a**   Find the coordinates of these two points. [4]

    **b**   Find the perpendicular bisector of the line $AB$. [4]

9   The line $y = mx + 1$ intersects the circle $x^2 + y^2 - 19x - 51 = 0$ at the point $P(5, 11)$.

    **a**   Find the coordinates of the point $Q$ where the line meets the curve again. [4]

    **b**   Find the equation of the perpendicular bisector of the line $PQ$. [3]

    **c**   Find the $x$-coordinates of the points where this perpendicular bisector intersects the circle.

       Give your answers in exact form. [4]

10

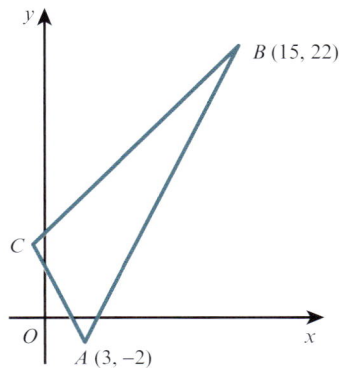

The diagram shows a triangle $ABC$ in which $A$ is $(3, -2)$ and $B$ is $(15, 22)$. The gradients of $AB$, $AC$ and $BC$ are $2m$, $-2m$ and $m$ respectively, where $m$ is a positive constant.

    **i**   Find the gradient of $AB$ and deduce the value of $m$. [2]

    **ii**   Find the coordinates of $C$. [4]

The perpendicular bisector of $AB$ meets $BC$ at $D$.

    **iii**   Find the coordinates of $D$. [4]

*Cambridge International AS & A Level Mathematics 9709 Paper 11 Q8 June 2010*

11   The point $A$ has coordinates $(-1, 6)$ and the point $B$ has coordinates $(7, 2)$.

    **i**   Find the equation of the perpendicular bisector of $AB$, giving your answer in the form $y = mx + c$. [4]

    **ii**   A point $C$ on the perpendicular bisector has coordinates $(p, q)$. The distance $OC$ is 2 units, where $O$ is the origin. Write down two equations involving $p$ and $q$ and hence find the coordinates of the possible positions of $C$. [5]

*Cambridge International AS & A Level Mathematics 9709 Paper 11 Q7 November 2013*

**12** The coordinates of $A$ are $(-3, 2)$ and the coordinates of $C$ are $(5, 6)$.

The mid-point of $AC$ is $M$ and the perpendicular bisector of $AC$ cuts the $x$-axis at $B$.

   **i**   Find the equation of $MB$ and the coordinates of $B$.   [5]

   **ii**  Show that $AB$ is perpendicular to $BC$.   [2]

   **iii** Given that $ABCD$ is a square, find the coordinates of $D$ and the length of $AD$.   [2]

*Cambridge International AS & A Level Mathematics 9709 Paper 11 Q9 June 2012*

**13** The points $A(1, -2)$ and $B(5, 4)$ lie on a circle with centre $C(6, p)$.

   **a** Find the equation of the perpendicular bisector of the line segment $AB$.   [4]

   **b** Use your answer to part **a** to find the value of $p$.   [1]

   **c** Find the equation of the circle.   [4]

**14**

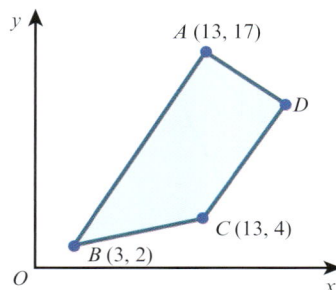

$ABCD$ is a trapezium with $AB$ parallel to $DC$ and angle $BAD = 90°$.

   **a** Calculate the coordinates of $D$.   [7]

   **b** Calculate the area of trapezium $ABCD$.   [2]

**15** The equation of a curve is $xy = 12$ and the equation of a line is $3x + y = k$, where $k$ is a constant.

   **a** In the case where $k = 20$, the line intersects the curve at the points $A$ and $B$.

   Find the midpoint of the line $AB$.   [4]

   **b** Find the set of values of $k$ for which the line $3x + y = k$ intersects the curve at two distinct points.   [4]

**16** $A$ is the point $(-3, 6)$ and $B$ is the point $(9, -10)$.

   **a** Find the equation of the line through $A$ and $B$.   [3]

   **b** Show that the perpendicular bisector of the line $AB$ is $3x - 4y = 17$.   [3]

   **c** A circle passes through $A$ and $B$ and has its centre on the line $x = 15$. Find the equation of this circle.   [4]

**17** The equation of a circle is $x^2 + y^2 - 8x + 4y + 4 = 0$.

   **a** Find the radius of the circle and the coordinates of its centre.   [4]

   **b** Find the $x$-coordinates of the points where the circle crosses the $x$-axis, giving your answers in exact form.   [4]

   **c** Show that the point $A(6, 2\sqrt{3} - 2)$ lies on the circle.   [2]

   **d** Show that the equation of the tangent to the circle at $A$ is $\sqrt{3}x + 3y = 12\sqrt{3} - 6$.   [4]

1   Solve the equation $\dfrac{4}{x^4} + 18 = \dfrac{17}{x^2}$.   [4]

2

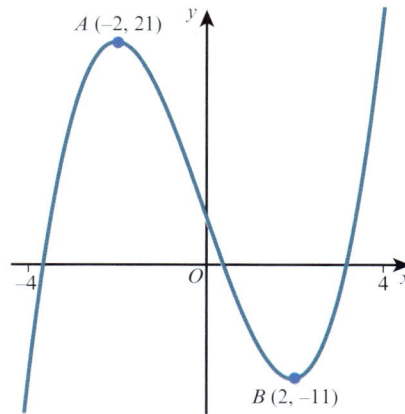

The diagram shows the graph of $y = f(x)$ for $-4 \leqslant x \leqslant 4$.

Sketch on separate diagrams, showing the coordinates of any turning points, the graphs of:

a   $y = f(x) + 5$   [2]

b   $y = -2f(x)$   [2]

3   The graph of $f(x) = ax + b$ is reflected in the $y$-axis and then translated by the vector $\begin{pmatrix} 0 \\ 3 \end{pmatrix}$.

The resulting function is $g(x) = 1 - 5x$. Find the value of $a$ and the value of $b$.   [4]

4   The graph of $y = (x + 1)^2$ is transformed by the composition of two transformations to the graph of $y = 2(x - 4)^2$. Find these two transformations.   [4]

5   The graph of $y = x^2 + 1$ is transformed by applying a reflection in the $x$-axis followed by a translation of $\begin{pmatrix} 3 \\ 2 \end{pmatrix}$. Find the equation of the resulting graph in the form $y = ax^2 + bx + c$.   [4]

6

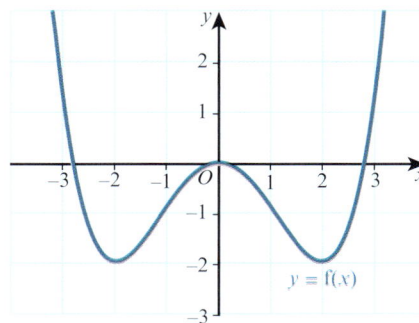

The diagram shows the graph of $y = f(x)$ for $-3 \leqslant x \leqslant 3$.

Sketch the graph of $y = 2 - f(x)$.   [4]

95

7 The function f is such that $f(x) = x^2 - 5x + 5$ for $x \in \mathbb{R}$.

    a Find the set of values of $x$ for which $f(x) \leqslant x$.    [3]

    b The line $y = mx - 11$ is a tangent to the curve $y = f(x)$.

      Find the two possible values of $m$.    [3]

8 The line $x + ky + k^2 = 0$, where $k$ is a constant, is a tangent to the curve $y^2 = 4x$ at the point $P$.

    Find, in terms of $k$, the coordinates of $P$.    [6]

9 $A$ is the point $(4, -6)$ and $B$ is the point $(12, 10)$. The perpendicular bisector of $AB$ intersects the $x$-axis at $C$ and the $y$-axis at $D$. Find the length of $CD$.    [6]

10 The points $A$, $B$ and $C$ have coordinates $A(2, 8)$, $B(9, 7)$ and $C(k, k - 2)$.

    a Given that $AB = BC$, show that a possible value of $k$ is 4 and find the other possible value of $k$.    [3]

    b For the case where $k = 4$, find the equation of the line that bisects angle $ABC$.    [4]

11 A curve has equation $xy = 12 + x$ and a line has equation $y = kx - 9$, where $k$ is a constant.

    a In the case where $k = 2$, find the coordinates of the points of intersection of the curve and the line.    [3]

    b Find the set of values of $k$ for which the line does not intersect the curve.    [4]

12 The function f is such that $f(x) = 2x - 3$ for $x \geqslant k$, where $k$ is a constant.

    The function g is such that $g(x) = x^2 - 4$ for $x \geqslant -4$.

    a Find the smallest value of $k$ for which the composite function gf can be formed.    [3]

    b Solve the inequality $gf(x) > 45$.    [4]

13 The functions f and g are defined by

$$f(x) = \frac{4}{x} - 2 \quad \text{for } x > 0,$$

$$g(x) = \frac{4}{5x + 2} \quad \text{for } x \geqslant 0.$$

    i Find and simplify an expression for $fg(x)$ and state the range of fg.    [3]

    ii Find an expression for $g^{-1}(x)$ and find the domain of $g^{-1}$.    [5]

*Cambridge International AS & A Level Mathematics 9709 Paper 11 Q8 November 2016*

14 The equation $x^2 + bx + c = 0$ has roots $-2$ and $7$.

    a Find the value of $b$ and the value of $c$.    [2]

    b Using these values of $b$ and $c$, find:

      i the coordinates of the vertex of the curve $y = x^2 + bx + c$    [3]

      ii the set of values of $x$ for which $x^2 + bx + c < 10$.    [3]

15  The line $L_1$ passes through the points $A(-6, 10)$ and $B(6, 2)$. The line $L_2$ is perpendicular to $L_1$ and passes through the point $C(-7, 2)$.

   a  Find the equation of the line $L_2$.                                                                                  [4]

   b  Find the coordinates of the point of intersection of lines $L_1$ and $L_2$.                                           [4]

16  A curve has equation $y = 12x - x^2$.

   a  Express $12x - x^2$ in the form $a - (x + b)^2$, where $a$ and $b$ are constants to be determined.                    [3]

   b  State the maximum value of $12x - x^2$.                                                                               [1]

   The function g is defined as  g: $x \mapsto 12x - x^2$, for $x \geq 6$.

   c  State the domain and range of $g^{-1}$.                                                                              [2]

   d  Find $g^{-1}(x)$.                                                                                                    [3]

17  a  Express $3x^2 + 12x - 1$ in the form $a(x + b)^2 + c$, where $a$, $b$ and $c$ are constants.                         [3]

   b  Write down the coordinates of the vertex of the curve  $y = 3x^2 + 12x - 1$.                                         [2]

   c  Find the set of values of $k$ for which $3x^2 + 12x - 1 = kx - 4$ has no real solutions.                             [4]

18  The function f is such that  $f(x) = 2x + 1$ for $x \in \mathbb{R}$.

   The function g is such that  $g(x) = 8 - ax - bx^2$ for $x \geq k$, where $a$, $b$ and $k$ are constants.

   The function fg is such that  $fg(x) = 17 - 24x - 4x^2$ for $x \geq k$.

   a  Find the value of $a$ and the value of $b$.                                                                          [3]

   b  Find the least possible value of $k$ for which g has an inverse.                                                     [4]

   c  For the value of $k$ found in part **b**, find $g^{-1}(x)$.                                                          [2]

19  A circle has centre $(8, 3)$ and passes through the point $P(13, 5)$.

   a  Find the equation of the circle.                                                                                    [4]

   b  Find the equation of the tangent to the circle at the point $P$.

      Give your answer in the form $ax + by = c$.                                                                         [5]

20  The function f is such that  $f(x) = 3x - 7$ for $x \in \mathbb{R}$.

   The function g is such that  $g(x) = \dfrac{18}{5 - x}$ for $x \in \mathbb{R}$, $x \neq 5$.

   a  Find the value of $x$ for which $fg(x) = 5$.                                                                         [3]

   b  Find $f^{-1}(x)$ and $g^{-1}(x)$.                                                                                    [3]

   c  Show that the equation $f^{-1}(x) = g^{-1}(x)$ has no real roots.                                                    [3]

21  A curve has equation $y = 2 - 3x - x^2$.

   a  Express $2 - 3x - x^2$ in the form $a - (x + b)^2$, where $a$ and $b$ are constants.                                 [2]

   b  Write down the coordinates of the maximum point on the curve.                                                       [1]

   c  Find the two values of $m$ for which the line $y = mx + 3$ is a tangent to the curve $y = 2 - 3x - x^2$.             [3]

   d  For each value of $m$ in part **c**, find the coordinates of the point where the line touches the curve.            [3]

97

22  A circle, $C$, has equation $x^2 + y^2 - 16x - 36 = 0$.

a  Find the coordinates of the centre of the circle. [2]

b  Find the radius of the circle. [2]

c  Find the coordinates of the points where the circle meets the $x$-axis. [2]

d  The point $P$ lies on the circle and the line $L$ is a tangent to $C$ at the point $P$. Given that the line $L$ has gradient $\dfrac{4}{3}$, find the equation of the perpendicular to the line $L$ at the point $P$. [3]

23  The function f is such that $f(x) = 3x - 2$ for $x \geqslant 0$.

The function g is such that $g(x) = 2x^2 - 8$ for $x \leqslant k$, where $k$ is a constant.

a  Find the greatest value of $k$ for which the composite function fg can be formed. [3]

b  For the case where $k = -3$:

i  find the range of fg [2]

ii  find $(fg)^{-1}(x)$ and state the domain and range of $(fg)^{-1}$. [4]

24  A curve has equation $xy = 20$ and a line has equation $x + 2y = k$, where $k$ is a constant.

a  In the case where $k = 14$, the line intersects the curve at the points $A$ and $B$.

Find:

i  the coordinates of the points $A$ and $B$ [3]

ii  the equation of the perpendicular bisector of the line $AB$. [4]

b  Find the values of $k$ for which the line is a tangent to the curve. [4]

# Chapter 4
# Circular measure

**In this section you will learn how to:**

■ understand the definition of a radian, and use the relationship between radians and degrees

■ use the formulae $s = r\theta$ and $A = \dfrac{1}{2} r^2 \theta$ to solve problems concerning the arc length and sector area of a circle.

**PREREQUISITE KNOWLEDGE**

| Where it comes from | What you should be able to do | Check your skills |
|---|---|---|
| IGCSE / O Level Mathematics | Find the perimeter and area of sectors. | 1 Find the perimeter and area of a sector of a circle with radius 6 cm and sector angle 30°. |
| IGCSE / O Level Mathematics | Use Pythagoras' theorem and trigonometry on right-angled triangles. | 2  <br> Find the value of $x$ and the value of $y$. |
| IGCSE / O Level Mathematics | Solve problems involving the sine and cosine rules for any triangle and the formula:<br><br>Area of triangle $= \dfrac{1}{2}ab\sin C$ | 3  <br> Find the value of $x$ and the area of the triangle. |

## Another measure for angles

At IGCSE / O Level, you will have always worked with angles that were measured in degrees. Have you ever wondered why there are 360° in one complete revolution? The original reason for choosing the degree as a unit of angular measure is unknown but there are a number of different theories.

- Ancient astronomers claimed that the Sun advanced in its path by one degree each day and that a solar year consisted of 360 days.

- The ancient Babylonians divided the circle into 6 equilateral triangles and then subdivided each angle at $O$ into 60 further parts, resulting in 360 divisions in one complete revolution.

- 360 has many factors that make division of the circle so much easier.

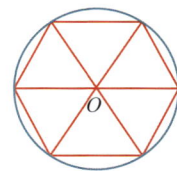

Degrees are not the only way in which we can measure angles. In this chapter you will learn how to use **radian** measure. This is sometimes referred to as the natural unit of angular measure and we use it extensively in mathematics because it can simplify many formulae and calculations.

## 4.1 Radians

In the diagram, the magnitude of angle $AOB$ is 1 radian.

1 radian is sometimes written as 1 rad, but often no symbol at all is used for angles measured in radians.

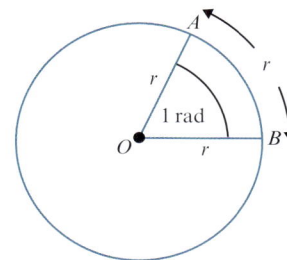

> ### KEY POINT 4.1
>
> An arc equal in length to the radius of a circle subtends an angle of 1 radian at the centre.

It follows that the circumference (an arc of length $2\pi r$) subtends an angle of $2\pi$ radians at the centre, therefore:

> ### KEY POINT 4.2
>
> $2\pi$ radians = $360°$
> $\pi$ radians = $180°$

When an angle is written in terms of $\pi$, we usually omit the word radian (or rad).

Hence, $\pi = 180°$.

## Converting from degrees to radians

Since $180° = \pi$, then $90° = \dfrac{\pi}{2}$, $45° = \dfrac{\pi}{4}$ etc.

We can convert angles that are not simple fractions of $180°$ using the following rule.

> ### KEY POINT 4.3
>
> To change from degrees to radians, multiply by $\dfrac{\pi}{180}$.

## Converting from radians to degrees

Since $\pi = 180°$, $\dfrac{\pi}{6} = 30°$, $\dfrac{\pi}{10} = 18°$ etc.

We can convert angles that are not simple fractions of $\pi$ using the following rule.

> ### KEY POINT 4.4
>
> To change from radians to degrees, multiply by $\dfrac{180}{\pi}$.

(It is useful to remember that 1 radian $= 1 \times \dfrac{180}{\pi} \approx 57°$.)

101

**WORKED EXAMPLE 4.1**

**a** Change $30°$ to radians, giving your answer in terms of $\pi$.

**b** Change $\dfrac{5\pi}{9}$ radians to degrees.

**Answer**

**a** **Method 1:**

$$180° = \pi \text{ radians}$$

$$\left(\dfrac{180}{6}\right)° = \dfrac{\pi}{6} \text{ radians}$$

$$30° = \dfrac{\pi}{6} \text{ radians}$$

**Method 2:**

$$30° = \left(30 \times \dfrac{\pi}{180}\right) \text{ radians}$$

$$30° = \dfrac{\pi}{6} \text{ radians}$$

**b** **Method 1:**

$$\pi \text{ radians} = 180°$$

$$\dfrac{\pi}{9} \text{ radians} = 20°$$

$$\dfrac{5\pi}{9} \text{ radians} = 100°$$

**Method 2:**

$$\dfrac{5\pi}{9} \text{ radians} = \left(\dfrac{5\pi}{9} \times \dfrac{180}{\pi}\right)°$$

$$\dfrac{5\pi}{9} \text{ radians} = 100°$$

In Worked example 4.1, we found that $30° = \dfrac{\pi}{6}$ radians.

There are other angles, which you should learn, that can be written as simple multiples of $\pi$.

There are:

| Degrees | 0° | 30° | 45° | 60° | 90° | 180° | 270° | 360° |
|---|---|---|---|---|---|---|---|---|
| Radians | 0 | $\dfrac{\pi}{6}$ | $\dfrac{\pi}{4}$ | $\dfrac{\pi}{3}$ | $\dfrac{\pi}{2}$ | $\pi$ | $\dfrac{3\pi}{2}$ | $2\pi$ |

We can quickly find other angles, such as $120°$, using these known angles.

**EXERCISE 4A**

**1** Change these angles to radians, giving your answers in terms of $\pi$.

   **a** $20°$      **b** $40°$      **c** $25°$      **d** $50°$      **e** $5°$

   **f** $150°$    **g** $135°$   **h** $210°$   **i** $225°$   **j** $300°$

   **k** $65°$     **l** $540°$   **m** $9°$     **n** $35°$    **o** $600°$

**2** Change these angles to degrees.

   **a** $\dfrac{\pi}{2}$    **b** $\dfrac{\pi}{3}$    **c** $\dfrac{\pi}{6}$    **d** $\dfrac{\pi}{12}$    **e** $\dfrac{4\pi}{3}$

   **f** $\dfrac{4\pi}{9}$   **g** $\dfrac{3\pi}{10}$   **h** $\dfrac{7\pi}{12}$   **i** $\dfrac{9\pi}{20}$   **j** $\dfrac{9\pi}{2}$

   **k** $\dfrac{7\pi}{5}$   **l** $\dfrac{4\pi}{15}$   **m** $\dfrac{5\pi}{4}$   **n** $\dfrac{7\pi}{3}$   **o** $\dfrac{9\pi}{8}$

3   Write each of these angles in radians, correct to 3 significant figures.

   a   28°         b   32°         c   47°         d   200°         e   320°

4   Write each of these angles in degrees, correct to 1 decimal place.

   a   1.2 rad      b   0.8 rad      c   1.34 rad      d   1.52 rad      e   0.79 rad

5   Copy and complete the tables, giving your answers in terms of $\pi$.

   a

| Degrees | 0 | 45 | 90 | 135 | 180 | 225 | 270 | 315 | 360 |
|---------|---|----|----|-----|-----|-----|-----|-----|-----|
| Radians | 0 |    |    |     | $\pi$ |   |     |     | $2\pi$ |

   b

| Degrees | 0 | 30 | 60 | 90 | 120 | 150 | 180 | 210 | 240 | 270 | 300 | 330 | 360 |
|---------|---|----|----|----|-----|-----|-----|-----|-----|-----|-----|-----|-----|
| Radians | 0 |    |    |    |     |     | $\pi$ |   |     |     |     |     | $2\pi$ |

6   Use your calculator to find:

   a   $\sin(0.7)$          b   $\tan(1.5)$          c   $\cos(0.9)$

   d   $\cos\dfrac{\pi}{2}$      e   $\sin\dfrac{\pi}{3}$      f   $\tan\dfrac{\pi}{5}$

> **TIP**
>
> You do not need to change the angle to degrees. You should set the angle mode on your calculator to radians.

7   Calculate the length of $QR$.

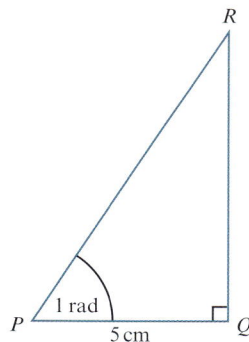

8   

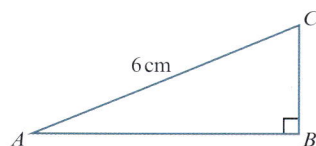

Robert is told the size of angle $BAC$ in degrees and he is then asked to calculate the length of the line $BC$. He uses his calculator but forgets that his calculator is in radian mode. Luckily he still manages to obtain the correct answer. Given that angle $BAC$ is between 10° and 15°, use graphing software to help you find the size of angle $BAC$, correct to 2 decimal places.

## EXPLORE 4.1

Discuss and explain, with the aid of diagrams, the meaning of each of these words.

$\boxed{\text{chord}}$ $\boxed{\text{arc}}$ $\boxed{\text{sector}}$ $\boxed{\text{segment}}$

Explain what is meant by:

- minor arc and major arc
- minor sector and major sector
- minor segment and major segment.

Given that the radius of a circle is $r$ cm and that the angle subtended at the centre of the circle by the chord $AB$ is $\theta°$, discuss and write down an expression, in terms of $r$ and $\theta$, for finding each of the following:

- length of minor arc $AB$
- perimeter of minor sector $AOB$
- perimeter of minor segment $AOB$
- length of chord $AB$
- area of minor sector $AOB$
- area of minor segment $AOB$.

What would the answers be if the angle $\theta$ was measured in radians instead?

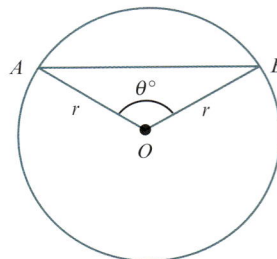

### DID YOU KNOW?

A geographical coordinate system is used to describe the location of any point on the Earth's surface. The coordinates used are longitude and latitude. 'Horizontal' circles and 'vertical' circles form the 'grid'. The horizontal circles are perpendicular to the axis of rotation of the Earth and are known as lines of latitude. The vertical circles pass through the North and South poles and are known as lines of longitude.

### WEB LINK

Try the *Where are you?* resource on the Underground Mathematics website.

## 4.2 Length of an arc

From the definition of a radian, an arc that subtends an angle of 1 radian at the centre of the circle is of length $r$. Hence, if an arc subtends an angle of $\theta$ radians at the centre, the length of the arc is $r\theta$.

### KEY POINT 4.5

Arc length $= r\theta$

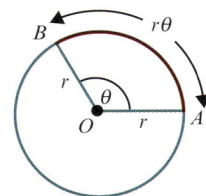

### WORKED EXAMPLE 4.2

An arc subtends an angle of $\dfrac{\pi}{3}$ radians at the centre of a circle with radius 15 cm.

Find the length of the arc in terms of $\pi$.

**Answer**

Arc length $= r\theta$

$= 15 \times \dfrac{\pi}{3}$

$= 5\pi \text{ cm}$

**WORKED EXAMPLE 4.3**

A sector has an angle of 1.5 radians and an arc length of 12 cm.

Find the radius of the sector.

**Answer**

Arc length $= r\theta$

$12 = r \times 1.5$

$r = 8\,\text{cm}$

**WORKED EXAMPLE 4.4**

Triangle $ABC$ is isosceles with $AC = CB = 8\,\text{cm}$.

$CD$ is an arc of a circle, centre $B$, and angle $ABC = 0.9$ radians.

Find:

   **a**  the length of arc $CD$

   **b**  the length of $AD$

   **c**  the perimeter of the shaded region.

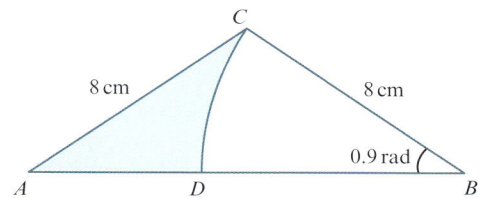

**Answer**

  **a**  Arc length $= r\theta$

             $= 8 \times 0.9$

             $= 7.2\,\text{cm}$

  **b**  $AB = 2 \times 8\cos 0.9 = 9.9457\ldots$

      $AD = AB - DB$

            $= 9.9457\ldots - 8$

            $= 1.95\,\text{cm}$ (to 3 significant figures)

  **c**  Perimeter $= DC + CA + AD$

                $= 7.2 + 8 + 1.945\ldots$

                $= 17.1\,\text{cm}$ (to 3 significant figures)

**EXERCISE 4B**

**1**  Find, in terms of $\pi$, the arc length of a sector of:

   **a**  radius 8 cm and angle $\dfrac{\pi}{4}$

   **b**  radius 7 cm and angle $\dfrac{3\pi}{7}$

   **c**  radius 16 cm and angle $\dfrac{3\pi}{8}$

   **d**  radius 24 cm and angle $\dfrac{7\pi}{6}$.

**2**  Find the arc length of a sector of:

   **a**  radius 10 cm and angle 1.3 radians

   **b**  radius 3.5 cm and angle 0.65 radians.

**3**  Find, in radians, the angle of a sector of:

   **a**  radius 10 cm and arc length 5 cm

   **b**  radius 12 cm and arc length 9.6 cm.

**4**  The High Roller Ferris wheel in the USA has a diameter of 158.5 metres.

   Calculate the distance travelled by a capsule as the wheel rotates through $\dfrac{\pi}{16}$ radians.

105

**5** Find the perimeter of each of these sectors.

**a**

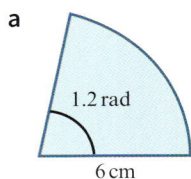

1.2 rad

6 cm

**b**

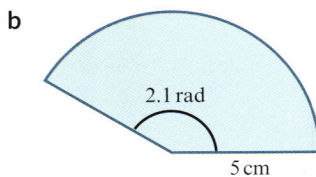

2.1 rad

5 cm

**c**

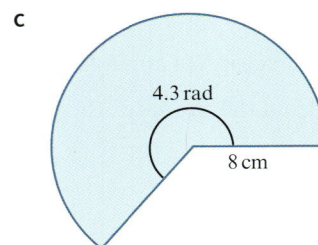

4.3 rad

8 cm

**6** The circle has radius 6 cm and centre $O$.

$PQ$ is a tangent to the circle at the point $P$.

$QRO$ is a straight line. Find:

  **a** angle $POQ$, in radians

  **b** the length of $QR$

  **c** the perimeter of the shaded area.

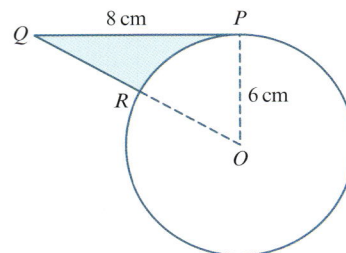

8 cm    P

Q

R    6 cm

O

**7** The circle has radius 7 cm and centre $O$.

$AB$ is a chord and angle $AOB = 2$ radians. Find:

  **a** the length of arc $AB$

  **b** the length of chord $AB$

  **c** the perimeter of the shaded segment.

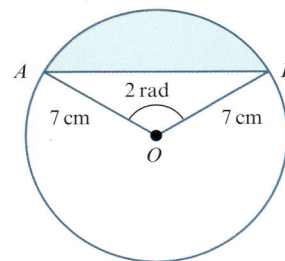

A    B

2 rad

7 cm    7 cm

O

**8** $ABCD$ is a rectangle with $AB = 5$ cm and $BC = 24$ cm.

$O$ is the midpoint of $BC$.

$OAED$ is a sector of a circle, centre $O$. Find:

  **a** the length of $AO$

  **b** angle $AOD$, in radians

  **c** the perimeter of the shaded region.

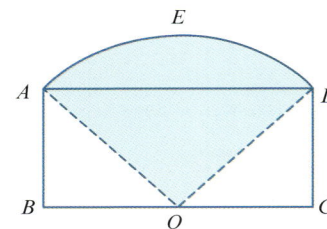

E

A    D

B    O    C

**9** The diagram shows a semicircle with radius 10 cm and centre $O$.
Angle $BOC = \theta$ radians. The perimeter of sector $AOC$ is twice the
perimeter of sector $BOC$.

  **a** Show that $\theta = \dfrac{\pi - 2}{3}$.

  **b** Find the perimeter of triangle $ABC$.

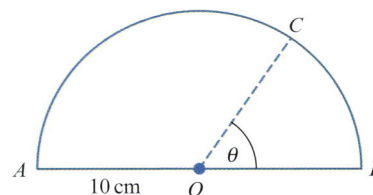

C

A    10 cm    O    $\theta$    B

**PS** **10** The diagram shows the cross-section of two cylindrical metal rods of radii
$x$ cm and $y$ cm. A thin band, of length $P$ cm, holds the two rods tightly
together.

Show that $P = 4\sqrt{xy} + \pi(x + y) + 2(x - y)\sin^{-1}\left(\dfrac{x - y}{x + y}\right)$.

[This question is based upon *Belt* on the Underground Mathematics website.]

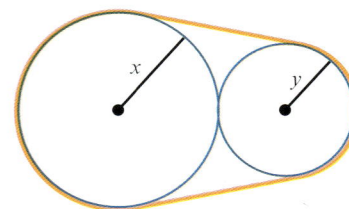

x    y

## 4.3 Area of a sector

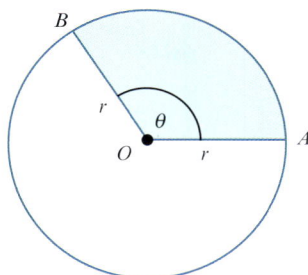

To find the formula for the area of a sector, we use the ratio:

$$\frac{\text{area of sector}}{\text{area of circle}} = \frac{\text{angle in the sector}}{\text{complete angle at the centre}}$$

When $\theta$ is measured in radians, the ratio becomes:

$$\frac{\text{area of sector}}{\pi r^2} = \frac{\theta}{2\pi}$$

$$\text{area of sector} = \frac{\theta}{2\pi} \times \pi r^2$$

**KEY POINT 4.6**

Area of sector $= \dfrac{1}{2} r^2 \theta$

107

**WORKED EXAMPLE 4.5**

Find the area of a sector of a circle with radius $9\,\text{cm}$ and angle $\dfrac{\pi}{6}$ radians.

Give your answer in terms of $\pi$.

**Answer**

$$\begin{aligned}
\text{Area of sector} &= \frac{1}{2} r^2 \theta \\
&= \frac{1}{2} \times 9^2 \times \frac{\pi}{6} \\
&= \frac{27\pi}{4} \text{ cm}^2
\end{aligned}$$

**WORKED EXAMPLE 4.6**

The circle has radius $6\,\text{cm}$ and centre $O$. $AB$ is a chord and angle $AOB = 1.2$ radians.
Find:

a   the area of sector $AOB$

b   the area of triangle $AOB$

c   the area of the shaded segment.

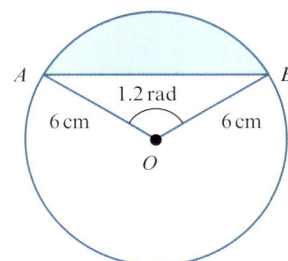

**Answer**

a   Area of sector $AOB = \dfrac{1}{2}\, r^2\theta$

$\qquad\qquad\qquad\;\; = \dfrac{1}{2} \times 6^2 \times 1.2$

$\qquad\qquad\qquad\;\; = 21.6\,\text{cm}^2$

b   Area of triangle $AOB = \dfrac{1}{2}\, ab\sin C$

$\qquad\qquad\qquad\qquad\;\; = \dfrac{1}{2} \times 6 \times 6 \times \sin 1.2$

$\qquad\qquad\qquad\qquad\;\; = 16.7767\ldots$

$\qquad\qquad\qquad\qquad\;\; = 16.8\,\text{cm}^2 \text{ (to 3 significant figures)}$

c   Area of shaded segment = area of sector $AOB$ − area of triangle $AOB$

$\qquad\qquad\qquad\qquad\qquad\quad = 21.6 - 16.7767\ldots$

$\qquad\qquad\qquad\qquad\qquad\quad = 4.82\,\text{cm}^2 \text{ (to 3 significant figures)}$

---

**WORKED EXAMPLE 4.7**

The diagram shows a circle inscribed inside a square of side length $10\,\text{cm}$.

A quarter circle, of radius $10\,\text{cm}$, is drawn with the vertex of the square as centre.

Find the shaded area.

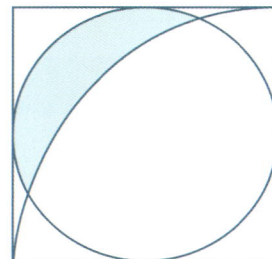

**Answer**

$OQ = 10\,\text{cm}$

Radius of inscribed circle = $5\,\text{cm}$

Pythagoras: $\dfrac{1}{2}$ (diagonal of square) $= \dfrac{1}{2}\left(\sqrt{10^2 + 10^2}\,\right) = 5\sqrt{2}\,\text{cm}$

Cosine rule: $\cos\alpha = \dfrac{5^2 + (5\sqrt{2})^2 - 10^2}{2 \times 5 \times 5\sqrt{2}}$

$\qquad\qquad\quad\; \alpha = 1.932\,\text{rad}$

Hence, $\beta = 2\pi - 2\alpha = 2.4189\,\text{rad}$

Sine rule: $\dfrac{\sin\theta}{5} = \dfrac{\sin\alpha}{10}$

$\qquad\qquad\; \theta = 0.4867\,\text{rad}$

Shaded area = area of segment $PQR$ − area of segment $PQS$

$\qquad\qquad\quad = \left(\dfrac{1}{2} \times 5^2 \times \beta - \dfrac{1}{2} \times 5^2 \times \sin\beta\right) - \left(\dfrac{1}{2} \times 10^2 \times 2\theta - \dfrac{1}{2} \times 10^2 \times \sin 2\theta\right)$

$\qquad\qquad\quad = 21.968 - 7.3296$

$\qquad\qquad\quad = 14.6\,\text{cm}^2 \text{ (to 3 significant figures)}$

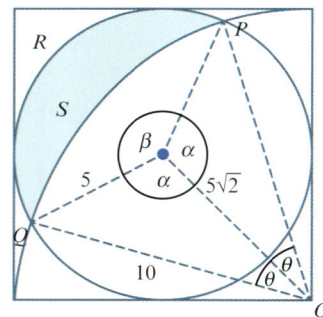

**EXERCISE 4C**

1  Find, in terms of $\pi$, the area of a sector of:

   **a**  radius 12 cm and angle $\dfrac{\pi}{6}$ radians

   **b**  radius 10 cm and angle $\dfrac{2\pi}{5}$ radians

   **c**  radius 4.5 cm and angle $\dfrac{2\pi}{9}$ radians

   **d**  radius 9 cm and angle $\dfrac{4\pi}{3}$ radians.

2  Find the area of a sector of:

   **a**  radius 34 cm and angle 1.5 radian

   **b**  radius 2.6 cm and angle 0.9 radians.

3  Find, in radians, the angle of a sector of:

   **a**  radius 4 cm and area 9 cm$^2$

   **b**  radius 6 cm and area 27 cm$^2$.

4  $AOB$ is a sector of a circle, centre $O$, with radius 8 cm.

   The length of arc $AB$ is 10 cm. Find:

   **a**  angle $AOB$, in radians

   **b**  the area of the sector $AOB$.

5  The diagram shows a sector, $POQ$, of a circle, centre $O$, with radius 4 cm. The length of arc $PQ$ is 7 cm. The lines $PX$ and $QX$ are tangents to the circle at $P$ and $Q$, respectively.

   **a**  Find angle $POQ$, in radians.

   **b**  Find the length of $PX$.

   **c**  Find the area of the shaded region.

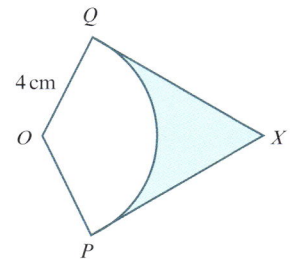

6  The diagram shows a sector, $POR$, of a circle, centre $O$, with radius 8 cm and sector angle $\dfrac{\pi}{3}$ radians. The lines $OR$ and $QR$ are perpendicular and $OPQ$ is a straight line.

   Find the exact area of the shaded region.

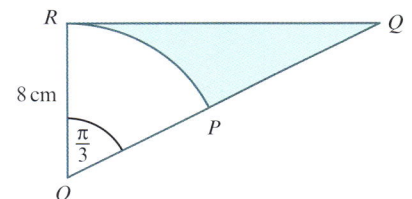

7  The diagram shows a sector, $AOB$, of a circle, centre $O$, with radius 5 cm and sector angle $\dfrac{\pi}{3}$ radians. The lines $AP$ and $BP$ are tangents to the circle at $A$ and $B$, respectively.

   **a**  Find the exact length of $AP$.

   **b**  Find the exact area of the shaded region.

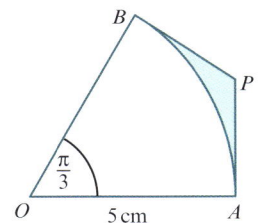

**PS**  8  The diagram shows three touching circles with radii 6 cm, 4 cm and 2 cm.

   Find the area of the shaded region.

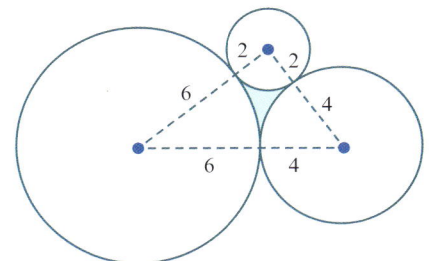

9  The diagram shows a semicircle, centre $O$, with radius $8\,\text{cm}$.

   $FH$ is the arc of a circle, centre $E$. Find the area of:

   a  triangle $EOF$          b  sector $FOG$

   c  sector $FEH$            d  the shaded region.

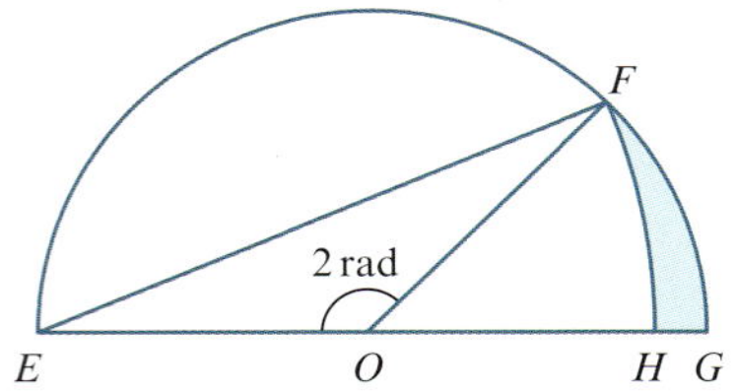

10  The diagram shows a sector, $EOG$, of a circle, centre $O$, with radius $r\,\text{cm}$.
    The line $GF$ is a tangent to the circle at $G$, and $E$ is the midpoint of $OF$.

    a  The perimeter of the shaded region is $P\,\text{cm}$.

       Show that $P = \dfrac{r}{3}(3 + 3\sqrt{3} + \pi)$.

    b  The area of the shaded region is $A\,\text{cm}^2$.

       Show that $A = \dfrac{r^2}{6}(3\sqrt{3} - \pi)$.

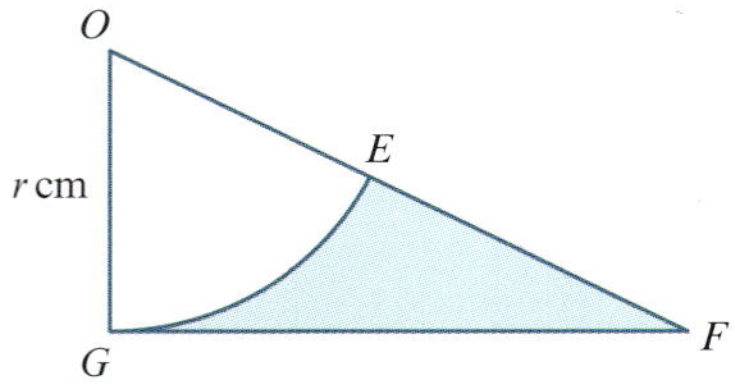

11  The diagram shows two circles with radius $r\,\text{cm}$.

    The centre of each circle lies on the circumference of the other circle.

    Find, in terms of $r$, the exact area of the shaded region.

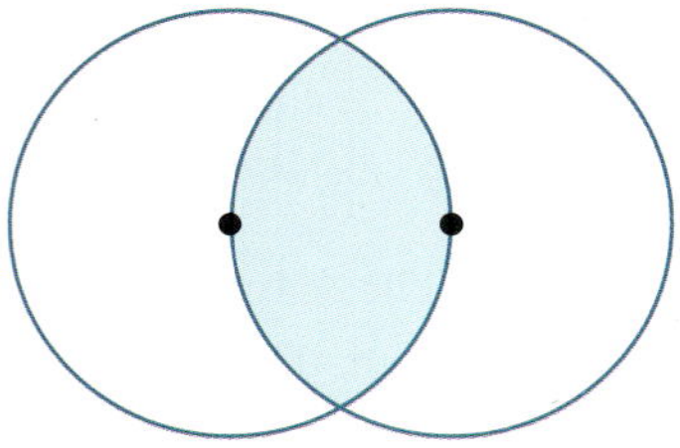

**PS** 12  The diagram shows a square of side length $10\,\text{cm}$.

    A quarter circle, of radius $10\,\text{cm}$, is drawn from each vertex of the square.
    Find the exact area of the shaded region.

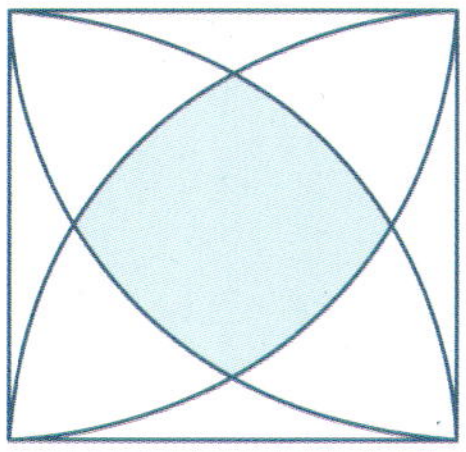

**PS** 13  The diagram shows a circle with radius $1\,\text{cm}$, centre $O$.

    Triangle $AOB$ is right angled and its hypotenuse $AB$ is a tangent to the circle at $P$.

    Angle $BAO = x$ radians.

    a  Find an expression for the length of $AB$ in terms of $\tan x$.

    b  Find the value of $x$ for which the two shaded areas are equal.

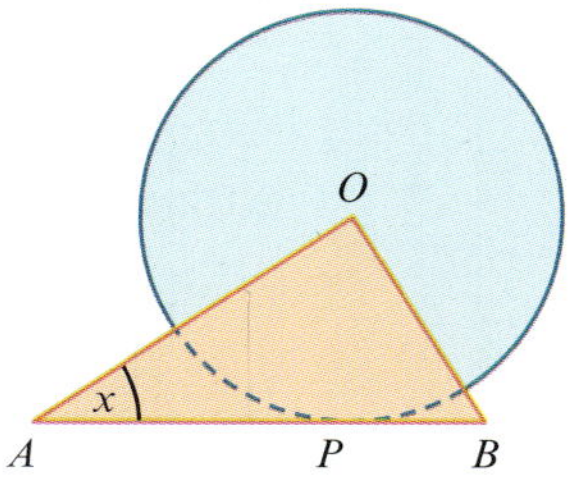

**P** 14  The diagram shows a sector, $AOB$, of a circle, centre $O$, with radius $R\,\text{cm}$ and
    sector angle $\dfrac{\pi}{3}$ radians.

    An inner circle of radius $r\,\text{cm}$ touches the three sides of the sector.

    a  Show that $R = 3r$.

    b  Show that $\dfrac{\text{area of inner circle}}{\text{area of sector}} = \dfrac{2}{3}$.

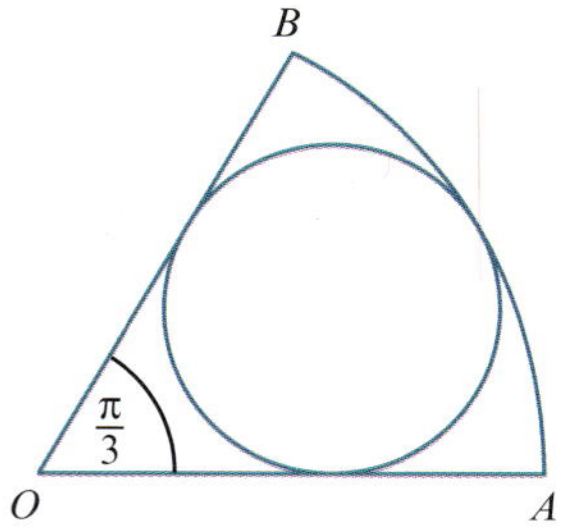

# Checklist of learning and understanding

**Radians and degrees**

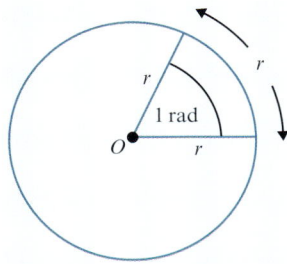

- One radian is the size of the angle subtended at the centre of a circle, radius $r$, by an arc of length $r$.
- $\pi$ radians $= 180°$
- To change from degrees to radians, multiply by $\dfrac{\pi}{180}$.
- To change from radians to degrees, multiply by $\dfrac{180}{\pi}$.

**Arc length and area of a sector**

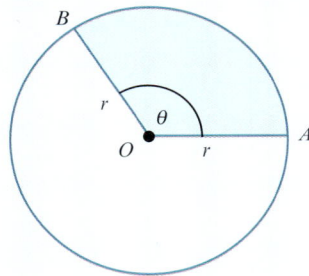

- When $\theta$ is measured in radians, the length of arc $AB$ is $r\theta$.
- When $\theta$ is measured in radians, the area of sector $AOB$ is $\dfrac{1}{2} r^2 \theta$.

**END-OF-CHAPTER REVIEW EXERCISE 4**

**1**

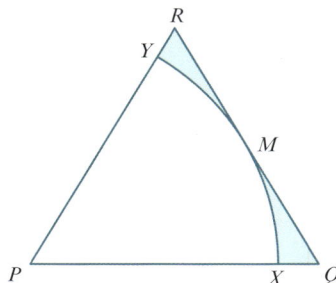

The diagram shows an equilateral triangle, $PQR$, with side length $5\,\text{cm}$. $M$ is the midpoint of the line $QR$. An arc of a circle, centre $P$, touches $QR$ at $M$ and meets $PQ$ at $X$ and $PR$ at $Y$. Find in terms of $\pi$ and $\sqrt{3}$:

a   the total perimeter of the shaded region   [5]

b   the total area of the shaded region.   [3]

**2**

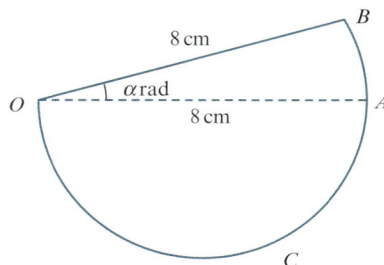

In the diagram, $OAB$ is a sector of a circle with centre $O$ and radius $8\,\text{cm}$. Angle $BOA$ is $\alpha$ radians. $OAC$ is a semicircle with diameter $OA$. The area of the semicircle $OAC$ is twice the area of the sector $OAB$.

i   Find $\alpha$ in terms of $\pi$.   [3]

ii   Find the perimeter of the complete figure in terms of $\pi$.   [2]

*Cambridge International AS & A Level Mathematics 9709 Paper 11 Q3 June 2013*

**3**

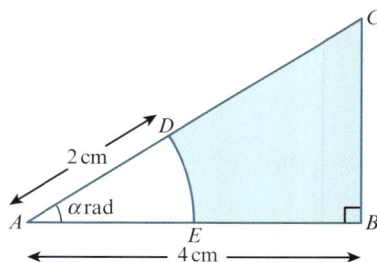

The diagram shows triangle $ABC$ in which $AB$ is perpendicular to $BC$. The length of $AB$ is $4\,\text{cm}$ and angle $CAB$ is $\alpha$ radians. The arc $DE$ with centre $A$ and radius $2\,\text{cm}$ meets $AC$ at $D$ and $AB$ at $E$. Find, in terms of $\alpha$,

i   the area of the shaded region,   [3]

ii   the perimeter of the shaded region.   [3]

*Cambridge International AS & A Level Mathematics 9709 Paper 11 Q6 June 2014*

**4**

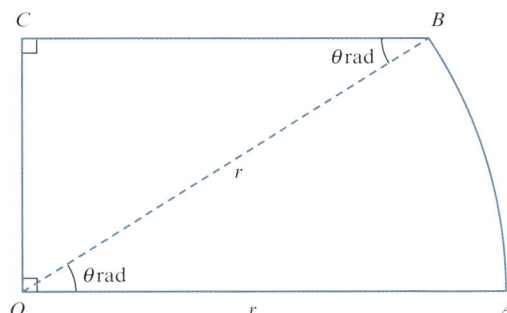

The diagram represents a metal plate $OABC$, consisting of a sector $OAB$ of a circle with centre $O$ and radius $r$, together with a triangle $OCB$ which is right-angled at $C$. Angle $AOB = \theta$ radians and $OC$ is perpendicular to $OA$.

i   Find an expression in terms of $r$ and $\theta$ for the perimeter of the plate.   [3]

ii  For the case where $r = 10$ and $\theta = \dfrac{1}{5}\pi$, find the area of the plate.   [3]

*Cambridge International AS & A Level Mathematics 9709 Paper 11 Q5 November 2011*

**5**

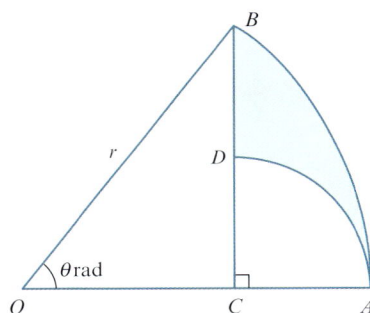

The diagram shows a sector $OAB$ of a circle with centre $O$ and radius $r$. Angle $AOB$ is $\theta$ radians. The point $C$ on $OA$ is such that $BC$ is perpendicular to $OA$. The point $D$ is on $BC$ and the circular arc $AD$ has centre $C$.

i   Find $AC$ in terms of $r$ and $\theta$.   [1]

ii  Find the perimeter of the shaded region $ABD$ when $\theta = \dfrac{1}{3}\pi$ and $r = 4$, giving your answer as an exact value.   [6]

*Cambridge International AS & A Level Mathematics 9709 Paper 11 Q6 November 2012*

**6**   A piece of wire of length 24 cm is bent to form the perimeter of a sector of a circle of radius $r$ cm.

i   Show that the area of the sector, $A$ cm$^2$, is given by $A = 12r - r^2$.   [3]

ii  Express $A$ in the form $a - (r - b)^2$, where $a$ and $b$ are constants.   [2]

iii Given that $r$ can vary, state the greatest value of $A$ and find the corresponding angle of the sector.   [2]

*Cambridge International AS & A Level Mathematics 9709 Paper 11 Q5 June 2015*

**7**

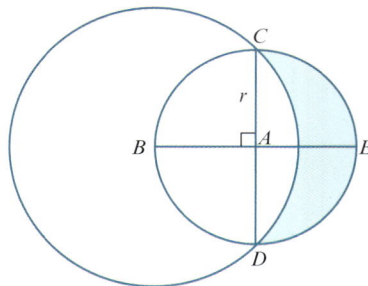

The diagram shows a circle with centre $A$ and radius $r$. Diameters $CAD$ and $BAE$ are perpendicular to each other. A larger circle has centre $B$ and passes through $C$ and $D$.

i   Show that the radius of the larger circle is $r\sqrt{2}$.   [1]

ii  Find the area of the shaded region in terms of $r$.   [6]

*Cambridge International AS & A Level Mathematics 9709 Paper 11 Q7 November 2015*

**8**

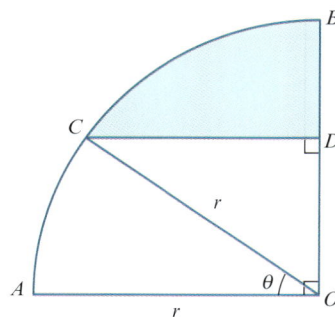

In the diagram, $AOB$ is a quarter circle with centre $O$ and radius $r$. The point $C$ lies on the arc $AB$ and the point $D$ lies on $OB$. The line $CD$ is parallel to $AO$ and angle $AOC = \theta$ radians.

i   Express the perimeter of the shaded region in terms of $r$, $\theta$ and $\pi$.   [4]

ii  For the case where $r = 5\,\text{cm}$ and $\theta = 0.6$, find the area of the shaded region.   [3]

*Cambridge International AS & A Level Mathematics 9709 Paper 11 Q7 June 2016*

**9**

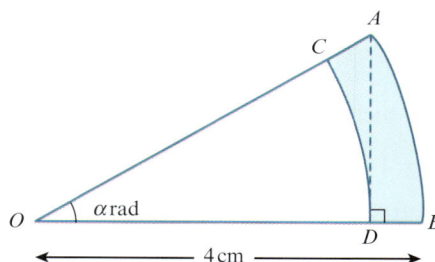

In the diagram, $AB$ is an arc of a circle with centre $O$ and radius $4\,\text{cm}$.
Angle $AOB$ is $\alpha$ radians. The point $D$ on $OB$ is such that $AD$ is perpendicular to $OB$.
The arc $DC$, with centre $O$, meets $OA$ at $C$.

i   Find an expression in terms of $\alpha$ for the perimeter of the shaded region $ABDC$.   [4]

ii   For the case where $\alpha = \dfrac{1}{6}\pi$, find the area of the shaded region $ABDC$, giving your answer in the form $k\pi$, where $k$ is a constant to be determined.   [4]

*Cambridge International AS & A Level Mathematics 9709 Paper 11 Q8 November 2014*

10

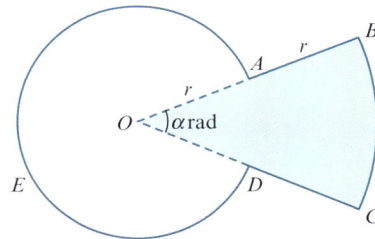

The diagram shows a metal plate made by fixing together two pieces, $OABCD$ (shaded) and $OAED$ (unshaded). The piece $OABCD$ is a minor sector of a circle with centre $O$ and radius $2r$. The piece $OAED$ is a major sector of a circle with centre $O$ and radius $r$. Angle $AOD$ is $\alpha$ radians. Simplifying your answers where possible, find, in terms of $\alpha$, $\pi$ and $r$,

i   the perimeter of the metal plate,   [3]

ii   the area of the metal plate.   [3]

It is now given that the shaded and unshaded pieces are equal in area.

iii   Find $\alpha$ in terms of $\pi$.   [2]

*Cambridge International AS & A Level Mathematics 9709 Paper 11 Q6 November 2013*

# Chapter 5
# Trigonometry

**In this section you will learn how to:**

- sketch and use graphs of the sine, cosine and tangent functions (for angles of any size, and using either degrees or radians)
- use the exact values of the sine, cosine and tangent of 30°, 45°, 60°, and related angles
- use the notations $\sin^{-1} x$, $\cos^{-1} x$, $\tan^{-1} x$ to denote the principal values of the inverse trigonometric relations
- use the identities $\dfrac{\sin \theta}{\cos \theta} = \tan \theta$ and $\sin^2 \theta + \cos^2 \theta = 1$
- find all the solutions of simple trigonometrical equations lying in a specified interval.

## PREREQUISITE KNOWLEDGE

| Where it comes from | What you should be able to do | Check your skills |
|---|---|---|
| IGCSE / O Level Mathematics | Use Pythagoras' theorem and trigonometry on right-angled triangles. | 1 Find each of the following in terms of $r$.<br>**a** $BC$<br>**b** $\sin\theta$<br>**c** $\cos\theta$<br>**d** $\tan\theta$ |
| Chapter 4 | Convert between degrees and radians. | 2 **a** Convert to radians.<br> **i** $45°$<br> **ii** $720°$<br> **iii** $150°$<br>**b** Convert to degrees.<br> **i** $\dfrac{\pi}{6}$<br> **ii** $\dfrac{7\pi}{2}$<br> **iii** $\dfrac{13\pi}{12}$ |
| IGCSE / O Level Mathematics | Solve quadratic equations. | 3 **a** Solve $x^2 - 5x = 0$.<br>**b** Solve $2x^2 + 7x - 15 = 0$. |

## Why do we study trigonometry?

You should already know how to calculate lengths and angles using the sine, cosine and tangent ratios. In this chapter you shall learn about some of the special rules connecting these trigonometric functions together with the special properties of their graphs. The graphs of $y = \sin x$ and $y = \cos x$ are sometimes referred to as waves.

Oscillations and waves occur in many situations in real life. A few examples of these are musical sound waves, light waves, water waves, electricity, vibrations of an aircraft wing and microwaves. Scientists and engineers represent these oscillations/waves using trigonometric functions.

**FAST FORWARD**

In the Pure Mathematics 2 & 3 Coursebook, Chapter 3 you shall learn about the secant, cosecant and cotangent functions, which are closely connected to the sine, cosine and tangent functions. You shall also learn many more rules involving these six trigonometric functions.

**WEB LINK**

Try the *Trigonometry: Triangles to functions* resource on the Underground Mathematics website.

117

## 5.1 Angles between 0° and 90°

You should already know the following trigonometric ratios.

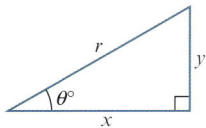

$$\sin \theta = \frac{\text{opposite}}{\text{hypotenuse}} \qquad \cos \theta = \frac{\text{adjacent}}{\text{hypotenuse}} \qquad \tan \theta = \frac{\text{opposite}}{\text{adjacent}}$$

$$\sin \theta = \frac{y}{r} \qquad \cos \theta = \frac{x}{r} \qquad \tan \theta = \frac{y}{x}$$

**WORKED EXAMPLE 5.1**

$\cos \theta = \dfrac{\sqrt{5}}{3}$, where $0° \leqslant \theta \leqslant 90°$.

**a** Find the exact values of:

    **i** $\cos^2 \theta$            **ii** $\sin \theta$            **iii** $\tan \theta$

**b** Show that $\dfrac{1 - \tan^2 \theta}{\cos \theta + \sin \theta} = \dfrac{3\sqrt{5} - 6}{5}$.

**Answer**

**a**  **i**  $\cos^2 \theta = \cos \theta \times \cos \theta$

$$= \left( \frac{\sqrt{5}}{3} \right)^2$$

$$= \frac{\sqrt{5}}{3} \times \frac{\sqrt{5}}{3}$$

$$= \frac{5}{9}$$

> **TIP**
>
> $\cos^2 \theta$ means $(\cos \theta)^2$

  **ii**  A right-angled triangle with angle $\theta$ is shown in this diagram.

Using Pythagoras' theorem:

$$x = \sqrt{3^2 - (\sqrt{5})^2} = 2$$

$$\therefore \sin \theta = \frac{2}{3}$$

  **iii**  From the triangle, $\tan \theta = \dfrac{2}{\sqrt{5}}$.

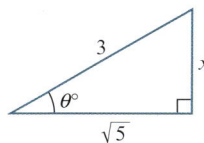

**b**  $\dfrac{1 - \tan^2 \theta}{\cos \theta + \sin \theta} = \dfrac{1 - \left( \dfrac{2}{\sqrt{5}} \right)^2}{\dfrac{\sqrt{5}}{3} + \dfrac{2}{3}}$     Simplify.

$$= \frac{\left( \dfrac{1}{5} \right)}{\left( \dfrac{\sqrt{5} + 2}{3} \right)}$$     Multiply the numerator and denominator by 15.

$$= \frac{3}{5 \left( \sqrt{5} + 2 \right)}$$     Multiply the numerator and denominator by $\sqrt{5} - 2$.

$$= \frac{3 \left( \sqrt{5} - 2 \right)}{5 \left( \sqrt{5} + 2 \right) \left( \sqrt{5} - 2 \right)}$$     $\left( \sqrt{5} + 2 \right)\left( \sqrt{5} - 2 \right)$ $= 5 - 2\sqrt{5} + 2\sqrt{5} - 4 = 1$

$$= \frac{3\sqrt{5} - 6}{5}$$

We can obtain exact values of the sine, cosine and tangent of 30°, 45° and 60° $\left(\text{or } \dfrac{\pi}{6}, \dfrac{\pi}{4} \text{ and } \dfrac{\pi}{3}\right)$ from the following two triangles.

**Triangle 1**

Consider a right-angled isosceles triangle whose two equal sides are of length 1 unit.

We find the third side using Pythagoras' theorem:

$$\sqrt{1^2 + 1^2} = \sqrt{2}$$

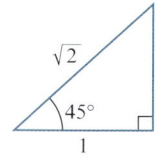

**Triangle 2**

Consider an equilateral triangle whose sides are of length 2 units.

The perpendicular bisector to the base splits the equilateral triangle into two congruent right-angled triangles.

We can find the height of the triangle using Pythagoras' theorem:

$$\sqrt{2^2 - 1^2} = \sqrt{3}$$

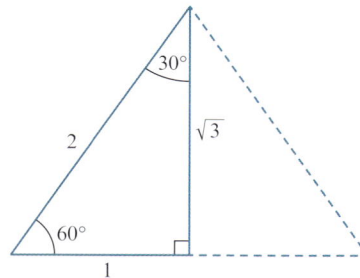

These two triangles give the important results:

| | $\sin\theta$ | $\cos\theta$ | $\tan\theta$ |
|---|---|---|---|
| $\theta = 30° = \dfrac{\pi}{6}$ | $\dfrac{1}{2}$ | $\dfrac{\sqrt{3}}{2}$ | $\dfrac{1}{\sqrt{3}}$ |
| $\theta = 45° = \dfrac{\pi}{4}$ | $\dfrac{1}{\sqrt{2}}$ | $\dfrac{1}{\sqrt{2}}$ | $1$ |
| $\theta = 60° = \dfrac{\pi}{3}$ | $\dfrac{\sqrt{3}}{2}$ | $\dfrac{1}{2}$ | $\sqrt{3}$ |

> **TIP**
>
> The value $\dfrac{1}{\sqrt{2}}$ can be written as $\dfrac{\sqrt{2}}{2}$.
>
> The value $\dfrac{1}{\sqrt{3}}$ can be written as $\dfrac{\sqrt{3}}{3}$.

**WORKED EXAMPLE 5.2**

Find the exact value of:

  a  $\sin 30° \cos 45°$     b  $\sin^2 \dfrac{\pi}{3}$     c  $\dfrac{2\cos\dfrac{\pi}{4}\sin\dfrac{\pi}{6}}{\cos^2\dfrac{\pi}{3} + \sin^2\dfrac{\pi}{3}}$

**Answer**

  a  $\sin 30° \cos 45° = \dfrac{1}{2} \times \dfrac{1}{\sqrt{2}}$

$\qquad\qquad\quad = \dfrac{1}{2\sqrt{2}}$  ............ Rationalise the denominator.

$\qquad\qquad\quad = \dfrac{1 \times \sqrt{2}}{2\sqrt{2} \times \sqrt{2}}$

$\qquad\qquad\quad = \dfrac{\sqrt{2}}{4}$

**b** $\sin^2 \dfrac{\pi}{3} = \sin \dfrac{\pi}{3} \times \sin \dfrac{\pi}{3}$ ............ $\sin^2 \dfrac{\pi}{3}$ means $\left( \sin \dfrac{\pi}{3} \right)^2$.

$= \dfrac{\sqrt{3}}{2} \times \dfrac{\sqrt{3}}{2}$

$= \dfrac{3}{4}$

**c** $\dfrac{2 \cos \dfrac{\pi}{4} \sin \dfrac{\pi}{6}}{\cos^2 \dfrac{\pi}{3} + \sin^2 \dfrac{\pi}{3}} = \dfrac{2 \times \dfrac{1}{\sqrt{2}} \times \dfrac{1}{2}}{\left( \dfrac{1}{2} \right)^2 + \left( \dfrac{\sqrt{3}}{2} \right)^2}$ ............ The denominator simplifies to $\dfrac{1}{4} + \dfrac{3}{4} = 1.$

$= \dfrac{1}{\sqrt{2}}$ ............ Rationalise the denominator.

$= \dfrac{\sqrt{2}}{\sqrt{2}\,\sqrt{2}}$

$= \dfrac{\sqrt{2}}{2}$

## EXERCISE 5A

**1** Given that $\cos\theta = \dfrac{4}{5}$ and that $\theta$ is acute, find the exact value of:

  **a** $\sin\theta$                        **b** $\tan\theta$                         **c** $2\sin\theta \cos\theta$

  **d** $\dfrac{5}{\tan\theta}$              **e** $\dfrac{1 - \sin^2\theta}{\cos\theta}$          **f** $\dfrac{3 - \sin\theta}{3 + \cos\theta}$

**2** Given that $\tan\theta = \dfrac{2}{\sqrt{5}}$ and that $\theta$ is acute, find the exact value of:

  **a** $\sin\theta$                        **b** $\cos\theta$                       **c** $\sin^2\theta + \cos^2\theta$

  **d** $\dfrac{\cos\theta}{\sin\theta}$           **e** $\dfrac{2}{\sin\theta + 1}$            **f** $\dfrac{5}{1 + \cos\theta}$

**3** Given that $\sin\theta = \dfrac{1}{4}$ and that $\theta$ is acute, find the exact value of:

  **a** $\cos\theta$                      **b** $\tan\theta$                       **c** $1 - \sin^2\theta$

  **d** $\dfrac{\sin\theta \cos\theta}{\tan\theta}$       **e** $\dfrac{1}{\tan\theta} + \dfrac{1}{\sin\theta}$      **f** $5 - \dfrac{\tan\theta}{\sin\theta}$

**4** Find the exact value of each of the following.

  **a** $\sin 30° \cos 60°$            **b** $\sin^2 45°$                **c** $\sin 45° + \cos 30°$

  **d** $\dfrac{\sin 60°}{\sin 30°}$          **e** $\dfrac{\sin^2 45°}{2 + \tan 60°}$      **f** $\dfrac{\sin^2 30° + \cos^2 30°}{2 \sin 45° \cos 45°}$

**5** Find the exact value of each of the following.

  **a** $\sin \dfrac{\pi}{4} \cos \dfrac{\pi}{4}$         **b** $\cos^2 \dfrac{\pi}{3}$             **c** $1 - 2 \sin^2 \dfrac{\pi}{6}$

  **d** $\dfrac{\sin \dfrac{\pi}{6} - \tan \dfrac{\pi}{3}}{\sin \dfrac{\pi}{4}}$      **e** $\dfrac{1}{\tan \dfrac{\pi}{4}} - \dfrac{1}{\cos \dfrac{\pi}{3}}$      **f** $\dfrac{\cos \dfrac{\pi}{3} + \tan \dfrac{\pi}{6}}{\sin \dfrac{\pi}{3}}$

**6** In the table, $0 \leqslant \theta \leqslant \dfrac{\pi}{2}$ and the missing function is from the list $\sin\theta$, $\tan\theta$, $\dfrac{1}{\cos\theta}$ and $\dfrac{1}{\tan\theta}$.

Without using a calculator, copy and complete the table.

|  | $\theta = \ldots$ | $\theta = \ldots$ | $\theta = \ldots$ |
|---|---|---|---|
| ...... | $1$ | $\sqrt{3}$ | $\dfrac{1}{\sqrt{3}}$ |
| $\cos\theta$ | $\dfrac{1}{\sqrt{2}}$ | $\dfrac{1}{2}$ | ...... |
| $\dfrac{1}{\sin\theta}$ | ...... | ...... | $2$ |

## 5.2 The general definition of an angle

We need to be able to use the three basic trigonometric functions for any angle.

To do this we need a general definition for an angle:

An angle is a measure of the rotation of a line segment $OP$ about a fixed point $O$. The angle is measured from the positive $x$-direction. An anticlockwise rotation is taken as positive and a clockwise rotation is taken as negative.

The Cartesian plane is divided into four **quadrants**, and the angle $\theta$ is said to be in the quadrant where $OP$ lies. In the previous diagram, $\theta$ is in the first quadrant.

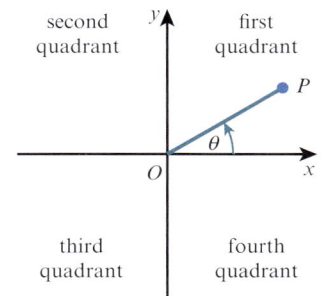

**WORKED EXAMPLE 5.3**

Draw a diagram showing the quadrant in which the rotating line $OP$ lies for each of the following angles. In each case, find the acute angle that the line $OP$ makes with the $x$-axis.

**a** $120°$      **b** $430°$      **c** $\dfrac{3\pi}{4}$      **d** $-\dfrac{2\pi}{3}$

**Answer**

**a** $120°$ is an anticlockwise rotation.

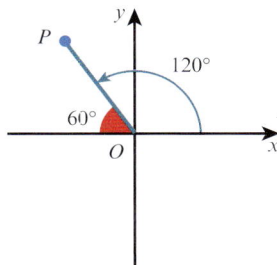

Acute angle made with $x$-axis $= 60°$

**b** $430°$ is an anticlockwise rotation.

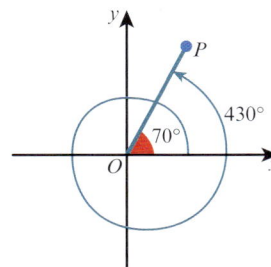

Acute angle made with $x$-axis $= 70°$

121

c $\dfrac{3\pi}{4}$ is an anticlockwise rotation.

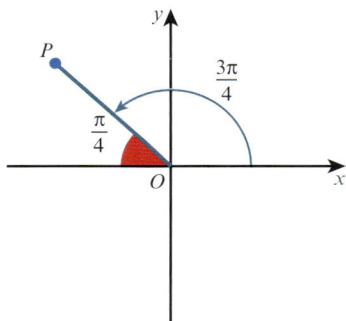

Acute angle made with $x$-axis $= \dfrac{\pi}{4}$

d $-\dfrac{2\pi}{3}$ is a clockwise rotation.

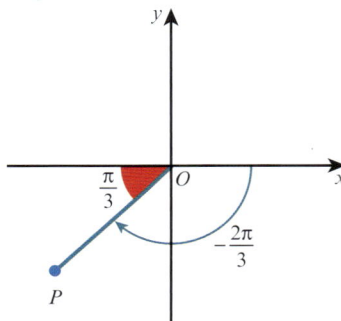

Acute angle made with $x$-axis $= \dfrac{\pi}{3}$

The acute angle made with the $x$-axis is sometimes called the **basic angle** or the **reference angle**.

## EXERCISE 5B

1 For each of the following diagrams, find the basic angle of $\theta$.

a

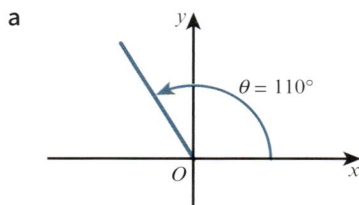

$\theta = 110°$

b

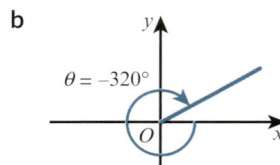

$\theta = -320°$

c

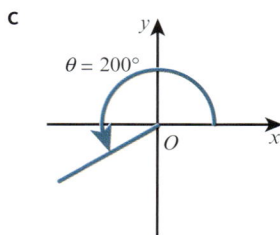

$\theta = 200°$

d

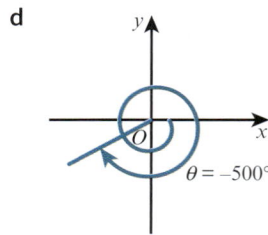

$\theta = -500°$

2 Draw a diagram showing the quadrant in which the rotating line $OP$ lies for each of the following angles. On each diagram, indicate clearly the direction of rotation and state the acute angle that the line $OP$ makes with the $x$-axis.

a $100°$

b $-100°$

c $310°$

d $-150°$

e $400°$

f $\dfrac{2\pi}{3}$

g $\dfrac{7\pi}{6}$

h $-\dfrac{5\pi}{3}$

i $\dfrac{13\pi}{9}$

j $-\dfrac{17\pi}{8}$

3   In each part of this question you are given the basic angle, $b$, the quadrant in which $\theta$ lies and the range in which $\theta$ lies. Find the value of $\theta$.

   a   $b = 55°$, second quadrant, $0° < \theta < 360°$

   b   $b = 20°$, third quadrant, $-180° < \theta < 0°$

   c   $b = 32°$, fourth quadrant, $360° < \theta < 720°$

   d   $b = \dfrac{\pi}{4}$, third quadrant, $0 < \theta < 2\pi$

   e   $b = \dfrac{\pi}{3}$, second quadrant, $2\pi < \theta < 4\pi$

   f   $b = \dfrac{\pi}{6}$, fourth quadrant, $-4\pi < \theta < -2\pi$

## 5.3 Trigonometric ratios of general angles

In general, trigonometric ratios of any angle $\theta$ in any quadrant are defined as:

### KEY POINT 5.1

$$\sin\theta = \frac{y}{r}, \ \cos\theta = \frac{x}{r}, \ \tan\theta = \frac{y}{x}, \ \text{when } x \ne 0$$

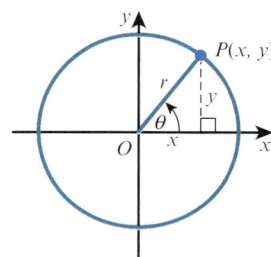

Where $x$ and $y$ are the coordinates of the point $P$ and $r$ is the length of $OP$, where $r = \sqrt{x^2 + y^2}$.

You need to know the signs of the three trigonometric ratios in each of the four quadrants.

### EXPLORE 5.1

$$\sin\theta = \frac{y}{r} \qquad\qquad \cos\theta = \frac{x}{r} \qquad\qquad \tan\theta = \frac{y}{x}$$

By considering whether $x$ and $y$ are positive or negative ($+$ or $-$) in each of the four quadrants, copy and complete the table. ($r$ is positive in all four quadrants.)

| | $\sin\theta$ | $\cos\theta$ | $\tan\theta$ |
|---|---|---|---|
| 1st quadrant | $\dfrac{y}{r} = \dfrac{+}{+} = +$ | $\dfrac{x}{r} = \dfrac{+}{+} = +$ | $\dfrac{y}{x} = \dfrac{+}{+} = +$ |
| 2nd quadrant | $\dfrac{y}{r} =$ | $\dfrac{x}{r} = \dfrac{-}{+} = -$ | $\dfrac{y}{x} =$ |
| 3rd quadrant | $\dfrac{y}{r} =$ | $\dfrac{x}{r} =$ | $\dfrac{y}{x} = \dfrac{-}{-} = +$ |
| 4th quadrant | $\dfrac{y}{r} = \dfrac{-}{+} = -$ | $\dfrac{x}{r} =$ | $\dfrac{y}{x} =$ |

On a copy of the diagram, record which ratios are positive in each quadrant.

The first quadrant has been completed for you.

(All three ratios are positive in the first quadrant.)

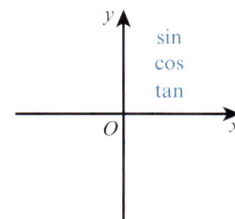

123

The diagram shows which trigonometric functions are positive in each quadrant.

We can memorise this diagram using a mnemonic such as '**A**ll **S**tudents **T**rust **C**ambridge'.

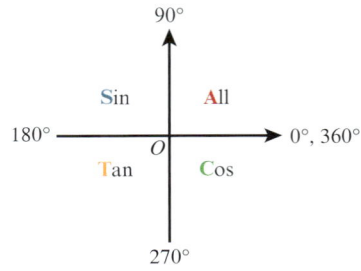

**WORKED EXAMPLE 5.4**

Express in terms of trigonometric ratios of acute angles:

  **a**  $\sin 140°$         **b**  $\cos(-130°)$

**Answer**

  **a**  The acute angle made with the $x$-axis is $40°$.

   In the second quadrant, sin is positive.

   $\sin 140° = \sin 40°$

  **b**  The acute angle made with the $x$-axis is $50°$.

   In the third quadrant only tan is positive, so cos is negative.

   $\cos(-130°) = -\cos 50°$

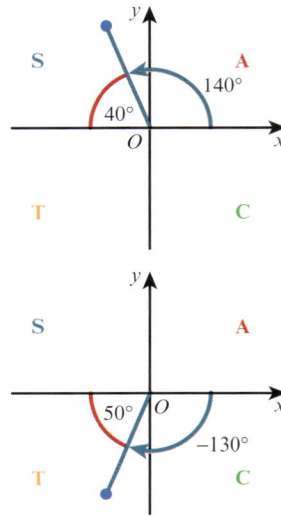

**WORKED EXAMPLE 5.5**

Given that $\cos\theta = -\dfrac{3}{5}$ and that $180° \leqslant \theta \leqslant 270°$, find the value of $\sin\theta$ and the value of $\tan\theta$.

**Answer**

$\theta$ is in the third quadrant.
sin is negative and tan is positive in this quadrant.

$y^2 + (-3)^2 = 5^2$

$\qquad y^2 = 25 - 9 = 16$

Since $y < 0$, $y = -4$.

$\therefore \sin\theta = \dfrac{-4}{5} = -\dfrac{4}{5}$ and $\tan\theta = \dfrac{-4}{-3} = \dfrac{4}{3}$.

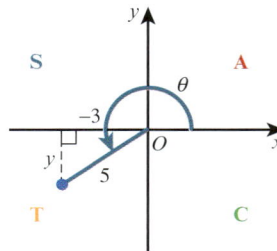

**WORKED EXAMPLE 5.6**

Without using a calculator, find the exact values of:

**a** $\sin 120°$

**b** $\cos \dfrac{7\pi}{6}$

**Answer**

**a** 120° lies in the second quadrant.

∴ $\sin 120°$ is positive.

Basic acute angle $= 180° - 120° = 60°$

∴ $\sin 120° = \sin 60° = \dfrac{\sqrt{3}}{2}$

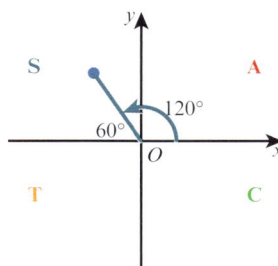

**b** $\dfrac{7\pi}{6}$ lies in the third quadrant.

∴ $\cos \dfrac{7\pi}{6}$ is negative.

Basic acute angle $= \dfrac{7\pi}{6} - \pi = \dfrac{\pi}{6}$

∴ $\cos \dfrac{7\pi}{6} = -\cos \dfrac{\pi}{6} = -\dfrac{\sqrt{3}}{2}$

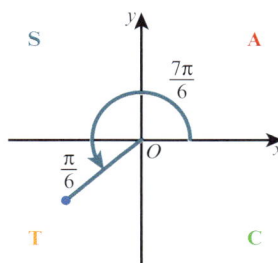

125

**WORKED EXAMPLE 5.7**

Given that $\sin 50° = b$, express each of the following in terms of $b$.

**a** $\sin 230°$        **b** $\cos 50°$        **c** $\tan 40°$        **d** $\tan 140°$

**Answer**

**a** 230° lies in the third quadrant.

∴ $\sin 230°$ is negative.

Basic acute angle $= 230° - 180° = 50°$

∴ $\sin 230° = -\sin 50° = -b$

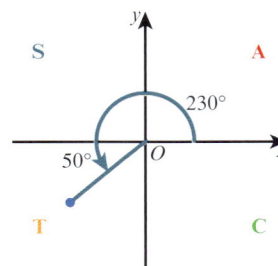

**b** Draw the right-angled triangle showing the angle $50°$:

$$\therefore \cos 50° = \frac{\sqrt{1-b^2}}{1} = \sqrt{1-b^2}$$

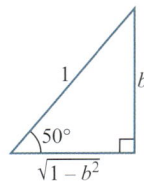

**c** Show $40°$ on the triangle:

$$\therefore \tan 40° = \frac{\sqrt{1-b^2}}{b}$$

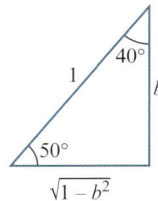

**d** $140°$ lies in the second quadrant.

$\therefore \tan 140°$ is negative.

Basic acute angle $= 180° - 140° = 40°$

$$\therefore \tan 140° = -\tan 40° = -\frac{\sqrt{1-b^2}}{b}$$

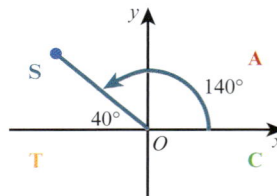

## EXERCISE 5C

**1** Express the following as trigonometric ratios of acute angles.

   **a** $\sin 190°$        **b** $\cos 305°$        **c** $\tan 125°$        **d** $\cos(-245°)$

   **e** $\cos \dfrac{4\pi}{5}$        **f** $\sin \dfrac{9\pi}{8}$        **g** $\cos\left(-\dfrac{7\pi}{10}\right)$        **h** $\tan \dfrac{11\pi}{9}$

**2** Without using a calculator, find the exact values of each of the following.

   **a** $\cos 120°$        **b** $\tan 330°$        **c** $\sin 225°$        **d** $\tan(-300°)$

   **e** $\sin \dfrac{4\pi}{3}$        **f** $\cos \dfrac{7\pi}{3}$        **g** $\tan\left(-\dfrac{\pi}{6}\right)$        **h** $\cos \dfrac{10\pi}{3}$

**3** Given that $\sin\theta < 0$ and $\tan\theta < 0$, name the quadrant in which angle $\theta$ lies.

**4** Given that $\sin\theta = \dfrac{2}{5}$ and that $\theta$ is obtuse, find the value of:

   **a** $\cos\theta$                **b** $\tan\theta$

**5** Given that $\cos\theta = -\dfrac{1}{\sqrt{3}}$ and that $180° \leqslant \theta \leqslant 270°$, find the value of:

   **a** $\sin\theta$                **b** $\tan\theta$

**6** Given that $\tan\theta = -\dfrac{5}{12}$ and that $180° \leqslant \theta \leqslant 360°$, find the value of:

   **a** $\sin\theta$                **b** $\cos\theta$

**7** Given that $\tan 25° = a$, express each of the following in terms of $a$.

   **a** $\tan 205°$        **b** $\sin 25°$        **c** $\cos 65°$        **d** $\cos 245°$

**8** Given that $\cos 77° = b$, express each of the following in terms of $b$.

    **a** $\sin 77°$     **b** $\tan 13°$     **c** $\sin 257°$     **d** $\cos 347°$

**9** Given that $\sin A = \dfrac{5}{13}$ and $\cos B = -\dfrac{4}{5}$, where $A$ and $B$ are in the same quadrant, find the value of:

    **a** $\cos A$     **b** $\tan A$     **c** $\sin B$     **d** $\tan B$

**10** Given that $\tan A = -\dfrac{2}{3}$ and $\cos B = \dfrac{3}{4}$, where $A$ and $B$ are in the same quadrant, find the value of:

    **a** $\sin A$     **b** $\cos A$     **c** $\sin B$     **d** $\tan B$

**11** In the table, $0° \leqslant \theta \leqslant 360°$ and the missing function is from the list $\cos\theta$, $\tan\theta$, $\dfrac{1}{\sin\theta}$ and $\dfrac{1}{\tan\theta}$.

Without using a calculator, copy and complete the table.

| | $\theta = 120°$ | $\theta = \ldots\ldots$ | $\theta = 210°$ |
|---|---|---|---|
| $\ldots\ldots$ | $\ldots\ldots$ | $-1$ | $\dfrac{1}{\sqrt{3}}$ |
| $\sin\theta$ | $\ldots\ldots$ | $\dfrac{1}{\sqrt{2}}$ | $-\dfrac{1}{2}$ |
| $\dfrac{1}{\cos\theta}$ | $-2$ | $-\sqrt{2}$ | $\ldots\ldots$ |

## 5.4 Graphs of trigonometric functions

**EXPLORE 5.2**

Consider taking a ride on a Ferris wheel with radius 50 metres that rotates at a constant speed.

You enter the ride from a platform that is level with the centre of the wheel and the wheel turns in an anticlockwise direction through one complete turn.

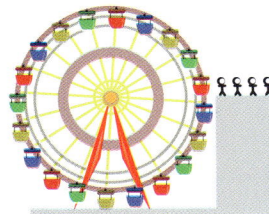

**1** Sketch the following two graphs and discuss their properties.

    **a** The graph of your *vertical displacement from the centre of the wheel* plotted against the *angle turned through*.

    **b** The graph of your *horizontal displacement from the centre of the wheel* plotted against the *angle turned through*.

**2** Discuss with your classmates what the two graphs would be like if you turned through two complete turns.

## The graphs of $y = \sin x$ and $y = \cos x$

Suppose that $OP$ makes an angle of $x$ with the positive horizontal axis and that $P$ moves around the unit circle, through one complete revolution.

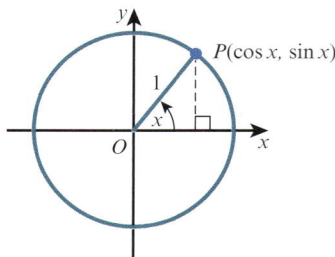

The coordinates of $P$ will be $(\cos x, \sin x)$.

The height of $P$ above the horizontal axis changes from $0 \rightarrow 1 \rightarrow 0 \rightarrow -1 \rightarrow 0$.

The graph of $\sin x$ against $x$ for $0° \leqslant x \leqslant 360°$ is therefore:

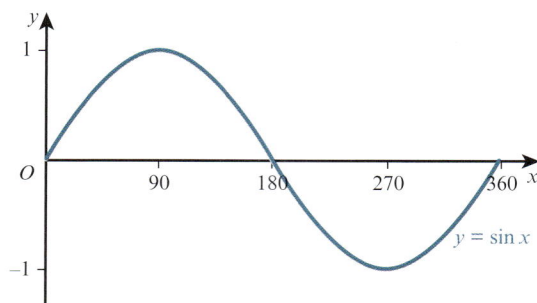

The displacement of $P$ from the vertical axis changes from $1 \rightarrow 0 \rightarrow -1 \rightarrow 0 \rightarrow 1$.

The graph of $\cos x$ against $x$ for $0° \leqslant x \leqslant 360°$ is therefore:

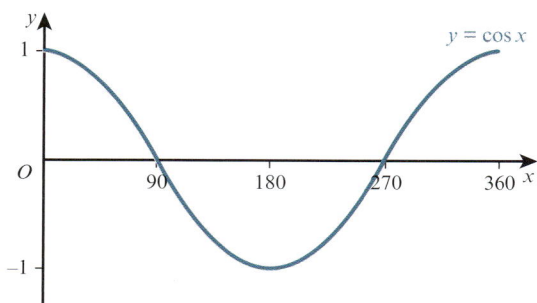

The graphs of $y = \sin x$ and $y = \cos x$ can be continued beyond $0° \leqslant x \leqslant 360°$:

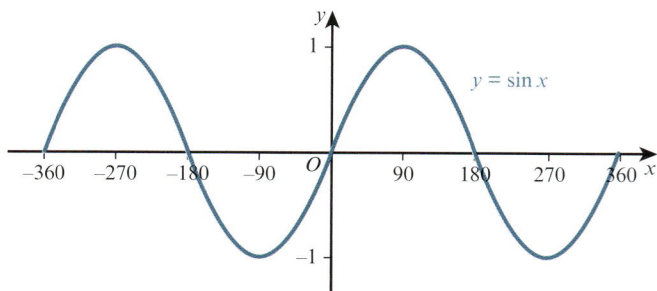

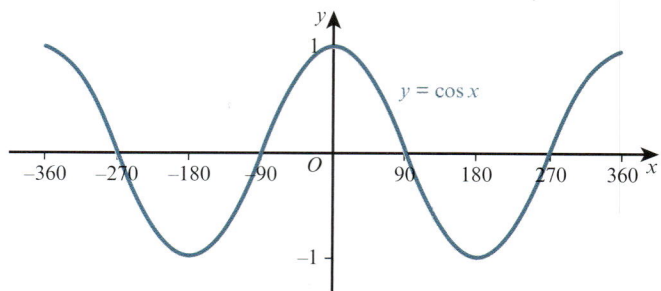

The sine and cosine functions are called **periodic functions** because they repeat themselves over and over again.

The **period** of a periodic function is defined as the length of one repetition or cycle.

The sine and cosine functions repeat every $360°$.

We say they have a period of $360°$ (or $2\pi$ radians).

The **amplitude** of a periodic function is defined as the distance between a maximum (or minimum) point and the principal axis.

The functions $y = \sin x$ and $y = \cos x$ both have amplitude 1.

The symmetry of the curve $y = \sin x$ shows these important relationships:

- $\sin(-x) = -\sin x$
- $\sin(180° - x) = \sin x$
- $\sin(180° + x) = -\sin x$
- $\sin(360° - x) = -\sin x$
- $\sin(360° + x) = \sin x$

**EXPLORE 5.3**

By considering the shape of the cosine curve, complete the following statements, giving your answers in terms of $\cos x$.

1  $\cos(-x) =$       2  $\cos(180° - x) =$       3  $\cos(180° + x) =$

4  $\cos(360° - x) =$       5  $\cos(360° + x) =$

## The graph of $y = \tan x$

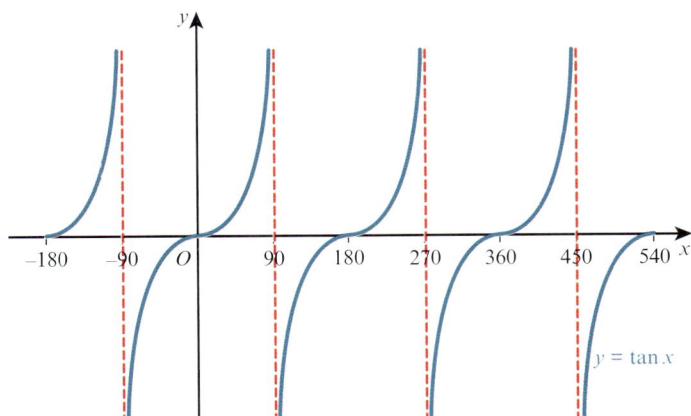

The tangent function behaves very differently to the sine and cosine functions.

The tangent function repeats its cycle every $180°$ so its period is $180°$ (or $\pi$ radians).

The red dashed lines at $x = \pm 90°$, $x = 270°$ and $x = 450°$ are called **asymptotes**. The branches of the graph get closer and closer to the asymptotes without ever reaching them.

The tangent function does not have an amplitude.

**EXPLORE 5.4**

By considering the shape of the tangent curve, complete the following statements, giving your answers in terms of $\tan x$.

1  $\tan(-x) =$     2  $\tan(180° - x) =$     3  $\tan(180° + x) =$

4  $\tan(360° - x) =$     5  $\tan(360° + x) =$

## Transformations of trigonometric functions

These rules for the transformations of the graph $y = f(x)$ can be used to transform graphs of trigonometric functions. These transformations include $y = a\,f(x)$, $y = f(ax)$, $y = f(x) + a$ and $y = f(x + a)$ and simple combinations of these.

### The graph of $y = a\sin x$

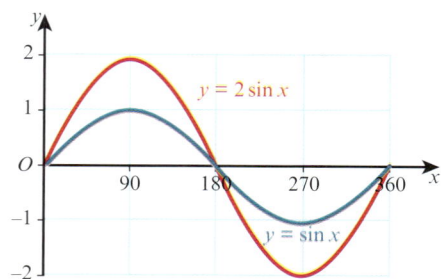

The graph of $y = 2\sin x$ is a stretch of the graph of $y = \sin x$.

It is a stretch, stretch factor 2, parallel to the $y$-axis.

The amplitude of $y = 2\sin x$ is 2 and the period is 360°.

### The graph of $y = \sin ax$

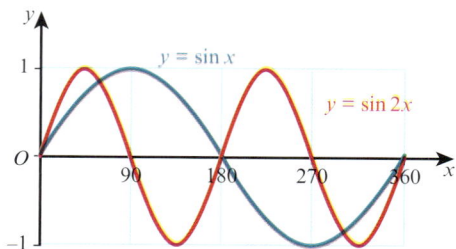

The graph of $y = \sin 2x$ is a stretch of the graph of $y = \sin x$.

It is a stretch, stretch factor $\dfrac{1}{2}$, parallel to the $x$-axis.

The amplitude of $y = \sin 2x$ is 1 and the period is 180°.

**REWIND**

In Section 2.6, you learnt some rules for the transformation of the graph $y = f(x)$. Here we will look at how these rules can be used to transform graphs of trigonometric functions.

## The graph of $y = a + \sin x$

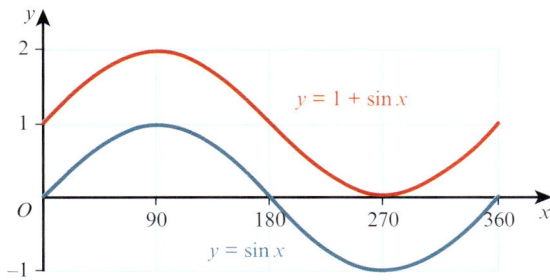

The graph of $y = 1 + \sin x$ is a translation of the graph of $y = \sin x$.

It is a translation of $\begin{pmatrix} 0 \\ 1 \end{pmatrix}$.

The amplitude of $y = 1 + \sin x$ is 1 and the period is $360°$.

## The graph of $y = \sin(x + a)$

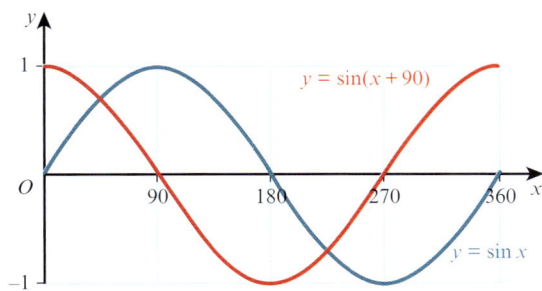

The graph of $y = \sin(x + 90)$ is a translation of the graph of $y = \sin x$.

It is a translation of $\begin{pmatrix} -90 \\ 0 \end{pmatrix}$.

The amplitude of $y = \sin(x + 90)$ is 1 and the period is $360°$.

**WORKED EXAMPLE 5.8**

On the same grid, sketch the graphs of $y = \sin x$ and $y = \sin(x - 90)$ for $0° \leqslant x \leqslant 360°$.

**Answer**

$y = \sin(x - 90)$ is a translation of the graph $y = \sin x$ by the vector $\begin{pmatrix} 90 \\ 0 \end{pmatrix}$.

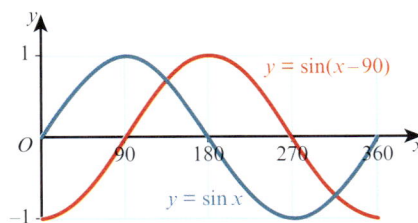

131

To sketch the graph of a trigonometric function, such as $y = 2\cos(x + 90) + 1$ for $0° \leqslant x \leqslant 360°$, we can build up the transformation in steps.

**Step 1:** Start with a sketch of $y = \cos x$.

Period $= 360°$

Amplitude $= 1$

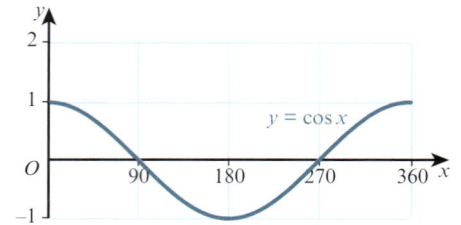

**Step 2:** Sketch the graph of $y = \cos(x + 90)$.

Translate $y = \cos x$ by the vector $\begin{pmatrix} -90 \\ 0 \end{pmatrix}$.

Period $= 360°$

Amplitude $= 1$

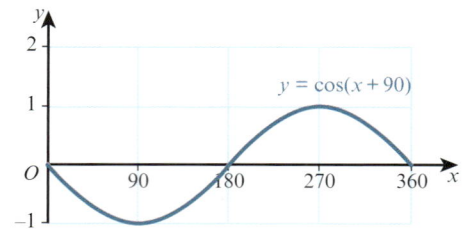

**Step 3:** Sketch the graph of $y = 2\cos(x + 90)$.

Stretch $y = \cos(x + 90)$ with stretch factor 2, parallel to the $y$-axis.

Period $= 360°$

Amplitude $= 2$

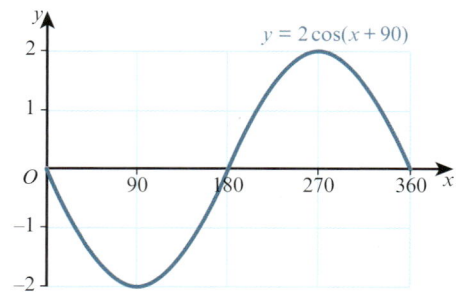

**Step 4:** Sketch the graph of $y = 2\cos(x + 90) + 1$.

Translate $y = 2\cos(x + 90)$ by the vector $\begin{pmatrix} 0 \\ 1 \end{pmatrix}$.

Period $= 360°$

Amplitude $= 2$

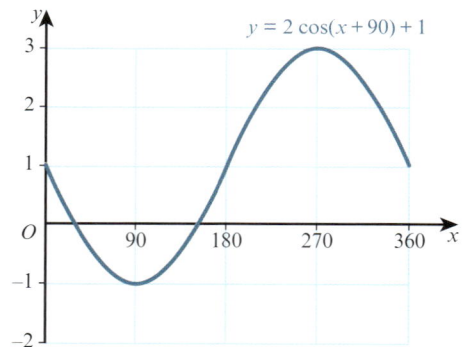

**WORKED EXAMPLE 5.9**

$f(x) = 3\cos 2x$ for $0° \leqslant x \leqslant 360°$.

a   Write down the period and amplitude of f.

b   Write down the coordinates of the maximum and minimum points on the curve $y = f(x)$.

c   Sketch the graph of $y = f(x)$.

d   Use your answer to part **c** to sketch the graph of $y = 1 + 3\cos 2x$.

**Answer**

**a** Period $= \dfrac{360°}{2} = 180°$

Amplitude $= 3$

**b** $y = \cos x$ has its maximum and minimum points at:

$(0°, 1), (180°, -1), (360°, 1), (540°, -1)$ and $(720°, 1)$

Hence, $f(x) = 3\cos 2x$ has its maximum and minimum points at:

$(0°, 3), (90°, -3), (180°, 3), (270°, -3)$ and $(360°, 3)$

**c**

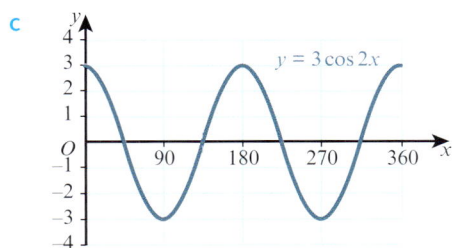

**d** $y = 1 + 3\cos 2x$ is a translation of the graph $y = 3\cos 2x$ by the vector $\begin{pmatrix} 0 \\ 1 \end{pmatrix}$.

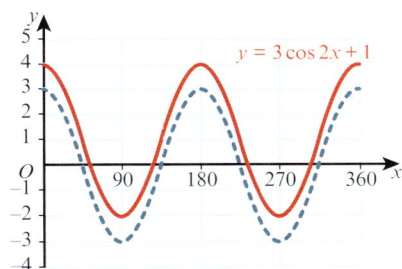

133

**WORKED EXAMPLE 5.10**

**a** On the same grid, sketch the graphs of $y = \sin 2x$ and $y = 1 + 3\cos 2x$ for $0° \leqslant x \leqslant 360°$.

**b** State the number of solutions of the equation $\sin 2x = 1 + 3\cos 2x$ for $0° \leqslant x \leqslant 360°$.

**Answer**

**a**

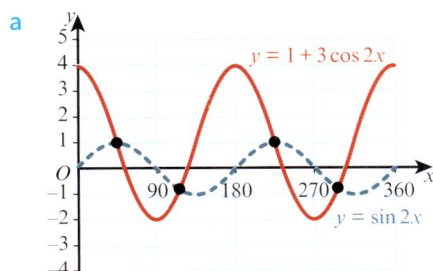

**b** The graphs of $y = \sin 2x$ and $y = 1 + 3\cos 2x$ intersect each other at four points in the interval.

Hence, the number of solutions of the equation $\sin 2x = 1 + 3\cos 2x$ is four.

**1** Write down the period of each of these functions.

    **a**  $y = \cos x°$
           **b**  $y = \sin 2x°$
           **c**  $y = 3\tan\dfrac{1}{2}x°$

    **d**  $y = 1 + 2\sin 3x°$
           **e**  $y = \tan(x - 30)°$
           **f**  $y = 5\cos(2x + 45)°$

**2** Write down the amplitude of each of these functions.

    **a**  $y = \sin x°$
           **b**  $y = 5\cos 2x°$
           **c**  $y = 7\sin\dfrac{1}{2}x°$

    **d**  $y = 2 - 3\cos 4x°$
           **e**  $y = 4\sin(2x + 60)°$
           **f**  $y = 2\sin(3x + 10)° + 5$

**3** Sketch the graph of each of these functions for $0° \leqslant x \leqslant 360°$.

    **a**  $y = 2\cos x$
           **b**  $y = \sin\dfrac{1}{2}x$
           **c**  $y = \tan 3x$

    **d**  $y = 3\cos 2x$
           **e**  $y = 1 + 3\cos x$
           **f**  $y = 2\sin 3x - 1$

    **g**  $y = \sin(x - 45)$
           **h**  $y = 2\cos(x + 60)$
           **i**  $y = \tan(x - 90)$

**4** **a** Sketch the graph of each of these functions for $0 \leqslant x \leqslant 2\pi$.

        **i**  $y = 2\sin x$
           **ii**  $y = \cos\left(x - \dfrac{\pi}{2}\right)$
           **iii**  $y = \sin\left(2x + \dfrac{\pi}{4}\right)$

    **b** Write down the coordinates of the turning points for your graph for part **a iii**.

**5** **a** On the same diagram, sketch the graphs of $y = \sin 2x$ and $y = 1 + \cos 2x$ for $0° \leqslant x \leqslant 360°$.

    **b** State the number of solutions of the equation $\sin 2x = 1 + \cos 2x$ for $0° \leqslant x \leqslant 360°$.

**6** **a** On the same diagram, sketch the graphs of $y = 2\sin x$ and $y = 2 + \cos 3x$ for $0 \leqslant x \leqslant 2\pi$.

    **b** Hence, state the number of solutions, in the interval $0 \leqslant x \leqslant 2\pi$, of the equation $2\sin x = 2 + \cos 3x$.

**7** **a** On the same diagram, sketch and label the graphs of $y = 3\sin x$ and $y = \cos 2x$ for the interval $0 \leqslant x \leqslant 2\pi$.

    **b** State the number of solutions of the equation $3\sin x = \cos 2x$ in the interval $0 \leqslant x \leqslant 2\pi$.

**8**

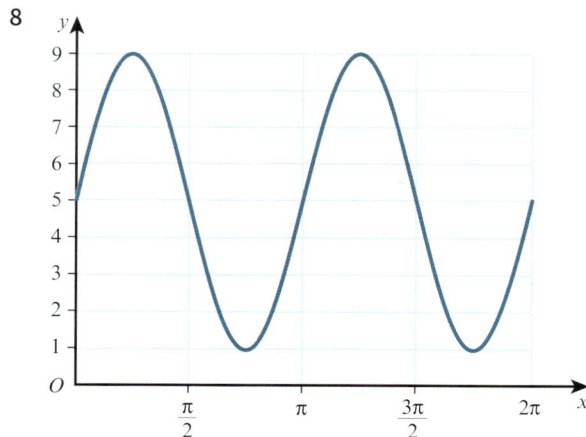

Part of the graph $y = a\sin bx + c$ is shown above.

Find the value of $a$, the value of $b$ and the value of $c$.

9

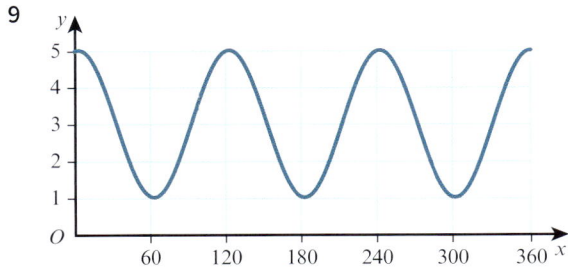

Part of the graph of $y = a + b\cos cx$ is shown above.

Write down the value of $a$, the value of $b$ and the value of $c$.

10 a  Sketch the graph of $y = 2\sin x$ for $-\pi \leqslant x \leqslant \pi$.

The straight line $y = kx$ intersects this curve at the maximum point.

b  Find the value of $k$. Give your answer in terms of $\pi$.

c  State the coordinates of the other points where the line intersects the curve.

11

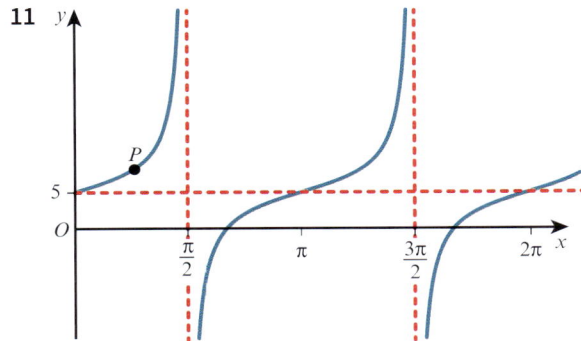

Part of the graph of $y = a\tan bx + c$ is shown above.

The graph passes through the point $P\left(\dfrac{\pi}{4}, 8\right)$.

Find the value of $a$, the value of $b$ and the value of $c$.

12  $f(x) = a + b\sin x$ for $0 \leqslant x \leqslant 2\pi$

Given that $f(0) = 3$ and that $f\left(\dfrac{7\pi}{6}\right) = 2$, find:

a  the value of $a$ and the value of $b$

b  the range of f.

13  $f(x) = a - b\cos x$ for $0° \leqslant x \leqslant 360°$, where $a$ and $b$ are positive constants.

The maximum value of $f(x)$ is 8 and the minimum value is −2.

a  Find the value of $a$ and the value of $b$.

b  Sketch the graph of $y = f(x)$.

14  $f(x) = a + b\sin cx$ for $0° \leqslant x \leqslant 360°$, where $a$ and $b$ are positive constants.

The maximum value of $f(x)$ is 9, the minimum value of $f(x)$ is 1 and the period is 120°.

Find the value of $a$, the value of $b$ and the value of $c$.

**15** $f(x) = A + 5\cos Bx$ for $0° \leqslant x \leqslant 120°$

The maximum value of $f(x)$ is 7 and the period is $60°$.

**a** Write down the value of $A$ and the value of $B$.

**b** Write down the amplitude of $f(x)$.

**c** Sketch the graph of $f(x)$.

**PS** **16** The graph of $y = \sin x$ is reflected in the line $x = \pi$ and then in the line $y = 1$.

Find the equation of the resulting function.

**PS** **17** The graph of $y = \cos x$ is reflected in the line $x = \dfrac{\pi}{2}$ and then in the line $y = 3$.

Find the equation of the resulting function.

## 5.5 Inverse trigonometric functions

The functions $y = \sin x$, $y = \cos x$ and $y = \tan x$ for $x \in \mathbb{R}$ are many-one functions. If, however, we suitably restrict the domain of each of these functions, it is possible to make the function one-one and hence we can define each inverse function.

The graphs of the suitably restricted functions $y = \sin x$, $y = \cos x$ and $y = \tan x$ and their inverse functions $y = \sin^{-1} x$, $y = \cos^{-1} x$ and $y = \tan^{-1} x$, together with their domains and ranges are:

**REWIND**

In Section 2.5 you learnt about the inverse of a function. Here we will look at the particular case of the inverse of a trigonometric function.

136

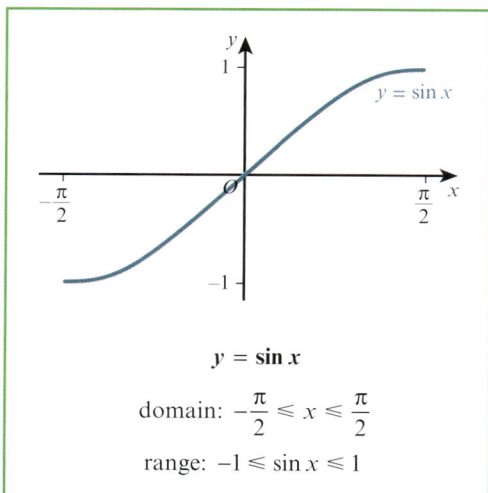

$y = \sin x$

domain: $-\dfrac{\pi}{2} \leqslant x \leqslant \dfrac{\pi}{2}$

range: $-1 \leqslant \sin x \leqslant 1$

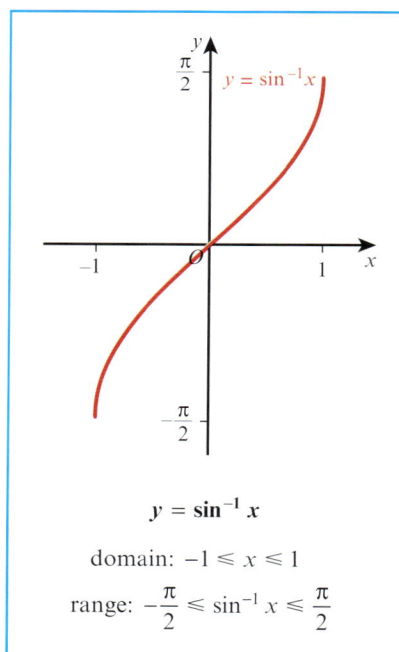

$y = \sin^{-1} x$

domain: $-1 \leqslant x \leqslant 1$

range: $-\dfrac{\pi}{2} \leqslant \sin^{-1} x \leqslant \dfrac{\pi}{2}$

**REWIND**

In Chapter 2 you learnt about functions and that only one-one functions can have an inverse function. You also learnt that if f and $f^{-1}$ are inverse functions, then the graph of $f^{-1}$ is a reflection of the graph of f in the line $y = x$.

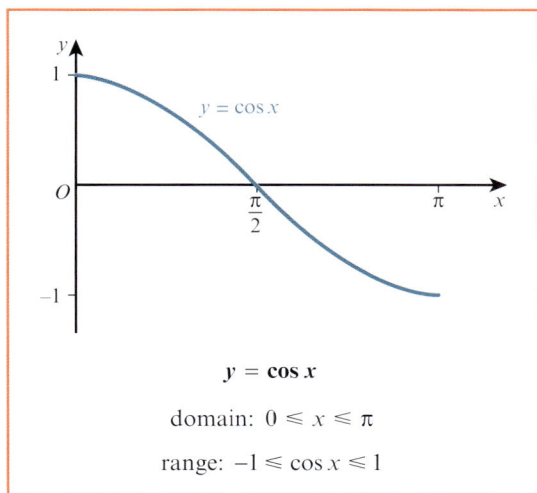

**y = cos x**

domain: $0 \leqslant x \leqslant \pi$

range: $-1 \leqslant \cos x \leqslant 1$

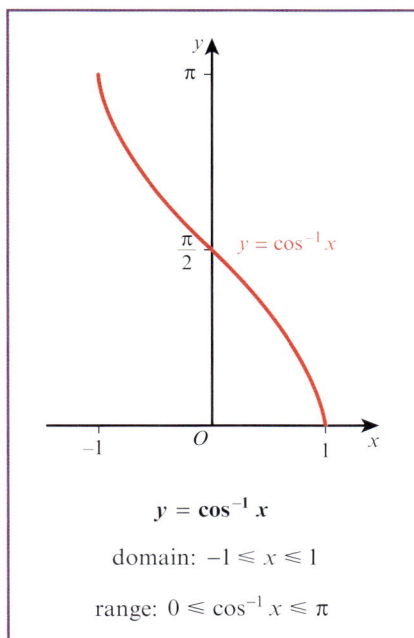

**y = cos⁻¹ x**

domain: $-1 \leqslant x \leqslant 1$

range: $0 \leqslant \cos^{-1} x \leqslant \pi$

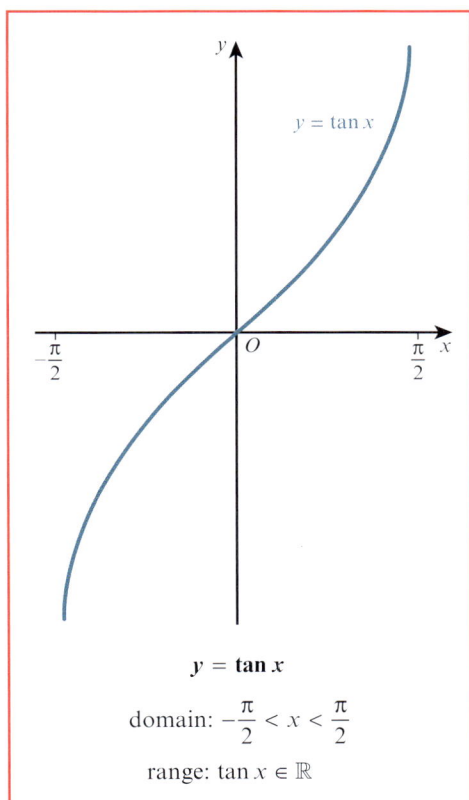

**y = tan x**

domain: $-\dfrac{\pi}{2} < x < \dfrac{\pi}{2}$

range: $\tan x \in \mathbb{R}$

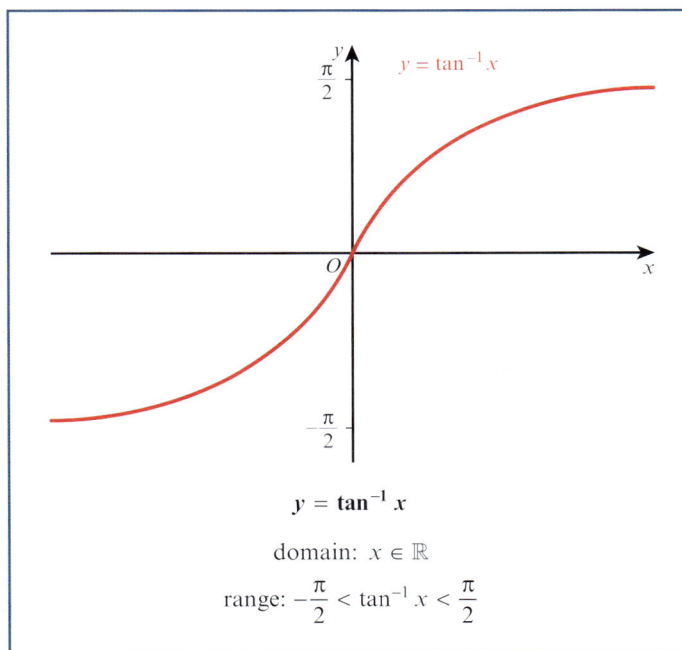

**y = tan⁻¹ x**

domain: $x \in \mathbb{R}$

range: $-\dfrac{\pi}{2} < \tan^{-1} x < \dfrac{\pi}{2}$

When solving the equation $\sin x = 0.5$ for $0 \leqslant x \leqslant \pi$, we can find one solution using the inverse functions:

$x = \sin^{-1} 0.5$

Using a calculator gives $x = \dfrac{\pi}{6}$.

The angle that the calculator gives is the one that lies in the range of the function $\sin^{-1}$.

(This is sometimes called the **principal angle**.)

The principal angle is the angle that lies in the range of the inverse trigonometric function.

There is a second angle, $x = \dfrac{5\pi}{6}$, that satisfies $\sin x = 0.5$ with $0 \leqslant x \leqslant \pi$. We can find this second angle either by using skills learnt earlier in this chapter or by using the symmetry of the curve $y = \sin x$.

**WORKED EXAMPLE 5.11**

🖩 The output of the $\sin^{-1}$, $\cos^{-1}$ and $\tan^{-1}$ functions can be given in degrees if that is needed.
Without using a calculator, write down, in degrees, the value of:

**a** $\sin^{-1} 0$

**b** $\cos^{-1}\left(\dfrac{\sqrt{3}}{2}\right)$

**c** $\tan^{-1}(-1)$

**Answer**

**a** $\sin^{-1} 0$ means the angle whose sine is 0, where $-90° \leqslant$ angle $\leqslant 90°$.

Hence, $\sin^{-1} 0 = 0°$.

**b** $\cos^{-1}\left(\dfrac{\sqrt{3}}{2}\right)$ means the angle whose cosine is $\dfrac{\sqrt{3}}{2}$, where $0° \leqslant$ angle $\leqslant 180°$.

Hence, $\cos^{-1}\left(\dfrac{\sqrt{3}}{2}\right) = 30°$.

**c** $\tan^{-1}(-1)$ means the angle whose tangent is $-1$, where $-90° \leqslant$ angle $\leqslant 90°$.

Hence, $\tan^{-1}(-1) = -45°$.

**WORKED EXAMPLE 5.12**

The function $f(x) = 3\sin\left(\dfrac{x}{2}\right) - 1$ is defined for the domain $-\pi \leqslant x \leqslant \pi$.

**a** Sketch the graph of $y = f(x)$ and explain why f has an inverse function.

**b** Find the range of f.

**c** Find $f^{-1}(x)$ and state its domain.

**Answer**

**a**

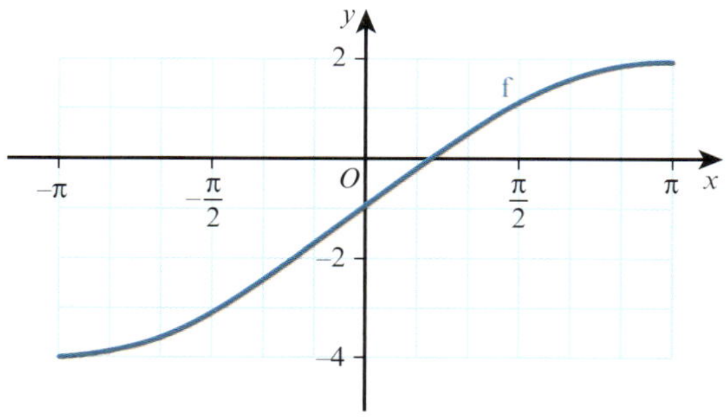

f has an inverse function because f is a one-one function with this domain.

**b** Range is $-4 \leqslant f(x) \leqslant 2$.

**c** $f(x) = 3\sin\left(\dfrac{x}{2}\right) - 1$

**Step 1:** Write the function as $y =$ $\longrightarrow$ $y = 3\sin\left(\dfrac{x}{2}\right) - 1$

**Step 2:** Interchange the $x$ and $y$ variables. $\longrightarrow$ $x = 3\sin\left(\dfrac{y}{2}\right) - 1$

**Step 3:** Rearrange to make $y$ the subject. $\longrightarrow$ $\dfrac{x+1}{3} = \sin\left(\dfrac{y}{2}\right)$

$$\dfrac{y}{2} = \sin^{-1}\left(\dfrac{x+1}{3}\right)$$

$$y = 2\sin^{-1}\left(\dfrac{x+1}{3}\right)$$

The inverse function is $f^{-1}(x) = 2\sin^{-1}\left(\dfrac{x+1}{3}\right)$ for $-4 \le x \le 2$.

## EXERCISE 5E

**1** Without using a calculator, write down, in degrees, the value of:

   **a**   $\cos^{-1} 1$           **b**   $\sin^{-1}\dfrac{1}{2}$           **c**   $\tan^{-1}\sqrt{3}$

   **d**   $\sin^{-1}(-1)$           **e**   $\tan^{-1}\left(-\sqrt{3}\right)$           **f**   $\cos^{-1}\left(-\dfrac{1}{\sqrt{2}}\right)$

**2** Without using a calculator, write down, in terms of $\pi$, the value of:

   **a**   $\sin^{-1} 0$           **b**   $\tan^{-1} 1$           **c**   $\cos^{-1}\left(\dfrac{1}{\sqrt{2}}\right)$

   **d**   $\tan^{-1}\left(-\dfrac{1}{\sqrt{3}}\right)$           **e**   $\cos^{-1}\left(-\dfrac{1}{2}\right)$           **f**   $\sin^{-1}\left(-\dfrac{\sqrt{3}}{2}\right)$

**3** Given that $\theta = \cos^{-1}\left(\dfrac{3}{5}\right)$, find the exact value of:

   **a**   $\sin^2\theta$           **b**   $\tan^2\theta$

**4** The function $f(x) = 3\sin x - 4$ is defined for the domain $-\dfrac{\pi}{2} \le x \le \dfrac{\pi}{2}$.

   **a**   Find the range of $f$.       **b**   Find $f^{-1}(x)$.

**5** The function $f(x) = 4 - 2\cos x$ is defined for the domain $0 \le x \le \pi$.

   **a**   Find the range of $f$ and sketch the graph of $y = f(x)$.

   **b**   Explain why $f$ has an inverse and find the equation of this inverse.

   **c**   Sketch the graph of $y = f^{-1}(x)$ on your graph for part **a**.

**6** The function $f(x) = 5 - 2\sin x$ is defined for the domain $\dfrac{\pi}{2} \le x \le p$.

   **a**   Find the largest value of $p$ for which $f$ has an inverse.

   **b**   For this value of $p$, find $f^{-1}(x)$ and state the domain of $f^{-1}$.

**7** The function $f(x) = 4\cos\left(\dfrac{x}{2}\right) - 5$ is defined for the domain $0 \le x \le 2\pi$.

   **a**   Find the range of $f$.       **b**   Find $f^{-1}(x)$ and state its range.

139

## 5.6 Trigonometric equations

Consider solving the equation $\sin x = 0.5$ for $-360° \leqslant x \leqslant 360°$.

One solution is given by $x = \sin^{-1}(0.5) = \dfrac{\pi}{6}$ (or $30°$).

There are, however, many more values of $x$ for which $\sin x = 0.5$.

The graph of $y = \sin x$ for $-360° \leqslant x \leqslant 360°$ is:

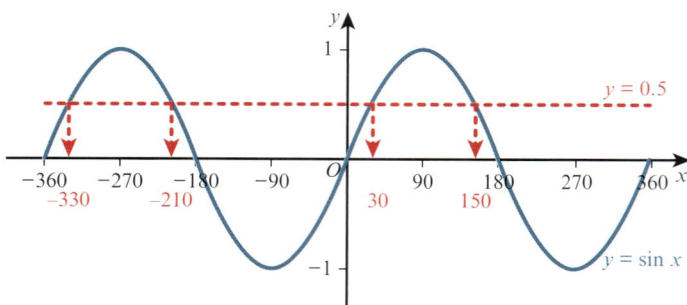

The graph shows there are four values of $x$, between $-360°$ and $360°$, for which $\sin x = 0.5$.

We can use the calculator value of $x = 30°$, together with the symmetry of the curve to find the remaining answers.

Hence, the solution of $\sin x = 0.5$ for $-360° \leqslant x \leqslant 360°$ is:

$$x = -330°, \ -210°, \ 30° \ \text{ or } \ 150°$$

**WORKED EXAMPLE 5.13**

Solve $\cos x = -0.7$ for $0° \leqslant x \leqslant 360°$.

**Answer**

$\cos x = -0.7$

One solution is $x = 134.4°$

> Use a calculator to find $\cos^{-1}(-0.7)$, correct to 1 decimal place.

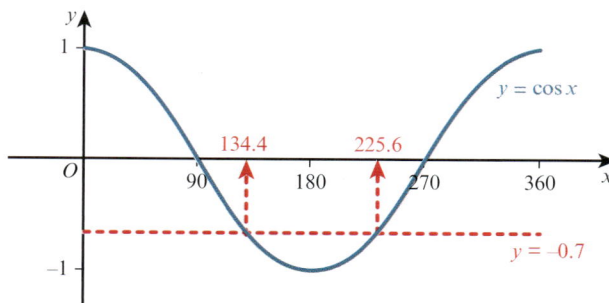

The sketch graph shows there are two values of $x$, between $-0°$ and $360°$, for which $\cos x = -0.7$.

Using the symmetry of the curve, the second value is $(360° - 134.4°) = 225.6°$.

Hence, the solution of $\cos x = -0.7$ for $0° \leqslant x \leqslant 360°$ is:

$$x = 134.4° \ \text{ or } \ 225.6° \quad \text{(correct to 1 decimal place)}$$

**WORKED EXAMPLE 5.14**

Solve $\tan 2A = -2.1$ for $0° \leqslant A \leqslant 180°$.

**Answer**

$$\tan 2A = -2.1 \qquad \text{Let } 2A = x.$$

$$\tan x = -2.1 \qquad \text{Use a calculator to find } \tan^{-1}(-2.1).$$

A solution is $x = -64.54°$.

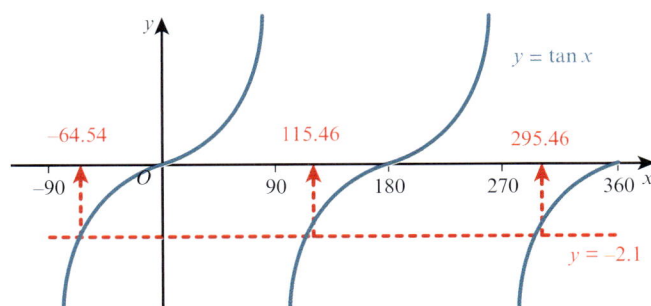

Using the symmetry of the curve:

$$x = -64.54° \qquad x = (-64.54° + 180°) \qquad x = (115.46° + 180°)$$
$$= 115.46° \qquad = 295.46°$$

Using $2A = x$:

$$2A = -64.54° \qquad 2A = 115.46° \qquad 2A = 295.46°$$
$$A = -32.3° \qquad A = 57.7° \qquad A = 147.7°$$

Hence, the solution of $\tan 2A = -2.1$ for $0° \leqslant A \leqslant 180°$ is:

$$A = 57.7° \text{ or } 147.7° \text{ (correct to 1 decimal place)}$$

141

**WORKED EXAMPLE 5.15**

Solve $\sin\left(2A + \dfrac{\pi}{6}\right) = 0.6$ for $0 \leqslant A \leqslant \pi$.

**Answer**

$$\sin\left(2A + \frac{\pi}{6}\right) = 0.6 \qquad \text{Let } 2A + \frac{\pi}{6} = x.$$

$$\sin x = 0.6 \qquad \text{Use a calculator to find } \sin^{-1}(0.6).$$

$$x = 0.6435 \text{ radians}$$

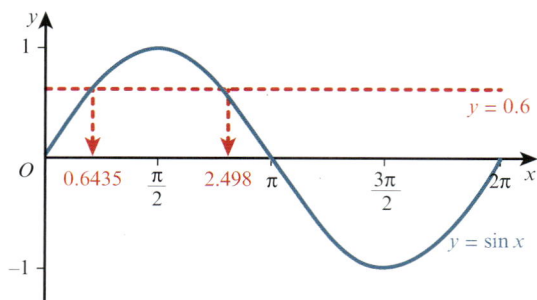

Using the symmetry of the curve:

$$x = 0.6435 \qquad\qquad x = \pi - 0.6435$$
$$= 2.498$$

Using $2A + \dfrac{\pi}{6} = x$:

$$2A + \dfrac{\pi}{6} = 0.6435 \qquad\qquad 2A + \dfrac{\pi}{6} = 2.498$$

$$A = \dfrac{1}{2}\left(0.6435 - \dfrac{\pi}{6}\right) \qquad\qquad A = \dfrac{1}{2}\left(2.498 - \dfrac{\pi}{6}\right)$$

$$A = 0.0600 \qquad\qquad A = 0.987$$

Hence, the solution of $\sin\left(2A + \dfrac{\pi}{6}\right) = 0.6$ for $0 \leqslant A \leqslant \pi$ is:

$$A = 0.0600 \text{ or } 0.987 \text{ radians (correct to 3 significant figures)}$$

Consider a right-angled triangle.

Two very important rules can be found using this triangle.

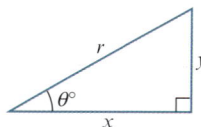

**Rule 1**

$$\tan\theta = \dfrac{y}{x} \qquad\qquad \text{Divide numerator and denominator by } r.$$

$$= \dfrac{\left(\dfrac{y}{r}\right)}{\left(\dfrac{x}{r}\right)} \qquad\qquad \text{Use } \sin\theta = \dfrac{y}{r} \text{ and } \cos\theta = \dfrac{x}{r}.$$

---

**KEY POINT 5.2**

$$\tan\theta = \dfrac{\sin\theta}{\cos\theta} \text{ for all } \theta \text{ with } \cos\theta \neq 0.$$

---

**Rule 2**

$$x^2 + y^2 = r^2 \qquad\qquad \text{Divide both sides by } r^2.$$

$$\left(\dfrac{x}{r}\right)^2 + \left(\dfrac{y}{r}\right)^2 = 1 \qquad\qquad \text{Use } \cos\theta = \dfrac{x}{r} \text{ and } \sin\theta = \dfrac{y}{r}.$$

> ## 🔍 KEY POINT 5.3
>
> $\cos^2\theta + \sin^2\theta = 1$ for all $\theta$.

If we use the unit circle definition of the trigonometric functions, we discover that these two important rules are true for all valid values of $\theta$. We can use them to help solve more complicated trigonometric equations.

## WORKED EXAMPLE 5.16

Solve $3\cos^2 x - \sin x \cos x = 0$ for $0° \leqslant x \leqslant 360°$.

**Answer**

$3\cos^2 x - \sin x \cos x = 0$        Factorise.

$\cos x(3\cos x - \sin x) = 0$

$\cos x = 0$   or   $3\cos x - \sin x = 0$

    $x = 90°, 270°$        $\sin x = 3\cos x$

                     $\tan x = 3$

                         $x = 71.6$   or   $180 + 71.6$

                         $x = 71.6°$   or   $251.6°$

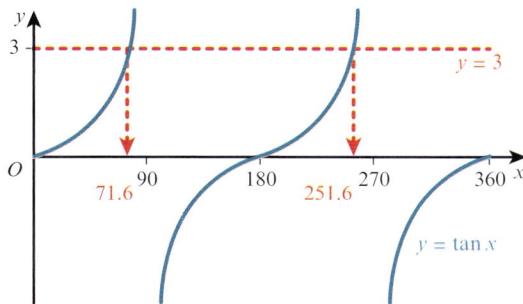

The solution of $3\cos^2 x - \sin x \cos x = 0$   for   $0° \leqslant x \leqslant 360°$ is:

$$x = 71.6°, 90°, 251.6° \text{ or } 270°$$

## WORKED EXAMPLE 5.17

Solve $2\sin^2 x + 3\cos x - 3 = 0$ for $0 \leqslant x \leqslant 2\pi$.

**Answer**

      $2\sin^2 x + 3\cos x - 3 = 0$        Replace $\sin^2 x$ with $1 - \cos^2 x$.

$2(1 - \cos^2 x) + 3\cos x - 3 = 0$        Expand brackets and collect terms.

      $2\cos^2 x - 3\cos x + 1 = 0$

    $(2\cos x - 1)(\cos x - 1) = 0$        Factorise.

$$\cos x = \frac{1}{2} \quad \text{or} \quad \cos x = 1$$

$$x = \frac{\pi}{3} \text{ or } 2\pi - \frac{\pi}{3} \qquad x = 0 \text{ or } 2\pi$$

$$x = \frac{\pi}{3} \text{ or } \frac{5\pi}{3}$$

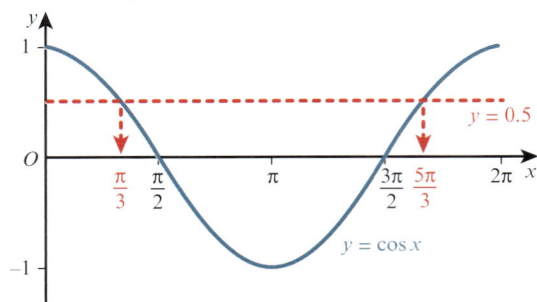

The solution of $2\sin^2 x + 3\cos x - 1 = 0$ for $0 \leqslant x \leqslant 2\pi$ is:

$$x = 0, \frac{\pi}{3}, \frac{5\pi}{3} \text{ or } 2\pi$$

## EXERCISE 5F

1   Solve each of these equations for $0° \leqslant x \leqslant 360°$.

    **a**   $\tan x = 1.5$        **b**   $\sin x = 0.4$        **c**   $\cos x = 0.7$        **d**   $\sin x = -0.3$

    **e**   $\cos x = -0.6$      **f**   $\tan x = -2$        **g**   $2\cos x - 1 = 0$        **h**   $5\sin x + 3 = 0$

2   Solve each of the these equations for $0 \leqslant x \leqslant 2\pi$.

    **a**   $\sin x = 0.3$        **b**   $\cos x = 0.5$        **c**   $\tan x = 3$        **d**   $\sin x = -0.7$

    **e**   $\tan x = -3$        **f**   $\cos x = -0.5$        **g**   $4\sin x = 3$        **h**   $5\tan x + 7 = 0$

3   Solve each of these equations for $0° \leqslant x \leqslant 180°$.

    **a**   $\cos 2x = 0.6$        **b**   $\sin 3x = 0.8$        **c**   $\tan 2x = 4$        **d**   $\sin 2x = -0.5$

    **e**   $3\cos 2x = 2$        **f**   $5\sin 2x = -4$        **g**   $4 + 2\tan 2x = 0$        **h**   $1 - 5\sin 2x = 0$

4   Solve each of these equations for the given domains.

    **a**   $\sin(x - 60°) = 0.5$    for $0° \leqslant x \leqslant 360°$       **b**   $\cos\left(x + \dfrac{\pi}{6}\right) = -0.5$  for $0 < x < 2\pi$

    **c**   $\cos(2x + 45°) = 0.8$  for $0° \leqslant x \leqslant 180°$       **d**   $3\sin(2x - 4) = 2$    for $0 < x < \pi$

    **e**   $2\tan\left(\dfrac{x}{2}\right) + \sqrt{3} = 0$  for $0° \leqslant x \leqslant 540°$       **f**   $\sqrt{2}\sin\left(\dfrac{x}{3} + \dfrac{\pi}{4}\right) = 1$  for $0 < x < 4\pi$

**5** Solve each of these equations for $0° \leqslant x \leqslant 360°$.

    **a**    $2\sin x = \cos x$                     **b**    $2\sin x - 3\cos x = 0$

    **c**    $4\sin x + 7\cos x = 0$            **d**    $3\cos 2x - 4\sin 2x = 0$

**6** Solve $4\sin(2x + 0.3) - 5\cos(2x + 0.3) = 0$ for $0 \leqslant x \leqslant \pi$.

**7** Solve each of these equations for $0° \leqslant x \leqslant 360°$.

    **a**    $\sin x \cos(x - 60) = 0$           **b**    $5\sin^2 x - 3\sin x = 0$

    **c**    $\tan^2 x = 5\tan x$                 **d**    $\sin^2 x + 2\sin x \cos x = 0$

    **e**    $2\sin x \cos x = \sin x$           **f**    $\sin x \tan x = 4\sin x$

**8** Solve each of these equations for $0° \leqslant x \leqslant 360°$.

    **a**    $4\cos^2 x = 1$                    **b**    $4\tan^2 x = 9$

**9** Solve each of these equations for $0° \leqslant x \leqslant 360°$.

    **a**    $2\sin^2 x + \sin x - 1 = 0$       **b**    $\tan^2 x + 2\tan x - 3 = 0$

    **c**    $3\cos^2 x - 2\cos x - 1 = 0$     **d**    $2\sin^2 x - \cos x - 1 = 0$

    **e**    $3\cos^2 x - 3 = \sin x$          **f**    $\cos x + 5 = 6\sin^2 x$

    **g**    $2\cos^2 x - \sin^2 x - 2\sin x - 1 = 0$     **h**    $1 + \tan x \cos x = 2\cos^2 x$

**10** Solve each of these equations for $0 \leqslant x \leqslant 2\pi$.

    **a**    $4\tan x = 3\cos x$              **b**    $2\cos^2 x + 5\sin x = 4$

**11** Solve $\sin^2 x + 3\sin x \cos x + 2\cos^2 x = 0$ for $0 \leqslant x \leqslant 2\pi$.

## 5.7 Trigonometric identities

$x + x = 2x$ is called an **identity** because it is true for all values of $x$.

When writing an identity, we often replace the $=$ symbol with a $\equiv$ symbol to emphasise that it is an identity.

Two commonly used trigonometric identities are:

$$\sin^2 x + \cos^2 x \equiv 1 \quad \text{and} \quad \tan x \equiv \frac{\sin x}{\cos x}$$

In this section you will learn how to use these two identities to simplify expressions and to prove other more complicated identities that involve $\sin x$, $\cos x$ and $\tan x$.

When proving an identity, it is usual to start with the more complicated side of the identity and prove that it simplifies to the less complicated side.

**WORKED EXAMPLE 5.18**

Express $4\cos^2 x - 3\sin^2 x$ in terms of $\cos x$.

**Answer**

$$4\cos^2 x - 3\sin^2 x \equiv 4\cos^2 x - 3(1 - \cos^2 x)$$

Replace $\sin^2 x$ with $1 - \cos^2 x$.

$$\equiv 4\cos^2 x - 3 + 3\cos^2 x$$

$$\equiv 7\cos^2 x - 3$$

**WORKED EXAMPLE 5.19**

> **TIP**
>
> LHS means left-hand side and RHS means right-hand side.

Prove the identity $\dfrac{1 + \sin x}{\cos x} + \dfrac{\cos x}{1 + \sin x} \equiv \dfrac{2}{\cos x}$.

**Answer**

$$\text{LHS} \equiv \frac{1 + \sin x}{\cos x} + \frac{\cos x}{1 + \sin x}$$

Add the two fractions.

$$\equiv \frac{(1 + \sin x)^2 + \cos^2 x}{\cos x(1 + \sin x)}$$

Expand the brackets in the numerator.

$$\equiv \frac{1 + 2\sin x + \sin^2 x + \cos^2 x}{\cos x(1 + \sin x)}$$

Use $\sin^2 x + \cos^2 x \equiv 1$.

$$\equiv \frac{2 + 2\sin x}{\cos x(1 + \sin x)}$$

Factorise the numerator.

$$\equiv \frac{2(1 + \sin x)}{\cos x(1 + \sin x)}$$

Divide numerator and denominator by $1 + \sin x$.

$$\equiv \frac{2}{\cos x}$$

$$\equiv \text{RHS}$$

**WORKED EXAMPLE 5.20**

Prove the identity $\dfrac{1 + \sin x}{1 - \sin x} \equiv \left( \tan x + \dfrac{1}{\cos x} \right)^2$.

**Answer**

$$\text{RHS} \equiv \left( \tan x + \frac{1}{\cos x} \right)^2$$

Use $\tan x = \dfrac{\sin x}{\cos x}$.

$$\equiv \left( \frac{\sin x}{\cos x} + \frac{1}{\cos x} \right)^2$$

Add the two fractions.

$$\equiv \left( \frac{\sin x + 1}{\cos x} \right)^2$$

146

$$\equiv \frac{(1+\sin x)^2}{\cos^2 x}$$ ............................................ Replace $\cos^2 x$ with $1-\sin^2 x$ in the denominator.

$$\equiv \frac{(1+\sin x)^2}{1-\sin^2 x}$$ ............................................ Use $1-\sin^2 x = (1+\sin x)(1-\sin x)$.

$$\equiv \frac{(1+\sin x)^2}{(1+\sin x)(1-\sin x)}$$ ............................................ Divide numerator and denominator by $1+\sin x$.

$$\equiv \frac{1+\sin x}{1-\sin x}$$

$$\equiv \text{LHS}$$

## EXPLORE 5.5

Equivalent trigonometric expressions:

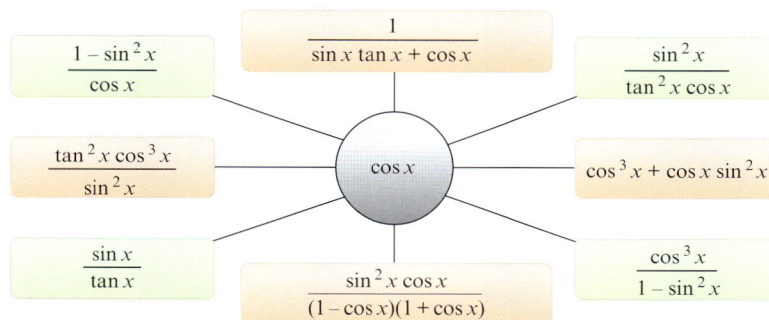

Discuss why each of the trigonometric expressions in the coloured boxes simplifies to $\cos x$.

Create trigonometric expressions of your own that simplify to $\sin x$.

(Your expressions must contain at least two different trigonometric ratios.)

Compare your answers with those of your classmates.

## EXERCISE 5G

1 Express $2\sin^2 x - 7\cos^2 x + 4$ in terms of $\sin x$.

2 Prove each of these identities.

a $\cos x \tan x \equiv \sin x$

b $\dfrac{1-\cos^2 x}{\sin x \cos x} \equiv \tan x$

c $\dfrac{\cos^2 x}{1-\sin x} \equiv 1+\sin x$

d $\dfrac{1+\sin x - \sin^2 x}{\cos x} \equiv \cos x + \tan x$

e $\dfrac{\cos^2 x - \sin^2 x}{\cos x + \sin x} + \sin x \equiv \cos x$

f $\cos^4 x + \sin^2 x \cos^2 x \equiv \cos^2 x$

147

**3** Prove each of these identities.

    **a** $(\sin x + \cos x)^2 \equiv 1 + 2\sin x \cos x$
        **b** $2(1 + \cos x) - (1 + \cos x)^2 \equiv \sin^2 x$

    **c** $2 - (\sin x + \cos x)^2 \equiv (\sin x - \cos x)^2$
        **d** $(\cos^2 x - 2)^2 - 3\sin^2 x \equiv \cos^4 x + \sin^2 x$

**4** Prove each of these identities.

    **a** $\cos^2 x - \sin^2 x \equiv 2\cos^2 x - 1$
        **b** $\cos^2 x - \sin^2 x \equiv 1 - 2\sin^2 x$

    **c** $\tan^2 x - \sin^2 x \equiv \tan^2 x \sin^2 x$
        **d** $\cos^4 x + \sin^2 x \equiv \sin^4 x + \cos^2 x$

**5** Prove each of these identities.

    **a** $\dfrac{\cos^2 x - \sin^2 x}{\cos x - \sin x} \equiv \cos x + \sin x$
        **b** $\sin^4 x - \cos^4 x \equiv 2\sin^2 x - 1$

    **c** $\dfrac{\cos^4 x - \sin^4 x}{\cos^2 x} \equiv 1 - \tan^2 x$
        **d** $\dfrac{\cos x}{\tan x(1 - \sin x)} \equiv 1 + \dfrac{1}{\sin x}$

    **e** $\dfrac{\sin x - \cos x}{\sin x + \cos x} \equiv \dfrac{\tan x - 1}{\tan x + 1}$
        **f** $\left(\dfrac{1}{\sin x} - \dfrac{1}{\tan x}\right)^2 \equiv \dfrac{1 - \cos x}{1 + \cos x}$

    **g** $\dfrac{\tan x + 1}{\sin x \tan x + \cos x} \equiv \sin x + \cos x$
        **h** $\dfrac{\sin^2 x(1 - \cos^2 x)}{\cos^2 x(1 - \sin^2 x)} \equiv \tan^4 x$

**6** Prove each of these identities.

    **a** $\dfrac{1}{\cos x} - \dfrac{\cos x}{1 + \sin x} \equiv \tan x$
        **b** $\tan x + \dfrac{1}{\tan x} \equiv \dfrac{1}{\sin x \cos x}$

    **c** $\dfrac{1}{\cos x} - \cos x \equiv \sin x \tan x$
        **d** $\dfrac{\sin x}{1 + \cos x} + \dfrac{1 + \cos x}{\sin x} \equiv \dfrac{2}{\sin x}$

    **e** $\dfrac{\sin x}{1 - \sin x} + \dfrac{\sin x}{1 + \sin x} \equiv \dfrac{2\tan x}{\cos x}$
        **f** $\dfrac{1 + \cos x}{1 - \cos x} - \dfrac{1 - \cos x}{1 + \cos x} \equiv \dfrac{4}{\sin x \tan x}$

**7** Show that $(1 + \cos x)^2 + (1 - \cos x)^2 + 2\sin^2 x$ has a constant value for all $x$ and state this value.

**8**   **a** Express $7\sin^2 x + 4\cos^2 x$ in the form $a + b\sin^2 x$.

    **b** State the range of the function $f(x) = 7\sin^2 x + 4\cos^2 x$, for the domain $0 \leqslant x \leqslant 2\pi$.

**9**   **a** Express $4\sin\theta - \cos^2\theta$ in the form $(\sin\theta + a)^2 + b$.

    **b** Hence, state the maximum and minimum values of $4\sin\theta - \cos^2\theta$, for the domain $0 \leqslant \theta \leqslant 2\pi$.

**PS** **P**  **10**  **a** Given that $a = \dfrac{1 - \sin\theta}{2\cos\theta}$, show that $\dfrac{1}{a} = \dfrac{2(1 + \sin\theta)}{\cos\theta}$.

    **b** Hence, find $\sin\theta$ and $\cos\theta$ in terms of $a$.

## 5.8 Further trigonometric equations

This section uses trigonometric identities to help solve some more complex trigonometric equations.

**WORKED EXAMPLE 5.21**

**a** Prove the identity $\dfrac{1 - \tan^2 \theta}{1 + \tan^2 \theta} \equiv 2\cos^2 \theta - 1$.

**b** Hence, solve the equation $\dfrac{1 - \tan^2 \theta}{1 + \tan^2 \theta} = 5\cos \theta - 3$ for $0° \leqslant \theta \leqslant 360°$.

**Answer**

**a** $\text{LHS} \equiv \dfrac{1 - \tan^2 \theta}{1 + \tan^2 \theta}$ 

Use $\tan \theta = \dfrac{\sin \theta}{\cos \theta}$.

$\equiv \dfrac{1 - \left(\dfrac{\sin \theta}{\cos \theta}\right)^2}{1 + \left(\dfrac{\sin \theta}{\cos \theta}\right)^2}$ 

Multiply numerator and denominator by $\cos^2 \theta$.

$\equiv \dfrac{\cos^2 \theta - \sin^2 \theta}{\cos^2 \theta + \sin^2 \theta}$ 

Use $\sin^2 \theta + \cos^2 \theta = 1$.

$\equiv \cos^2 \theta - \sin^2 \theta$ 

Replace $\sin^2 \theta$ with $1 - \cos^2 \theta$.

$\equiv \cos^2 \theta - (1 - \cos^2 \theta)$ 

Simplify.

$\equiv 2\cos^2 \theta - 1$

$\equiv \text{RHS}$

**b** $\dfrac{1 - \tan^2 \theta}{1 + \tan^2 \theta} = 5\cos \theta - 3$ 

Use the result from part **a**.

$2\cos^2 \theta - 1 = 5\cos \theta - 3$ 

Rearrange.

$2\cos^2 \theta - 5\cos \theta + 2 = 0$ 

Factorise.

$(2\cos \theta - 1)(\cos \theta - 2) = 0$

$\cos \theta = \dfrac{1}{2}$ or $\cos \theta = 2$ 

$\cos \theta = 2$ has no solutions.

$\theta = \cos^{-1}\left(\dfrac{1}{2}\right)$

$\theta = 60°$ or $\theta = 360° - 60°$

Solution is $\theta = 60°$ or $\theta = 300°$.

**EXERCISE 5H**

1   a   Show that the equation $\cos\theta + \sin\theta = 5\cos\theta$ can be written in the form $\tan\theta = k$.

   b   Hence, solve the equation $\cos\theta + \sin\theta = 5\cos\theta$ for $0° \leqslant \theta \leqslant 360°$.

2   a   Show that the equation $3\sin^2\theta + 5\sin\theta\cos\theta = 2\cos^2\theta$ can be written in the form
       $3\tan^2\theta + 5\tan\theta - 2 = 0$.

   b   Hence, solve the equation $3\sin^2\theta + 5\sin\theta\cos\theta = 2\cos^2\theta$ for $0° \leqslant \theta \leqslant 180°$.

3   a   Show that the equation $8\sin^2\theta + 2\cos^2\theta - \cos\theta = 6$ can be written in the form $6\cos^2\theta + \cos\theta - 2 = 0$.

   b   Hence, solve the equation $8\sin^2\theta + 2\cos^2\theta - \cos\theta = 6$ for $0° \leqslant \theta \leqslant 360°$.

4   a   Show that the equation $4\sin^4\theta + 14 = 19\cos^2\theta$ can be written in the form $4x^2 + 19x - 5 = 0$, where
       $x = \sin^2\theta$.

   b   Hence, solve the equation $4\sin^4\theta + 14 = 19\cos^2\theta$ for $0° \leqslant \theta \leqslant 360°$.

5   a   Show that the equation $\sin\theta\tan\theta = 3$ can be written in the form $\cos^2\theta + 3\cos\theta - 1 = 0$.

   b   Hence, solve the equation $\sin\theta\tan\theta = 3$ for $0° \leqslant \theta \leqslant 360°$.

6   a   Show that the equation $5(2\sin\theta - \cos\theta) = 4(\sin\theta + 2\cos\theta)$ can be written in the form $\tan\theta = \dfrac{13}{6}$.

   b   Hence, solve the equation $5(2\sin\theta - \cos\theta) = 4(\sin\theta + 2\cos\theta)$ for $0° \leqslant \theta \leqslant 360°$.

7   a   Prove the identity $\dfrac{\sin\theta}{1+\cos\theta} + \dfrac{1+\cos\theta}{\sin\theta} \equiv \dfrac{2}{\sin\theta}$.

   b   Hence, solve the equation $\dfrac{\sin\theta}{1+\cos\theta} + \dfrac{1+\cos\theta}{\sin\theta} = 1 + 3\sin\theta$ for $0° \leqslant \theta \leqslant 360°$.

8   a   Prove the identity $\dfrac{\cos\theta}{\tan\theta\,(1+\sin\theta)} \equiv \dfrac{1}{\sin\theta} - 1$.

   b   Hence, solve the equation $\dfrac{\cos\theta}{\tan\theta\,(1+\sin\theta)} = 1$ for $0° \leqslant \theta \leqslant 360°$.

9   a   Prove the identity $\dfrac{1}{1+\sin\theta} + \dfrac{1}{1-\sin\theta} \equiv \dfrac{2}{\cos^2\theta}$.

   b   Hence, solve the equation $\cos\theta\left(\dfrac{1}{1+\sin\theta} + \dfrac{1}{1-\sin\theta}\right) = 5$ for $0° \leqslant \theta \leqslant 360°$.

10  a   Prove the identity $\left(\dfrac{1}{\sin\theta} + \dfrac{1}{\tan\theta}\right)^2 \equiv \dfrac{1+\cos\theta}{1-\cos\theta}$.

   b   Hence, solve the equation $\left(\dfrac{1}{\sin\theta} + \dfrac{1}{\tan\theta}\right)^2 = 2$ for $0° \leqslant \theta \leqslant 360°$.

11  a   Prove the identity $\cos^4\theta - \sin^4\theta \equiv 2\cos^2\theta - 1$.

   b   Hence, solve the equation $\cos^4\theta - \sin^4\theta = \dfrac{1}{2}$ for $0° \leqslant \theta \leqslant 360°$.

# Checklist of learning and understanding

**Exact values of trigonometric functions**

| | $\sin \theta$ | $\cos \theta$ | $\tan \theta$ |
|---|---|---|---|
| $\theta = 30° = \dfrac{\pi}{6}$ | $\dfrac{1}{2}$ | $\dfrac{\sqrt{3}}{2}$ | $\dfrac{1}{\sqrt{3}}$ |
| $\theta = 45° = \dfrac{\pi}{4}$ | $\dfrac{1}{\sqrt{2}}$ | $\dfrac{1}{\sqrt{2}}$ | $1$ |
| $\theta = 60° = \dfrac{\pi}{3}$ | $\dfrac{\sqrt{3}}{2}$ | $\dfrac{1}{2}$ | $\sqrt{3}$ |

**Positive and negative angles**

- Angles measured anticlockwise from the positive $x$-direction are positive.

- Angles measured clockwise from the positive $x$-direction are negative.

**Diagram showing where sin, cos and tan are positive**

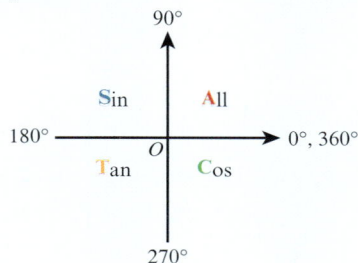

- Useful mnemonic: '**A**ll **S**tudents **T**rust **C**ambridge'.

**Graphs of trigonometric functions**

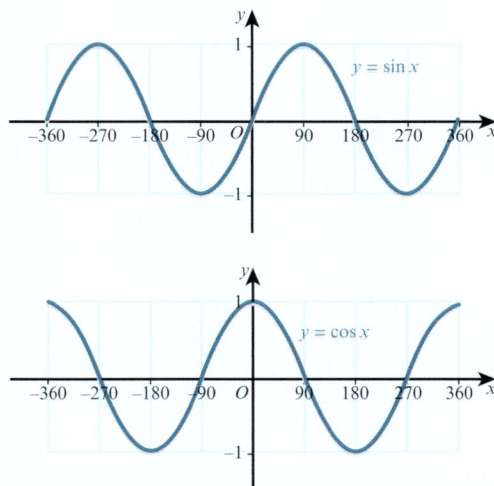

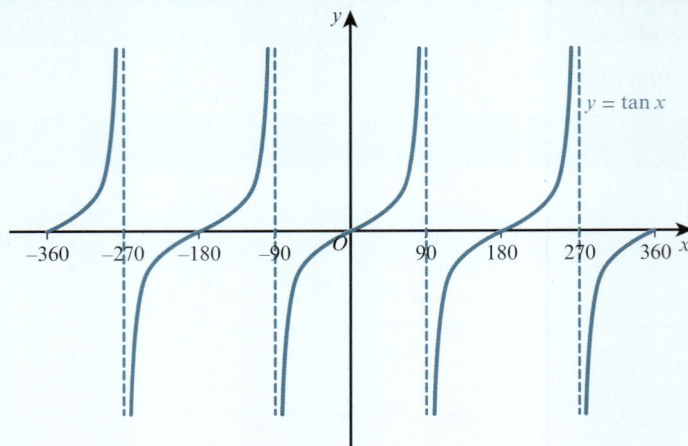

- The graph of $y = a\sin x$ is a stretch of $y = \sin x$, stretch factor $a$, parallel to the $y$-axis.

- The graph of $y = \sin(ax)$ is a stretch of $y = \sin x$, stretch factor $\dfrac{1}{a}$, parallel to the $x$-axis.

- The graph of $y = a + \sin x$ is a translation of $y = \sin x$ by the vector $\begin{pmatrix} 0 \\ a \end{pmatrix}$.

- The graph of $y = \sin(x + a)$ is a translation of $y = \sin x$ by the vector $\begin{pmatrix} -a \\ 0 \end{pmatrix}$.

### Inverse trigonometric functions

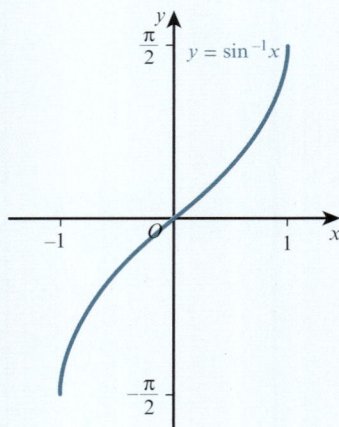

$y = \sin^{-1} x$
domain: $-1 \leqslant x \leqslant 1$
range: $-\dfrac{\pi}{2} \leqslant \sin^{-1} x \leqslant \dfrac{\pi}{2}$

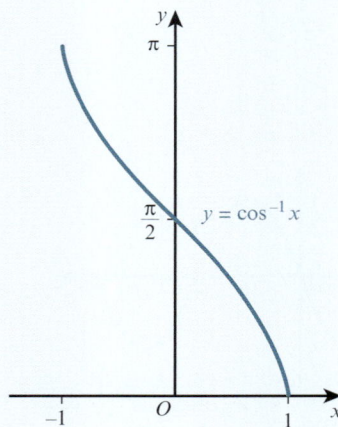

$y = \cos^{-1} x$
domain: $-1 \leqslant x \leqslant 1$
range: $0 \leqslant \cos^{-1} x \leqslant \pi$

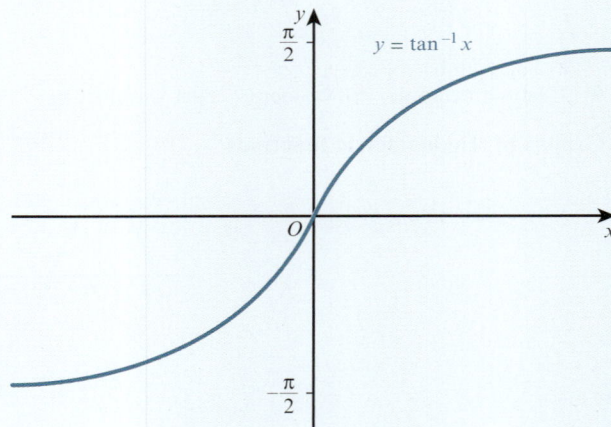

$y = \tan^{-1} x$
domain: $x \in \mathbb{R}$
range: $-\dfrac{\pi}{2} \leqslant \tan^{-1} x \leqslant \dfrac{\pi}{2}$

### Trigonometric identities

- $\tan x \equiv \dfrac{\sin x}{\cos x}$

- $\sin^2 x + \cos^2 x \equiv 1$

152

**END-OF-CHAPTER REVIEW EXERCISE 5**

1

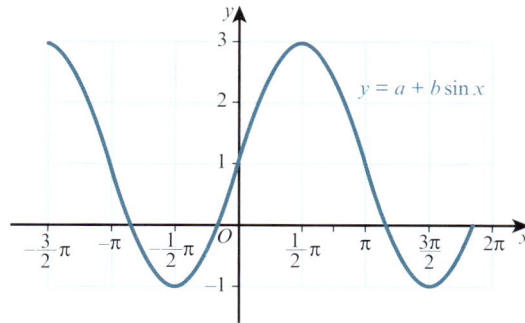

The diagram shows part of the graph of $y = a + b\sin x$.

State the values of the constants $a$ and $b$. [2]

*Cambridge International AS & A Level Mathematics 9709 Paper 11 Q1 June 2014*

2  Find the value of $x$ satisfying the equation $\sin^{-1}(x-1) = \tan^{-1}(3)$. [3]

*Cambridge International AS & A Level Mathematics 9709 Paper 11 Q2 November 2014*

3  Given that $\theta$ is an acute angle measured in radians and that $\cos\theta = k$, find, in terms of $k$, an expression for:

a  $\sin\theta$ [1]

b  $\tan\theta$ [1]

c  $\cos(\pi - \theta)$ [1]

4  Solve the equation $\cos^{-1}(8x^4 + 14x^2 - 16) = \pi$. [4]

5  Solve the equation $\sin 2x = 5\cos 2x$, for $0° \le x \le 180°$. [4]

6  Solve the equation $\dfrac{13\sin^2\theta}{2 + \cos\theta} + \cos\theta = 2$ for $0° \le \theta \le 180°$. [4]

*Cambridge International AS & A Level Mathematics 9709 Paper 11 Q3 November 2014*

7  Solve the equation $2\cos^2 x = 5\sin x - 1$ for $0° \le x \le 360°$. [4]

8  i  Sketch, on a single diagram, the graphs of $y = \cos 2\theta$ and $y = \dfrac{1}{2}$ for $0 \le \theta \le 2\pi$. [3]

ii  Write down the number of roots of the equation $2\cos 2\theta - 1 = 0$ in the interval $0 \le \theta \le 2\pi$. [1]

iii  Deduce the number of roots of the equation $2\cos 2\theta - 1 = 0$ in the interval $10\pi \le \theta \le 20\pi$. [1]

*Cambridge International AS & A Level Mathematics 9709 Paper 11 Q3 November 2011*

9  i  Show that the equation $2\tan^2\theta \sin^2\theta = 1$ can be written in the form $2\sin^4\theta + \sin^2\theta - 1 = 0$. [2]

ii  Hence solve the equation $2\tan^2\theta \sin^2\theta = 1$ for $0° \le \theta \le 360°$. [4]

*Cambridge International AS & A Level Mathematics 9709 Paper 11 Q5 June 2011*

10 i Solve the equation $4\sin^2 x + 8\cos x - 7 = 0$ for $0° \leqslant x \leqslant 360°$. [4]

ii Hence find the solution of the equation $4\sin^2\left(\dfrac{1}{2}\theta\right) + 8\cos\left(\dfrac{1}{2}\theta\right) - 7 = 0$ for $0° \leqslant \theta \leqslant 360°$. [2]

*Cambridge International AS & A Level Mathematics 9709 Paper 11 Q4 November 2013*

11 i Prove the identity $\dfrac{\sin x \tan x}{1 - \cos x} \equiv 1 + \dfrac{1}{\cos x}$. [3]

ii Hence solve the equation $\dfrac{\sin x \tan x}{1 - \cos x} + 2 = 0$, for $0° \leqslant x \leqslant 360°$. [3]

*Cambridge International AS & A Level Mathematics 9709 Paper 11 Q4 November 2010*

12 a Solve the equation $\dfrac{2 - \sin x}{1 + 2\sin x} = \dfrac{3}{4}$ for $0 \leqslant x \leqslant 2\pi$. [3]

b Solve the equation $\sin x - 2\cos x = 2(2\sin x - 3\cos x)$ for $-\pi \leqslant x \leqslant \pi$. [4]

13 A function f is defined by $f : x \rightarrow 3 - 2\tan\left(\dfrac{1}{2}x\right)$ for $0 \leqslant x \leqslant \pi$.

i State the range of f. [1]

ii State the exact value of $f\left(\dfrac{2}{3}\pi\right)$. [1]

iii Sketch the graph of $y = f(x)$. [2]

iv Obtain an expression, in terms of $x$, for $f^{-1}(x)$. [3]

*Cambridge International AS & A Level Mathematics 9709 Paper 11 Q7 November 2010*

14 i Solve the equation $2\cos^2\theta = 3\sin\theta$, for $0° \leqslant \theta \leqslant 360°$. [4]

ii The smallest positive solution of the equation $2\cos^2(n\theta) = 3\sin(n\theta)$, where $n$ is a positive integer, is $10°$. State the value of $n$ and hence find the largest solution of this equation in the interval $0° \leqslant \theta \leqslant 360°$. [3]

*Cambridge International AS & A Level Mathematics 9709 Paper 11 Q7 November 2012*

15 i Show that $\dfrac{\sin\theta}{\sin\theta + \cos\theta} + \dfrac{\cos\theta}{\sin\theta - \cos\theta} \equiv \dfrac{1}{\sin^2\theta - \cos^2\theta}$. [3]

ii Hence solve the equation $\dfrac{\sin\theta}{\sin\theta + \cos\theta} + \dfrac{\cos\theta}{\sin\theta - \cos\theta} = 3$, for $0° \leqslant \theta \leqslant 360°$. [4]

*Cambridge International AS & A Level Mathematics 9709 Paper 11 Q5 June 2013*

16 i Show that the equation $\dfrac{4\cos\theta}{\tan\theta} + 15 = 0$ can be expressed as $4\sin^2\theta - 15\sin\theta - 4 = 0$. [3]

ii Hence solve the equation $\dfrac{4\cos\theta}{\tan\theta} + 15 = 0$ for $0° \leqslant \theta \leqslant 360°$. [3]

*Cambridge International AS & A Level Mathematics 9709 Paper 11 Q4 November 2015*

17 The function $f : x \rightarrow 5 + 3\cos\left(\dfrac{1}{2}x\right)$ is defined for $0 \leqslant x \leqslant 2\pi$.

i Solve the equation $f(x) = 7$, giving your answer correct to 2 decimal places. [3]

ii Sketch the graph of $y = f(x)$. [2]

iii Explain why f has an inverse. [1]

iv Obtain an expression for $f^{-1}(x)$. [3]

*Cambridge International AS & A Level Mathematics 9709 Paper 11 Q8 June 2015*

# Chapter 6
# Series

**In this chapter you will learn how to:**

■ use the expansion of $(a + b)^n$, where $n$ is a positive integer

■ recognise arithmetic and geometric progressions

■ use the formulae for the $n$th term and for the sum of the first $n$ terms to solve problems involving arithmetic or geometric progressions

■ use the condition for the convergence of a geometric progression, and the formula for the sum to infinity of a convergent geometric progression.

## PREREQUISITE KNOWLEDGE

| Where it comes from | What you should be able to do | Check your skills |
|---|---|---|
| IGCSE / O Level Mathematics | Expand brackets. | 1  Expand:<br>a  $(2x+3)^2$<br>b  $(1-3x)(1+2x-3x^2)$ |
| IGCSE / O Level Mathematics | Simplify indices. | 2  Simplify:<br>a  $(5x^2)^3$<br>b  $(-2x^3)^5$ |
| IGCSE / O Level Mathematics | Find the $n$th term of a linear sequence. | 3  Find the $n$th term of these linear sequences.<br>a  $5, 7, 9, 11, 13, \dots$<br>b  $8, 5, 2, -1, -4, \dots$ |

## Why study series?

At IGCSE / O Level you learnt how to expand expressions such as $(1+x)^2$. In this chapter you will learn how to expand expressions of the form $(1+x)^n$, where $n$ can be any positive integer. Expansions of this type are called binomial expansions.

This chapter also covers arithmetic and geometric progressions. Both the mathematical and the real world are full of number sequences that have particular special properties. You will learn how to find the sum of the numbers in these progressions. Some fractal patterns can generate these types of sequences.

### FAST FORWARD

In the Pure Mathematics 2 and 3 Coursebook, Chapter 7, you will learn how to expand these expressions for any real value of $n$.

### FAST FORWARD

Properties of binomial expansions are also used in probability theory, which you will learn about if you go on to study the Probability and Statistics 1 Coursebook, Chapter 7.

### WEB LINK

Try the *Sequences* and *Counting and binomials* resources on the Underground Mathematics website.

## 6.1 Binomial expansion of $(a+b)^n$

**Binomial** means 'two terms'.

The word is used in algebra for expressions such as $x+3$ and $5x-2y$.

You should already know that $(a+b)^2 = a^2 + 2ab + b^2$.

The expansion of $(a+b)^2$ can be used to expand $(a+b)^3$:

$(a+b)^3 = (a+b)(a^2+2ab+b^2)$
$\quad = a^3 + 2a^2b + ab^2 + a^2b + 2ab^2 + b^3$
$\quad = a^3 + 3a^2b + 3ab^2 + b^3$

Similarly, it can be shown that $(a+b)^4 = a^4 + 4a^3b + 6a^2b^2 + 4ab^3 + b^4$.

Writing the expansions of $(a+b)^n$ in full in order:

$(a+b)^0 = 1$
$(a+b)^1 = 1a + 1b$
$(a+b)^2 = 1a^2 + 2ab + 1b^2$
$(a+b)^3 = 1a^3 + 3a^2b + 3ab^2 + 1b^3$
$(a+b)^4 = 1a^4 + 4a^3b + 6a^2b^2 + 4ab^3 + 1b^4$

If you look at the expansion of $(a + b)^4$, you should notice that the powers of $a$ and $b$ form a pattern.

- The first term is $a^4$ and then the power of $a$ decreases by 1 while the power of $b$ increases by 1 in each successive term.
- All of the terms have a total index of 4 ($a^4$, $a^3b$, $a^2b^2$, $ab^3$ and $b^4$).

There is a similar pattern in the other expansions.

The coefficients also form a pattern that is known as **Pascal's triangle**.

$n = 0$:               1

$n = 1$:            1    1

$n = 2$:         1    2    1

$n = 3$:      1    3    3    1

$n = 4$:   1    4    6    4    1

The next row is then:

$n = 5$: 1   5   10   10   5   1

This row can then be used to write down the expansion of $(a + b)^5$:

$$(a + b)^5 = 1a^5 + 5a^4b + 10a^3b^2 + 10a^2b^3 + 5ab^4 + 1b^5$$

> **TIP**
>
> Each row always starts and finishes with a 1.
>
> Each number is the sum of the two numbers in the row above it.

**EXPLORE 6.1**

```
              1
           1     1
        1     2     1
     1     3     3     1
  1     4     6     4     1
1     5    10    10    5     1
1   6   15   20   15   6   1
```

There are many number patterns to be found in Pascal's triangle.

For example, the numbers 1, 4, 10 and 20 have been highlighted.

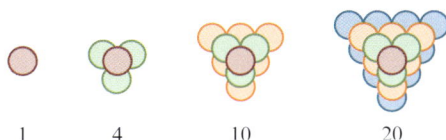

|   1   |   4   |   10  |   20  |

These numbers are called tetrahedral numbers.

1. What do you notice if you find the total of each row in Pascal's triangle? Can you explain your findings?

2. Can you find the Fibonacci sequence (1, 1, 2, 3, 5, 8, 13, …) in Pascal's triangle? You may want to add terms together.

3. Pascal's triangle has many other number patterns. Which number patterns can you find?

> **DID YOU KNOW?**
>
> Pascal's triangle is named after the French mathematician Blaise Pascal (1623–1662).

**WORKED EXAMPLE 6.1**

Use Pascal's triangle to find the expansion of:

**a** $(3x + 2)^3$        **b** $(5 - 2x)^4$

**Answer**

**a** $(3x + 2)^3$

The index is 3 so use the row for $n = 3$ in Pascal's triangle $(1, 3, 3, 1)$.

$(3x + 2)^3 = 1(3x)^3 + 3(3x)^2(2) + 3(3x)(2)^2 + 1(2)^3$

$\qquad\qquad = 27x^3 + 54x^2 + 36x + 8$

**b** $(5 - 2x)^4$

The index is 4 so use the row for $n = 4$ in Pascal's triangle $(1, 4, 6, 4, 1)$.

$(5 - 2x)^4 = 1(5)^4 + 4(5)^3(-2x) + 6(5)^2(-2x)^2 + 4(5)(-2x)^3 + 1(-2x)^4$

$\qquad\qquad = 625 - 1000x + 600x^2 - 160x^3 + 16x^4$

**WORKED EXAMPLE 6.2**

**a** Use Pascal's triangle to expand $(1 - 2x)^5$.

**b** Find the coefficient of $x^3$ in the expansion of $(3 + 5x)(1 - 2x)^5$.

**Answer**

**a** $(1 - 2x)^5$

The index is 5 so use the row for $n = 5$ in Pascal's triangle $(1, 5, 10, 10, 5, 1)$.

$(1 - 2x)^5 = 1(1)^5 + 5(1)^4(-2x) + 10(1)^3(-2x)^2 + 10(1)^2(-2x)^3 + 5(1)(-2x)^4 + 1(-2x)^5$

$\qquad\qquad = 1 - 10x + 40x^2 - 80x^3 + 80x^4 - 32x^5$

**b** $(3 + 5x)(1 - 2x)^5 = (3 + 5x)(1 - 10x + 40x^2 - 80x^3 + 80x^4 - 32x^5)$

The term in $x^3$ comes from the products:

$(3 + 5x)(1 - 10x + 40x^2 - 80x^3 + 80x^4 - 32x^5)$

$3 \times (-80x^3) = -240x^3$ and $5x \times 40x^2 = 200x^3$

Coefficient of $x^3 = -240 + 200 = -40$.

**EXERCISE 6A**

**1** Use Pascal's triangle to find the expansions of:

**a** $(x + 2)^3$     **b** $(1 - x)^4$     **c** $(x + y)^3$     **d** $(2 - x)^3$

**e** $(x - y)^4$     **f** $(2x + 3y)^3$     **g** $(2x - 3)^4$     **h** $\left(x^2 + \dfrac{3}{2x^3}\right)^3$

158

**2** Find the coefficient of $x^3$ in the expansions of:

  **a** $(x+3)^4$    **b** $(1+x)^5$    **c** $(3-x)^5$    **d** $(4+x)^4$

  **e** $(x-2)^5$    **f** $(2x-1)^4$    **g** $(4x+3)^4$    **h** $\left(5-\dfrac{x}{2}\right)^4$

**3** $(3+x)^5 + (3-x)^5 \equiv A + Bx^2 + Cx^4$

  Find the value of $A$, the value of $B$ and the value of $C$.

**4** The coefficient of $x^2$ in the expansion of $(3+ax)^4$ is 216.

  Find the possible values of the constant $a$.

**5 a** Expand $(2+x)^4$.

  **b** Use your answer to part **a** to express $(2+\sqrt{3})^4$ in the form $a+b\sqrt{3}$.

**6 a** Expand $(1+x)^3$.

  **b** Use your answer to part **a** to express:

   **i** $(1+\sqrt{5})^3$ in the form $a+b\sqrt{5}$

   **ii** $(1-\sqrt{5})^3$ in the form $c+d\sqrt{5}$.

  **c** Use your answers to part **b** to simplify $(1+\sqrt{5})^3 + (1-\sqrt{5})^3$.

**7** Expand $(1+x)(2+3x)^4$.

**8 a** Expand $(x^2-1)^4$.

  **b** Find the coefficient of $x^6$ in the expansion of $(1-2x^2)(x^2-1)^4$.

**9** Find the coefficient of $x^2$ in the expansion of $\left(3x-\dfrac{2}{x}\right)^4$.

**10** Find the term independent of $x$ in the expansion of $\left(x^2-\dfrac{3}{x^2}\right)^4$.

**11 a** Find the first three terms, in ascending powers of $y$, in the expansion of $(1+y)^4$.

  **b** By replacing $y$ with $5x-2x^2$, find the coefficient of $x^2$ in the expansion of $(1+5x-2x^2)^4$.

**12** The coefficient of $x^2$ in the expansion of $(1+ax)^4$ is 30 times the coefficient of $x$ in the expansion of $\left(1+\dfrac{ax}{3}\right)^3$. Find the value of $a$.

**13** Find the power of $x$ that has the greatest coefficient in the expansion of $\left(3x^4+\dfrac{1}{x}\right)^4$.

**14 a** Write down the expansion of $(x+y)^5$.

  **b** Without using a calculator and using your result from part **a**, find the value of $\left(10\dfrac{1}{4}\right)^5$, correct to the nearest hundred.

**15 a** Given that $\left(x^2+\dfrac{1}{x}\right)^4 - \left(x^2-\dfrac{1}{x}\right)^4 = px^5 + \dfrac{q}{x}$, find the value of $p$ and the value of $q$.

  **b** Hence, without using a calculator, find the exact value of $\left(2+\dfrac{1}{\sqrt{2}}\right)^4 - \left(2-\dfrac{1}{\sqrt{2}}\right)^4$.

**(PS)** **16** $y = x + \dfrac{1}{x}$

    **a** Express $x^3 + \dfrac{1}{x^3}$ in terms of $y$.

    **b** Express $x^5 + \dfrac{1}{x^5}$ in terms of $y$.

## 6.2 Binomial coefficients

Pascal's triangle can be used to expand $(a + b)^n$ for any positive integer $n$, but if $n$ is large it can take a long time to write out all the rows in the triangle. Hence, we need a more efficient method to find the coefficients in the expansions. The coefficients in the binomial expansion of $(1 + x)^n$ are known as **binomial coefficients**.

---

### EXPLORE 6.2

Consider the expansion:

$$(1+x)^5 = 1 + 5x + 10x^2 + 10x^3 + 5x^4 + x^5$$

    The coefficients are: 1   5   10     10    5    1

🖩 Find the $nCr$ function on your calculator. On some calculators this may be $^nC_r$ or $\begin{pmatrix} n \\ r \end{pmatrix}$.

🖩 **1** Use your calculator to find the values of:

$$\begin{pmatrix} 5 \\ 0 \end{pmatrix}, \begin{pmatrix} 5 \\ 1 \end{pmatrix}, \begin{pmatrix} 5 \\ 2 \end{pmatrix}, \begin{pmatrix} 5 \\ 3 \end{pmatrix}, \begin{pmatrix} 5 \\ 4 \end{pmatrix} \text{ and } \begin{pmatrix} 5 \\ 5 \end{pmatrix}.$$

**2** What do you notice about your answers to question 1?

**3** Complete the following four statements.

The coefficient of $x^2$ in the expansion of $(1 + x)^5$ is $\begin{pmatrix} 5 \\ \cdots \end{pmatrix}$.

The coefficient of $x^r$ in the expansion of $(1 + x)^n$ is $\begin{pmatrix} \cdots \\ \cdots \end{pmatrix}$.

The coefficient of the 4th term in the expansion of $(1 + x)^5$ is $\begin{pmatrix} 5 \\ \cdots \end{pmatrix}$.

The coefficient of the $(r + 1)$th term in the expansion of $(1 + x)^n$ is $\begin{pmatrix} \cdots \\ \cdots \end{pmatrix}$.

---

> **TIP**
>
> To find $\begin{pmatrix} 5 \\ 2 \end{pmatrix}$, key in
>
> $\boxed{5}\ \boxed{nCr}\ \boxed{2}$.

We write the binomial expansion of $(1 + x)^n$, where $n$ is a positive integer as:

### 🔍 KEY POINT 6.1

If $n$ is a positive integer, then $(1 + x)^n = \begin{pmatrix} n \\ 0 \end{pmatrix} + \begin{pmatrix} n \\ 1 \end{pmatrix} x + \begin{pmatrix} n \\ 2 \end{pmatrix} x^2 + \cdots + \begin{pmatrix} n \\ n \end{pmatrix} x^n.$

We can therefore write the expansion of $(1 + x)^n$ using binomial coefficients; the result is known as the **Binomial theorem**.

We can use the Binomial theorem to expand $(a + b)^n$, too. We can write $(a + b)^n = a^n \left( 1 + \dfrac{b}{a} \right)^n$ (assuming that $a \neq 0$).

**🔍 KEY POINT 6.2**

$$(a + b)^n = \binom{n}{0} a^n + \binom{n}{1} a^{n-1}b^1 + \binom{n}{2} a^{n-2}b^2 + \cdots + \binom{n}{n} b^n$$

**WORKED EXAMPLE 6.3**

Find, in ascending powers of $x$, the first four terms in the expansion of:

a $(1 + x)^{15}$            b $(2 - 3x)^{10}$

**Answer**

a $(1 + x)^{15} = \binom{15}{0} + \binom{15}{1} x + \binom{15}{2} x^2 + \binom{15}{3} x^3 + \cdots$

$= 1 + 15x + 105x^2 + 455x^3 + \cdots$

b $(2 - 3x)^{10} = \binom{10}{0} 2^{10} + \binom{10}{1} 2^9(-3x)^1 + \binom{10}{2} 2^8(-3x)^2 + \binom{10}{3} 2^7(-3x)^3 + \cdots$

$= 1024 - 15\,360x + 103\,680x^2 - 414\,720x^3 + \cdots$

You should also know how to work out the binomial coefficients without using a calculator.

From Pascal's triangle, we know that $\binom{5}{0} = 1$ and $\binom{5}{5} = 1$.

In general, we can write this as:

**🔍 KEY POINT 6.3**

$$\binom{n}{0} = 1 \quad \text{and} \quad \binom{n}{n} = 1$$

We write $\binom{5}{1}, \binom{5}{2}, \binom{5}{3}$ and $\binom{5}{4}$ as:

$$\binom{5}{1} = \frac{5}{1} = 5 \qquad \binom{5}{2} = \frac{5 \times 4}{2 \times 1} = 10 \qquad \binom{5}{3} = \frac{5 \times 4 \times 3}{3 \times 2 \times 1} = 10 \qquad \binom{5}{4} = \frac{5 \times 4 \times 3 \times 2}{4 \times 3 \times 2 \times 1} = 5$$

In general, if $r$ is a positive integer less than $n$, then:

> **KEY POINT 6.4**
>
> $$\binom{n}{r} = \frac{n \times (n-1) \times (n-2) \times \cdots \times (n-r+1)}{r \times (r-1) \times (r-2) \times \cdots \times 3 \times 2 \times 1}$$

**WORKED EXAMPLE 6.4**

**a** Without using a calculator, find the value of $\binom{8}{4}$.

**b** Find an expression, in terms of $n$, for $\binom{n}{4}$.

**Answer**

**a** $\binom{8}{4} = \dfrac{8 \times 7 \times 6 \times 5}{4 \times 3 \times 2 \times 1} = 70$

**b** $\binom{n}{4} = \dfrac{n \times (n-1) \times (n-2) \times (n-3)}{4 \times 3 \times 2 \times 1} = \dfrac{n(n-1)(n-2)(n-3)}{24}$

**WORKED EXAMPLE 6.5**

When $\left(1 - \dfrac{x}{3}\right)^n$ is expanded in ascending powers of $x$, the coefficient of $x^2$ is 4. Given that $n$ is the positive integer, find the value of $n$.

**Answer**

$$\text{Term in } x^2 = \binom{n}{2}\left(-\frac{x}{3}\right)^2 = \frac{n \times (n-1)}{2 \times 1} \times \frac{x^2}{9} = \frac{n \times (n-1)}{18}x^2$$

$$\frac{n \times (n-1)}{18} = 4$$

$$n(n-1) = 72$$

$$n^2 - n - 72 = 0$$

$$(n-9)(n+8) = 0$$

$$n = 9 \text{ or } n = -8$$

As $n$ is a positive integer, $n = 9$.

**WORKED EXAMPLE 6.6**

When $(2 + kx)^8$ is expanded, the coefficient of $x^5$ is two times the coefficient of $x^4$. Given that $k > 0$, find the value of $k$.

**Answer**

Term in $x^5 = \begin{pmatrix} 8 \\ 5 \end{pmatrix} (2)^3 (kx)^5 = 448k^5x^5$

Term in $x^4 = \begin{pmatrix} 8 \\ 4 \end{pmatrix} (2)^4 (kx)^4 = 1120k^4x^4$

Coefficient of $x^5 = 2 \times$ coefficient of $x^4$

$$448k^5 = 2 \times 1120k^4$$

$$448k^5 - 2240k^4 = 0$$

$$448k^4(k - 5) = 0$$

$$k = 0 \text{ or } k = 5$$

As $k$ is a positive integer, $k = 5$.

**WORKED EXAMPLE 6.7**

**a**  Obtain the first three terms in the expansion of $(2 - x)(1 + 2x)^9$.

**b**  Use your answer to part **a** to estimate the value of $1.99 \times 1.02^9$.

**Answer**

**a**  $(2 - x)(1 + 2x)^9 = (2 - x)\left[ \begin{pmatrix} 9 \\ 0 \end{pmatrix} + \begin{pmatrix} 9 \\ 1 \end{pmatrix} (2x)^1 + \begin{pmatrix} 9 \\ 2 \end{pmatrix} (2x)^2 + \cdots \right]$

$$= (2 - x)(1 + 18x + 144x^2 + \cdots)$$

$$= 2(1 + 18x + 144x^2 + \cdots) - x(1 + 18x + 144x^2 + \cdots)$$

$$= 2 + (2 \times 18 - 1)x + (2 \times 144 - 18)x^2 + \cdots$$

$$= 2 + 35x + 270x^2 + \cdots$$

**b**  $(2 - x)(1 + 2x)^9 = 2 + 35x + 270x^2 + \cdots$ 　　　　　　Let $x = 0.01$.

$$1.99 \times 1.02^9 \approx 2 + 35(0.01) + 270(0.01)^2$$

$$1.99 \times 1.02^9 \approx 2.377$$

There is an alternative formula for calculating $\begin{pmatrix} n \\ r \end{pmatrix}$. To be able to understand and apply the alternative formula, we need to first know about **factorial** notation.

We write 6! to mean $6 \times 5 \times 4 \times 3 \times 2 \times 1$, and call it '6 factorial'.

In general, if $n$ is a positive integer, then:

> **KEY POINT 6.5**
>
> $n! = n \times (n-1) \times (n-2) \times (n-3) \times \cdots \times 3 \times 2 \times 1$

The formula for $\begin{pmatrix} n \\ r \end{pmatrix}$ then becomes:

> **KEY POINT 6.6**
>
> $\begin{pmatrix} n \\ r \end{pmatrix} = \dfrac{n!}{r!\,(n-r)!}$

**WORKED EXAMPLE 6.8**

Use the formula $\begin{pmatrix} n \\ r \end{pmatrix} = \dfrac{n!}{r!\,(n-r)!}$ to find the value of:

a $\begin{pmatrix} 8 \\ 4 \end{pmatrix}$ 

b $\begin{pmatrix} 9 \\ 3 \end{pmatrix}$

**Answer**

a $\begin{pmatrix} 8 \\ 4 \end{pmatrix} = \dfrac{8!}{4!\,(8-4)!} = \dfrac{8!}{4!\,4!} = 70$

b $\begin{pmatrix} 9 \\ 3 \end{pmatrix} = \dfrac{9!}{3!\,(9-3)!} = \dfrac{9!}{3!\,6!} = 84$

**WORKED EXAMPLE 6.9**

Find the term independent of $x$ in the expansion of $\left( x + \dfrac{5}{x^2} \right)^9$.

**Answer**

$\left( x + \dfrac{5}{x^2} \right)^9 = \begin{pmatrix} 9 \\ 0 \end{pmatrix} x^9 + \begin{pmatrix} 9 \\ 1 \end{pmatrix} x^8 \left( \dfrac{5}{x^2} \right)^1 + \begin{pmatrix} 9 \\ 2 \end{pmatrix} x^7 \left( \dfrac{5}{x^2} \right)^2 + \begin{pmatrix} 9 \\ 3 \end{pmatrix} x^6 \left( \dfrac{5}{x^2} \right)^3 + \cdots$

The term that is independent of $x$ is the term that when simplified does not involve $x$.

The $x$ terms cancel each other out when the power of $x$ is double the power of $\dfrac{5}{x^2}$.

Also, the sum of these powers must be 9.

Hence, we are looking for powers of 6 and 3, respectively, and the corresponding binomial coefficient is $\begin{pmatrix} 9 \\ 3 \end{pmatrix}$.

The term independent of $x$ is:

$\begin{pmatrix} 9 \\ 3 \end{pmatrix} x^6 \left( \dfrac{5}{x^2} \right)^3 = 84 \times x^6 \times \dfrac{125}{x^6} = 10\,500$

## EXERCISE 6B

**1** Without using a calculator, find the value of each of the following.

a $\begin{pmatrix} 7 \\ 3 \end{pmatrix}$

b $\begin{pmatrix} 9 \\ 6 \end{pmatrix}$

c $\begin{pmatrix} 12 \\ 4 \end{pmatrix}$

d $\begin{pmatrix} 15 \\ 6 \end{pmatrix}$

**2** Express each of the following in terms of $n$.

a $\begin{pmatrix} n \\ 2 \end{pmatrix}$

b $\begin{pmatrix} n \\ 1 \end{pmatrix}$

c $\begin{pmatrix} n \\ 3 \end{pmatrix}$

**3** Use the formula $\begin{pmatrix} n \\ r \end{pmatrix} = \dfrac{n!}{r!\,(n-r)!}$ to find the value of of each of the following.

a $\begin{pmatrix} 10 \\ 2 \end{pmatrix}$

b $\begin{pmatrix} 8 \\ 5 \end{pmatrix}$

c $\begin{pmatrix} 14 \\ 3 \end{pmatrix}$

d $\begin{pmatrix} 12 \\ 7 \end{pmatrix}$

**4** Find, in ascending powers of $x$, the first three terms in each of the following expansions.

a $(1+2x)^8$

b $(1-3x)^{10}$

c $\left(1+\dfrac{x}{2}\right)^7$

d $(1+x^2)^{12}$

e $\left(3+\dfrac{x}{2}\right)^7$

f $(2-x)^{13}$

g $(2+x^2)^8$

h $\left(2+\dfrac{x^2}{2}\right)^9$

**5** Find the coefficient of $x^3$ in each of the following expansions.

a $(1-x)^9$

b $(1+3x)^{12}$

c $\left(2+\dfrac{x}{4}\right)^7$

d $\left(3-\dfrac{x}{3}\right)^{10}$

**6** Find the coefficient of $x^4$ in the expansion of $(2x+1)^{12}$.

**7** Find the term in $x^5$ in the expansion of $(5-2x)^8$.

**8** Find the coefficient of $x^8 y^5$ in the expansion of $(x-2y)^{13}$.

**9** Find the term independent of $x$ in the expansion of $\left(x-\dfrac{3}{x^2}\right)^{12}$.

**10** Find, in ascending powers of $x$, the first three terms of each of the following expansions.

a $(1-x)(2+x)^7$

b $(1+2x)(1-3x)^{10}$

c $(1+x)\left(1-\dfrac{x}{2}\right)^8$

**11 a** Find, in ascending powers of $x$, the first three terms in the expansion of $(2+x)^{10}$.

**b** By replacing $x$ with $2y-3y^2$, find the first three terms in the expansion of $(2+2y-3y^2)^{10}$.

**12 a** Find, in ascending powers of $x$, the first three terms in the expansion of $\left(1-\dfrac{x}{2}\right)^8$.

**b** Hence, obtain the coefficient of $x^2$ in the expansion of $(2+3x-x^2)\left(1-\dfrac{x}{2}\right)^8$.

**13** Find the first three terms, in ascending powers of $x$, in the expansion of $(2-3x)^4(1+2x)^{10}$.

**14** The first four terms, in ascending powers of $x$, in the expansion of $(1+ax+bx^2)^7$ are $1-14x+91x^2+px^3$. Find the values of $a$, $b$ and $p$.

**15** The first two terms, in ascending powers of $x$, in the expansion of $(1+x)\left(2-\dfrac{x}{4}\right)^n$ are $p+qx^2$. Find the values of $n$, $p$ and $q$.

165

## 6.3 Arithmetic progressions

At IGCSE / O Level you learnt that a number sequence is a list of numbers and that the numbers in the sequence are called the **terms** of the sequence.

A linear sequence such as 5, 8, 11, 14, 17, ... is also called an **arithmetic progression**. Each term differs from the term before by a constant. This constant is called the **common difference**.

The notation used for arithmetic progressions is:

$a$ = first term          $d$ = common difference          $l$ = last term

The common difference is also allowed to be zero or negative. For example, 10, 6, 2, –2, ... and 5, 5, 5, 5, ... are both arithmetic progressions.

The first five terms of an arithmetic progression whose first term is $a$ and whose common difference is $d$ are:

| $a$ | $a + d$ | $a + 2d$ | $a + 3d$ | $a + 4d$ |
|---|---|---|---|---|
| term 1 | term 2 | term 3 | term 4 | term 5 |

From this pattern, you can see that the formula for the $n$th term is given by:

> **KEY POINT 6.7**
>
> $n$th term $= a + (n - 1)d$

**WORKED EXAMPLE 6.10**

Find the number of terms in the arithmetic progression –3, 1, 5, 9, 13, ... , 237.

**Answer**

$n$th term $= a + (n - 1)d$ .................................................... Use $a = -3$, $d = 4$ and $n$th term $= 237$.

$237 = -3 + 4(n - 1)$ .................................................... Solve.

$n - 1 = 60$

$n = 61$

**WORKED EXAMPLE 6.11**

The fourth term of an arithmetic progression is 7 and the tenth term is 16. Find the first term and the common difference.

**Answer**

4th term $= 7$      $\Rightarrow a + 3d = 7$ --------- (1)

10th term $= 16$     $\Rightarrow a + 9d = 16$ ------- (2)

$(2) - (1)$ gives $6d = 9$

$d = 1.5$

Substituting into (1) gives $a + 4.5 = 7$

$a = 2.5$

First term $= 2.5$, common difference $= 1.5$

**WORKED EXAMPLE 6.12**

The $n$th term of an arithmetic progression is $5 - 6n$. Find the first term and the common difference.

**Answer**

1st term $= 5 - 6(1) = -1$ .................................................... Substitute $n = 1$ into $n$th term $= 5 - 6n$.

2nd term $= 5 - 6(2) = -7$ .................................................... Substitute $n = 2$ into $n$th term $= 5 - 6n$.

Common difference $=$ 2nd term $-$ 1st term $= -6$

## The sum of an arithmetic progression

When the terms in a sequence are added together we call the resulting sum a **series**.

**EXPLORE 6.3**

$$1 + 2 + 3 + 4 + \cdots + 97 + 98 + 99 + 100 = ?$$

It is said that, at the age of seven or eight, the famous mathematician Carl Gauss was asked to find the sum of the numbers from 1 to 100. His teacher expected this task to keep him occupied for some time but Gauss surprised him by writing down the correct answer almost immediately. His method involved adding the numbers in pairs: $1 + 100 = 101$, $2 + 99 = 101$, $3 + 98 = 101, \ldots$

1   Can you complete Gauss's method to find the answer?

2   Use Gauss's method to find the sum of:

    **a**  $2 + 4 + 6 + 8 + \cdots + 494 + 496 + 498 + 500$

    **b**  $5 + 10 + 15 + 20 + \cdots + 185 + 190 + 195 + 200$

    **c**  $6 + 9 + 12 + 15 + \cdots + 93 + 96 + 99 + 102$

3   Use Gauss's method to find an expression, in terms of $n$, for the sum:
    $1 + 2 + 3 + 4 + \cdots + (n - 3) + (n - 2) + (n - 1) + n$

The sum of an arithmetic progression, $S_n$, can be written as:

**🔍 KEY POINT 6.8**

$$S_n = \frac{n}{2}(a + l) \qquad \text{or} \qquad S_n = \frac{n}{2}[2a + (n - 1)d]$$

We can prove this result as follows, by writing out the series in full.

$$S_n = a + (a+d) + (a+2d) + \cdots + (l-2d) + (l-d) + l$$

Reversing:
$$S_n = l + (l-d) + (l-2d) + \cdots + (a+2d) + (a+d) + a$$

Adding:
$$2S_n = (a+l) + (a+l) + (a+l) + \cdots + (a+l) + (a+l) + (a+l)$$
$$2S_n = n(a+l), \text{ as there are } n \text{ terms in the series}$$

So $S_n = \dfrac{n}{2}(a+l)$.

Using $l = a + (n-1)d$, this can be rewritten as $S_n = \dfrac{n}{2}[2a + (n-1)d]$.

It is useful to remember the following rule that applies for all sequences.

**KEY POINT 6.9**

$n$th term $= S_n - S_{n-1}$

**WORKED EXAMPLE 6.13**

In an arithmetic progression, the 1st term is $-12$, the 17th term is 12 and the last term is 45.
Find the sum of all the terms in the progression.

**Answer**

We start by working out the common difference.

$n$th term $= a + (n-1)d$ — Use $n$th term $= 12$ when $n = 17$ and $a = -12$.

$12 = -12 + 16d$ — Solve.

$d = \dfrac{3}{2}$

We now determine the number of terms in the whole sequence.

$n$th term $= a + (n-1)d$ — Use $n$th term $= 45$ when $a = -12$ and $d = \dfrac{3}{2}$.

$45 = -12 + \dfrac{3}{2}(n-1)$ — Solve.

$n - 1 = 38$

$n = 39$

Finally, we can work out the sum of all the terms.

$S_n = \dfrac{n}{2}(a+l)$ — Use $a = -12$, $l = 45$ and $n = 39$.

$S_{39} = \dfrac{39}{2}(-12 + 45)$

$= 643\tfrac{1}{2}$

**WORKED EXAMPLE 6.14**

The 10th term in an arithmetic progression is 14 and the sum of the first 7 terms is 42.

Find the first term of the progression and the common difference.

**Answer**

$n$th term $= a + (n - 1)d$ — — — — — — Use $n$th term $= 14$ when $n = 10$.

$\qquad 14 = a + 9d$ ----- (1)

$\qquad S_n = \dfrac{n}{2}[2a + (n - 1)d]$ — — — — Use $n = 7$ and $S_7 = 42$.

$\qquad 42 = \dfrac{7}{2}(2a + 6d)$

$\qquad 6 = a + 3d$ ----- (2)

(1) − (2) gives $6d = 8$

$\qquad\qquad d = \dfrac{4}{3}$

Substituting $d = \dfrac{4}{3}$ into equation (1) gives $a = 2$.

First term $= 2$, common difference $= \dfrac{4}{3}$

**WORKED EXAMPLE 6.15**

The sum of the first $n$ terms, $S_n$, of a particular arithmetic progression is given by $S_n = 4n^2 + n$.

   **a** Find the first term and the common difference.

   **b** Find an expression for the $n$th term.

**Answer**

   **a** $S_1 = 4(1)^2 + 1 = 5$ — — — — — — First term $= 5$

      $S_2 = 4(2)^2 + 2 = 18$ — — — — — First term + second term $= 18$

      Second term $= 18 - 5 = 13$

      First term $= 5$, common difference $= 8$

   **b Method 1:**

      $n$th term $= a + (n - 1)d$ — — — — Use $a = 5$, $d = 8$.

      $\qquad\quad = 5 + 8(n - 1)$

      $\qquad\quad = 8n - 3$

   **Method 2:**

      $n$th term $= S_n - S_{n-1} = 4n^2 + n - [4(n - 1)^2 + (n - 1)]$

      $\qquad\quad = 4n^2 + n - (4n^2 - 8n + 4 + n - 1)$

      $\qquad\quad = 8n - 3$

1 The first term in an arithmetic progression is $a$ and the common difference is $d$.

Write down expressions, in terms of $a$ and $d$, for the seventh term and the 19th term.

2 Find the number of terms and the sum of each of these arithmetic series.

   **a** $13 + 17 + 21 + \cdots + 97$                    **b** $152 + 149 + 146 + \cdots + 50$

3 Find the sum of each of these arithmetic series.

   **a** $5 + 12 + 19 + \cdots$ (17 terms)          **b** $4 + 1 + (-2) + \cdots$ (38 terms)

   **c** $\dfrac{1}{3} + \dfrac{1}{2} + \dfrac{2}{3} + \cdots$ (20 terms)        **d** $-x - 5x - 9x - \cdots$ (40 terms)

4 The first term of an arithmetic progression is 15 and the sum of the first 20 terms is 1630. Find the common difference.

5 In an arithmetic progression, the first term is $-27$, the 16th term is 78 and the last term is 169.

   **a** Find the common difference and the number of terms.

   **b** Find the sum of the terms in this progression.

6 The first two terms in an arithmetic progression are 146 and 139. The last term is $-43$. Find the sum of all the terms in this progression.

7 The first two terms in an arithmetic progression are 2 and 9. The last term in the progression is the only number that is greater than 150. Find the sum of all the terms in the progression.

8 The first term of an arithmetic progression is 15 and the last term is 27. The sum of the first five terms is 79. Find the number of terms in this progression.

9 Find the sum of all the integers between 100 and 300 that are multiples of 7.

10 The first term of an arithmetic progression is 2 and the 11th term is 17. The sum of all the terms in the progression is 500. Find the number of terms in the progression.

11 Robert buys a car for $8000 in total (including interest). He pays for the car by making monthly payments that are in arithmetic progression. The first payment that he makes is $200 and the debt is fully repaid after 16 payments. Find the fifth payment.

12 The sixth term of an arithmetic progression is $-3$ and the sum of the first ten terms is $-10$.

   **a** Find the first term and the common difference.

   **b** Given that the $n$th term of this progression is $-59$, find the value of $n$.

13 The sum of the first $n$ terms, $S_n$, of a particular arithmetic progression is given by $S_n = 4n^2 + 3n$. Find the first term and the common difference.

14 The sum of the first $n$ terms, $S_n$, of a particular arithmetic progression is given by $S_n = 12n - 2n^2$. Find the first term and the common difference.

**15** The sum of the first $n$ terms, $S_n$, of a particular arithmetic progression is given by $S_n = \dfrac{1}{4}(5n^2 - 17n)$.

Find an expression for the $n$th term.

**16** A circle is divided into ten sectors. The sizes of the angles of the sectors are in arithmetic progression. The angle of the largest sector is seven times the angle of the smallest sector. Find the angle of the smallest sector.

**17** An arithmetic sequence has first term $a$ and common difference $d$. The sum of the first 20 terms is seven times the sum of the first five terms.

 **a** Find $d$ in terms of $a$.        **b** Find the 65th term in terms of $a$.

**18** The tenth term in an arithmetic progression is three times the third term. Show that the sum of the first ten terms is eight times the sum of the first three terms.

**(P)** **19** The first term of an arithmetic progression is $\sin^2 x$ and the second term is 1.

 **a** Write down an expression, in terms of $\sin x$, for the fifth term of this progression.

 **b** Show that the sum of the first ten terms of this progression is $10 + 35\cos^2 x$.

**(PS)** **20** The sum of the digits in the number 67 is 13 (as $6 + 7 = 13$).

 **a** Show that the sum of the digits of the integers from 19 to 21 is 15.

 **b** Find the sum of the digits of the integers from 1 to 99.

## 6.4 Geometric progressions

The sequence $2, 6, 18, 54, \ldots$ is called a **geometric progression**. Each term is three times the preceding term. The constant multiplier, 3, is called the **common ratio**.

The notation used for a geometric progression is:

 $a = $ first term   $r = $ common ratio

The first five terms of a geometric progression whose first term is $a$ and whose common ratio is $r$ are:

| $a$ | $ar$ | $ar^2$ | $ar^3$ | $ar^4$ |
|---|---|---|---|---|
| term 1 | term 2 | term 3 | term 4 | term 5 |

This leads to the formula for the $n$th term of a geometric progression:

> **KEY POINT 6.10**
>
> $n$th term $= ar^{n-1}$

**WORKED EXAMPLE 6.16**

The fifth term of a geometric progression is 1 and the common ratio is $\dfrac{1}{2}$.
Find the eighth term and an expression for the $n$th term.

**Answer**

$n$th term $= ar^{n-1}$       Use $n$th term $= 1$ when $n = 5$ and $r = \dfrac{1}{2}$.

$$1 = a\left(\frac{1}{2}\right)^4$$

$$a = 16$$

$$\text{8th term} = 16\left(\frac{1}{2}\right)^7 = \frac{1}{8}$$

$$n\text{th term} = ar^{n-1} = 16\left(\frac{1}{2}\right)^{n-1}$$

**WORKED EXAMPLE 6.17**

The second and fifth terms in a geometric progression are 12 and 40.5, respectively. Find the first term and the common ratio. Hence, write down an expression for the $n$th term.

**Answer**

$12 = ar$ - - - - - - -(1)

$40.5 = ar^4$ - - - - -(2)

(2) ÷ (1) gives $\dfrac{ar^4}{ar} = \dfrac{40.5}{12}$

$$r^3 = \frac{27}{8}$$

$$r = \frac{3}{2}$$

Substituting $r = \dfrac{3}{2}$ into equation (1) gives $a = 8$.

First term $= 8$, common ratio $= \dfrac{3}{2}$, $n$th term $= 8\left(\dfrac{3}{2}\right)^{n-1}$.

**WORKED EXAMPLE 6.18**

The $n$th term of a geometric progression is $9\left(-\dfrac{2}{3}\right)^n$. Find the first term and the common ratio.

**Answer**

$$\text{1st term} = 9\left(-\frac{2}{3}\right)^1 = -6$$

$$\text{2nd term} = 9\left(-\frac{2}{3}\right)^2 = 4$$

Common ratio $= \dfrac{\text{2nd term}}{\text{1st term}} = \dfrac{4}{-6} = -\dfrac{2}{3}$

This is also clear from the formula directly: each term is $\left(-\dfrac{2}{3}\right)$ times the previous one.

First term $= -6$, common ratio $= -\dfrac{2}{3}$.

172

In this Explore activity you are *not* allowed to use a calculator.

1 Consider the sum of the first 10 terms, $S_{10}$, of a geometric progression with $a = 1$ and $r = 3$.

$$S_{10} = 1 + 3 + 3^2 + 3^3 + \cdots + 3^7 + 3^8 + 3^9$$

a Multiply both sides of the previous equation by the common ratio, 3, and complete the following statement.

$$3S_{10} = 3 + 3^2 + 3^{\cdots} + 3^{\cdots} + \cdots + 3^{\cdots} + 3^{\cdots} + 3^{\cdots}$$

b How does this compare to the original expression? Can you use this to find a simpler way of expressing the sum $S_{10}$?

2 Use the method from **question 1** to find an alternative way of expressing each of the following.

a $1 + r + r^2 + \cdots$ (10 terms)

b $a + ar + ar^2 + \cdots$ (10 terms)

c $a + ar + ar^2 + \cdots$ ($n$ terms)

You will have discovered in Explore 6.4 that the sum of a geometric progression, $S_n$, can be written as:

173

**KEY POINT 6.11**

$$S_n = \frac{a(1 - r^n)}{1 - r} \quad \text{or} \quad S_n = \frac{a(r^n - 1)}{r - 1}$$

**TIP**

These formulae are not defined when $r = 1$.

Either formula can be used but it is usually easier to:

● Use the first formula when $-1 < r < 1$.
● Use the second formula when $r > 1$ or when $r \leqslant -1$.

This is the proof of the formulae in Key point 6.11.

$$S_n = a + ar + ar^2 + \cdots + ar^{n-3} + ar^{n-2} + ar^{n-1} \quad \text{------------(1)}$$

$r \times (1)$: $\quad rS_n = \quad ar + ar^2 + \cdots + ar^{n-3} + ar^{n-2} + ar^{n-1} + ar^n \quad \text{--------(2)}$

$(2) - (1)$: $\quad rS_n - S_n = ar^n - a$

$$(r - 1)S_n = a(r^n - 1)$$

$$S_n = \frac{a(r^n - 1)}{r - 1}$$

Multiplying the numerator and the denominator by $-1$ gives the alternative formula

$$S_n = \frac{a(1 - r^n)}{1 - r}.$$

Can you see why this formula does not work when $r = 1$?

**WORKED EXAMPLE 6.19**

Find the sum of the first 12 terms of the geometric series $3 + 6 + 12 + 24 + \cdots$.

**Answer**

$S_n = \dfrac{a(r^n - 1)}{r - 1}$ ⋯⋯⋯⋯⋯⋯ Use $a = 3$, $r = 2$ and $n = 12$.

$S_{12} = \dfrac{3(2^{12} - 1)}{2 - 1}$ ⋯⋯⋯⋯⋯⋯ Simplify.

$= 12\,285$

**WORKED EXAMPLE 6.20**

The third term of a geometric progression is nine times the first term. The sum of the first six terms is $k$ times the sum of the first two terms. Find the value of $k$.

**Answer**

3rd term $= 9 \times$ first term

$ar^2 = 9a$ ⋯⋯⋯⋯ Divide both sides by $a$ (which we assume is non-zero) and solve.

$r = \pm 3$

Use $S_6 = kS_2$

$\dfrac{a(r^6 - 1)}{r - 1} = \dfrac{ka(r^2 - 1)}{r - 1}$ ⋯⋯⋯⋯ Rearrange to make $k$ the subject.

$k = \dfrac{r^6 - 1}{r^2 - 1}$

When $r = 3$, $k = 91$ and when $r = -3$, $k = 91$.

Hence, $k = 91$.

**EXERCISE 6D**

1 Identify whether the following sequences are geometric.
If they are geometric, write down the common ratio and the eighth term.

a $2, 4, 8, 14, \ldots$

b $7, 21, 63, 189, \ldots$

c $81, -27, 9, -3, \ldots$

d $\dfrac{1}{9}, \dfrac{2}{9}, \dfrac{4}{9}, \dfrac{7}{9}, \ldots$

e $1, 0.4, 0.16, 0.64, \ldots$

f $1, -1, 1, -1, \ldots$

2 The first term in a geometric progression is $a$ and the common ratio is $r$. Write down expressions, in terms of $a$ and $r$, for the sixth term and the 15th term.

3 The first term of a geometric progression is 270 and the fourth term is 80. Find the common ratio.

4 The first term of a geometric progression is 50 and the second term is $-30$. Find the fourth term.

5 The second term of a geometric progression is 12 and the fourth term is 27. Given that all the terms are positive, find the common ratio and the first term.

6 The sum of the second and third terms in a geometric progression is 84. The second term is 16 less than the first term. Given that all the terms in the progression are positive, find the first term.

7   Three consecutive terms of a geometric progression are $x$, 4 and $x + 6$. Find the possible values of $x$.

8   Find the sum of the first eight terms of each of these geometric series.

    **a**   $3 + 6 + 12 + 24 + \cdots$                **b**   $128 + 64 + 32 + 16 + \cdots$

    **c**   $1 - 2 + 4 - 8 + \cdots$                   **d**   $243 + 162 + 108 + 72 + \cdots$

9   The first four terms of a geometric progression are $0.5$, $1$, $2$ and $4$. Find the smallest number of terms that will give a sum greater than $1\,000\,000$.

10  A ball is thrown vertically upwards from the ground. The ball rises to a height of 8 m and then falls and bounces. After each bounce it rises to $\dfrac{3}{4}$ of the height of the previous bounce.

    **a**   Write down an expression for the height that the ball rises after the $n$th impact with the ground.

    **b**   Find the total distance that the ball travels from the first throw to the fifth impact with the ground.

11  The second term of a geometric progression is 24 and the third term is $12(x + 1)$.

    **a**   Find, in terms of $x$, the first term of the progression.

    **b**   Given that the sum of the first three terms is 76, find the possible values of $x$.

12  The third term of a geometric progression is nine times the first term. The sum of the first four terms is $k$ times the first term. Find the possible values of $k$.

13  A company makes a donation to charity each year. The value of the donation increases exponentially by 10% each year. The value of the donation in 2010 was $10\,000$.

    **a**   Find the value of the donation in 2016.

    **b**   Find the total value of the donations made during the years 2010 to 2016, inclusive.

**P**   14  A geometric progression has first term $a$, common ratio $r$ and sum to $n$ terms $S_n$.

    Show that $\dfrac{S_{3n} - S_{2n}}{S_n} = r^{2n}$.

**P**   15  Consider the sequence $1, 1, 3, \dfrac{1}{3}, 9, \dfrac{1}{9}, 27, \dfrac{1}{27}, 81, \dfrac{1}{81}, \ldots$.

    Show that the sum of the first $2n$ terms of the sequence is $\dfrac{1}{2}(2 + 3^n - 3^{1-n})$.

**P**   16  Let $S_n = 1 + 11 + 111 + 1111 + 11111 + \cdots$ to $n$ terms.

    Show that $S_n = \dfrac{10^{n+1} - 10 - 9n}{81}$.

## 6.5 Infinite geometric series

An infinite sequence is a sequence whose terms continue forever.

Consider the infinite geometric progression where $a = 2$ and $r = \dfrac{1}{2}$, so it begins

$2, 1, \dfrac{1}{2}, \dfrac{1}{4}, \dfrac{1}{8}, \ldots$. We can work out the sum of the first $n$ terms of this:

$S_1 = 2$, $S_2 = 3$, $S_3 = 3\dfrac{1}{2}$, $S_4 = 3\dfrac{3}{4}$, $S_5 = 3\dfrac{7}{8}$, and so on.

These sums are getting closer and closer to 4.

The diagram of the 2 by 2 square is a visual representation of this series. If the pattern of rectangles inside the square is continued, the total area of the rectangles approximates the area of the whole square (which is 4) increasingly well as more rectangles are included.

We therefore say that the sum of the infinite geometric series $2 + 1 + \dfrac{1}{2} + \dfrac{1}{4} + \dfrac{1}{8} + \cdots$ is 4, because the sum of the first $n$ terms gets as close to 4 as we like as $n$ gets larger. We write $2 + 1 + \dfrac{1}{2} + \dfrac{1}{4} + \dfrac{1}{8} + \cdots = 4$. We also say that the sum to infinity of this series is 4, and that the series converges to 4. A series that converges is also known as a **convergent series**.

You might be wondering why we can say this, as no matter how many terms we add up, the answer is always less than 4. The simplest answer is because it works. Mathematicians and philosophers have struggled with the idea of infinity for thousands of years, and whether something like '$2 + 1 + \dfrac{1}{2} + \dfrac{1}{4} + \dfrac{1}{8} + \cdots$' even makes sense. But over the past few hundred years, we have worked out that writing '$2 + 1 + \dfrac{1}{2} + \dfrac{1}{4} + \dfrac{1}{8} + \cdots = 4$' turns out to be very useful, and gives us answers that work consistently when we try to do more mathematics with them.

You are probably also familiar with a very important example of an infinite geometric series without realising it! What do we mean by the recurring decimal 0.3333...?

We can write this as a series: $0.3333\ldots = \dfrac{3}{10} + \dfrac{3}{100} + \dfrac{3}{1000} + \cdots$. If we work out the sum of the first $n$ terms of this geometric series, we find $S_1 = \dfrac{3}{10} = 0.3$, $S_2 = \dfrac{33}{100} = 0.33$, $S_3 = \dfrac{333}{1000} = 0.333$ and so on. These sums are getting as close as we like to $\dfrac{1}{3}$, so we say that the sum of the infinite series is equal to $\dfrac{1}{3}$, and we write $\dfrac{1}{3} = 0.3333\ldots$. This justifies what you have been writing for many years. Using the formula we will be working out shortly, we can easily write any recurring decimal as an exact fraction.

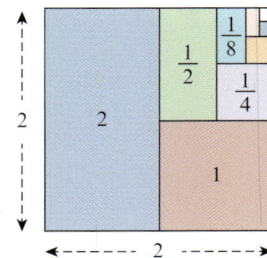

## DID YOU KNOW?

The first person to introduce infinite decimal numbers was Simon Stevin in 1585. He was an influential mathematician who popularised the use of decimals more generally as well, through a publication called *De Thiende* ('The tenth').

## EXPLORE 6.5

1 Investigate whether these infinite geometric series converge or not. You could use a spreadsheet to help with the calculations. If they converge, state their sum to infinity.

   a $\quad a = \dfrac{3}{5}, r = -2$

   b $\quad a = 3, r = -\dfrac{1}{5}$

   c $\quad a = 6, r = \dfrac{2}{3}$

   d $\quad a = -\dfrac{1}{2}, r = -2$

2 Find other convergent geometric series of your own.
   In each case, find the sum to infinity.

3 Can you find a condition for $r$ for which a geometric series is convergent?

Consider the geometric series $a + ar + ar^2 + ar^3 + \cdots + ar^{n-1}$.

The sum, $S_n$, is given by the formula $S_n = \dfrac{a(1 - r^n)}{1 - r}$.

If $-1 < r < 1$, then as $n$ gets larger and larger, $r^n$ gets closer and closer to 0.

We say that as $n$ tends to infinity, $r^n$ tends to zero, and we write 'as $n \to \infty$, $r^n \to 0$'.

Hence, as $n \to \infty$, $\dfrac{a(1 - r^n)}{1 - r} \to \dfrac{a(1 - 0)}{1 - r} = \dfrac{a}{1 - r}$.

This gives the result:

## KEY POINT 6.12

$S_\infty = \dfrac{a}{1 - r}$ provided that $-1 < r < 1$.

If $r \geqslant 1$ or $r \leqslant -1$, then $r^n$ does not converge, and so the series itself does not converge. So an infinite geometric series converges when and only when $-1 < r < 1$.

## WORKED EXAMPLE 6.21

The first four terms of a geometric progression are 5, 4, 3.2 and 2.56.

   **a**  Write down the common ratio.

   **b**  Find the sum to infinity.

**Answer**

   **a**  Common ratio $= \dfrac{\text{second term}}{\text{first term}} = \dfrac{4}{5}$

   **b**  $S_\infty = \dfrac{a}{1 - r}$               Use $a = 5$ and $r = \dfrac{4}{5}$.

         $= \dfrac{5}{1 - \left(\dfrac{4}{5}\right)}$

         $= 25$

## WORKED EXAMPLE 6.22

A geometric progression has a common ratio of $-\dfrac{2}{3}$ and the sum of the first three terms is 63.

   **a**  Find the first term of the progression.

   **b**  Find the sum to infinity.

**Answer**

   **a**  $S_3 = \dfrac{a(1 - r^3)}{1 - r}$           Use $S_3 = 63$ and $r = -\dfrac{2}{3}$.

        $63 = \dfrac{a\left(1 - \left(-\dfrac{2}{3}\right)^3\right)}{1 - \left(-\dfrac{2}{3}\right)}$       Simplify.

$$63 = \frac{a \times \frac{35}{27}}{\frac{5}{3}}$$

⋯⋯⋯⋯⋯⋯⋯⋯⋯⋯⋯⋯ Solve.

$$a = 81$$

**b** $S_\infty = \dfrac{a}{1-r}$ ⋯⋯⋯⋯⋯⋯⋯⋯⋯ Use $a = 81$ and $r = -\dfrac{2}{3}$.

$$= \frac{81}{1 - \left(-\frac{2}{3}\right)}$$

$$= 48\frac{3}{5}$$

**EXERCISE 6E**

1   Find the sum to infinity of each of the following geometric series.

   **a**   $2 + \dfrac{2}{3} + \dfrac{2}{9} + \dfrac{2}{27} + \cdots$          **b**   $1 + 0.1 + 0.01 + 0.001 + \cdots$

   **c**   $40 - 20 + 10 - 5 + \cdots$          **d**   $-64 + 48 - 36 + 27 - \cdots$

2   The first four terms of a geometric progression are $1$, $0.5^2$, $0.5^4$ and $0.5^6$. Find the sum to infinity.

3   The first term of a geometric progression is 8 and the second term is 6. Find the sum to infinity.

4   The first term of a geometric progression is 270 and the fourth term is 80. Find the common ratio and the sum to infinity.

5   **a**   Write the recurring decimal $0.\dot{5}\dot{7}$ as the sum of a geometric progression.

   **b**   Use your answer to part **a** to show that $0.\dot{5}\dot{7}$ can be written as $\dfrac{19}{33}$.

6   The first term of a geometric progression is 150 and the sum to infinity is 200. Find the common ratio and the sum of the first four terms.

7   The second term of a geometric progression is 4.5 and the sum to infinity is 18. Find the common ratio and the first term.

8   Write the recurring decimal $0.315151515\ldots$ as a fraction.

9   The second term of a geometric progression is 9 and the fourth term is 4. Given that the common ratio is positive, find:

   **a**   the common ratio and the first term

   **b**   the sum to infinity.

10   The third term of a geometric progression is 16 and the sixth term is $-\dfrac{1}{4}$.

   **a**   Find the common ratio and the first term.

   **b**   Find the sum to infinity.

**11** The first three terms of a geometric progression are 135, $k$ and 60. Given that all the terms in the progression are positive, find:

   **a** the value of $k$

   **b** the sum to infinity.

**12** The first three terms of a geometric progression are $k + 12$, $k$ and $k - 9$, respectively.

   **a** Find the value of $k$.

   **b** Find the sum to infinity.

**13** The fourth term of a geometric progression is 48 and the sum to infinity is five times the first term. Find the first term.

**14** A geometric progression has first term $a$ and common ratio $r$. The sum of the first three terms is 3.92 and the sum to infinity is 5. Find the value of $a$ and the value of $r$.

**15** The first term of a geometric progression is 1 and the second term is $2\cos x$, where $0 < x < \dfrac{\pi}{2}$. Find the set of values of $x$ for which this progression is convergent.

**(PS)**   **16** A circle of radius 1 cm is drawn touching the three edges of an equilateral triangle.

   Three smaller circles are then drawn at each corner to touch the original circle and two edges of the triangle.

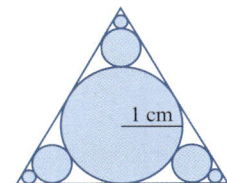

   This process is then repeated an infinite number of times, as shown in the diagram.

   **a** Find the sum of the circumferences of all the circles.

   **b** Find the sum of the areas of all the circles.

**(P)**   **17**

pattern 1        pattern 2        pattern 3        pattern 4

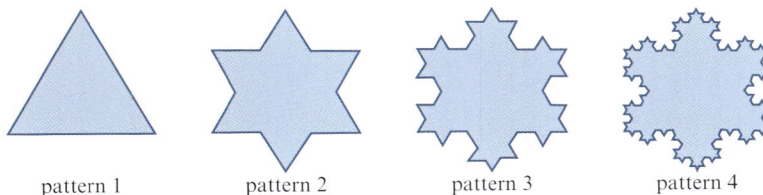

We can construct a Koch snowflake as follows.
Starting with an equilateral triangle (pattern 1), we perform the following steps to produce pattern 2.

**Step 1:** Divide each line segment into three equal segments.

**Step 2:** Draw an equilateral triangle, pointing outwards, that has the middle segment from step 1 as its base.

**Step 3:** Remove the line segments that were used as the base of the equilateral triangles in step 2.

These three steps are then repeated to produce the next pattern.

   **a** Let $p_n$ be the perimeter of pattern $n$. Show that the sequence $p_1, p_2, p_3, \ldots$ tends to infinity.

   **b** Let $A_n$ be the area of pattern $n$. Show that the sequence $A_1, A_2, A_3, \ldots$ tends to $\dfrac{8}{5}$ times the area of the original triangle.

   **c** The Koch snowflake is the limit of the patterns. It has infinite perimeter but an area of $\dfrac{8}{5}$ of the original triangle, as you have shown. This snowflake pattern is an example of a fractal. Use the internet to find out about the Sierpinski triangle fractal.

## 6.6 Further arithmetic and geometric series

**EXPLORE 6.6**

$$a, b, c, \ldots$$

1   Given that $a$, $b$ and $c$ are in arithmetic progression, find an equation connecting $a$, $b$ and $c$.

2   Given that $a$, $b$ and $c$ are in geometric progression, find an equation connecting $a$, $b$ and $c$.

**WORKED EXAMPLE 6.23**

The first, second and third terms of an arithmetic series are $x$, $y$ and $x^2$. The first, second and third terms of a geometric series are $x$, $x^2$ and $y$. Given that $x < 0$, find:

a   the value of $x$ and the value of $y$

b   the sum to infinity of the geometric series

c   the sum of the first 20 terms of the arithmetic series.

**Answer**

a   Arithmetic series is: $x + y + x^2 + \cdots$ ............ Use common differences.

$$y - x = x^2 - y$$

$$2y = x^2 + x \text{ -------- (1)}$$

Geometric series is: $x + x^2 + y + \cdots$ ............ Use common ratios.

$$\frac{y}{x^2} = \frac{x^2}{x}$$

$$y = x^3 \text{ ------------ (2)}$$

(1) and (2) give $2x^3 = x^2 + x$ ............ Divide by $x$ (since $x \neq 0$) and rearrange.

$$2x^2 - x - 1 = 0$$ ............ Factorise and solve.

$$(2x + 1)(x - 1) = 0$$

$$x = -\frac{1}{2} \text{ or } x = 1$$ ............ $x \neq 1$ since $x < 0$.

Hence, $x = -\frac{1}{2}$ and $y = -\frac{1}{8}$.

b   $S_\infty = \dfrac{a}{1-r}$ ............ Use $a = -\dfrac{1}{2}$ and $r = -\dfrac{1}{2}$.

$$S_\infty = \frac{-\frac{1}{2}}{1 - \left(-\frac{1}{2}\right)} = -\frac{1}{3}$$

c   $S_n = \dfrac{n}{2}[2a + (n-1)d]$ ............ Use $n = 20$, $a = -\dfrac{1}{2}$, $d = -\dfrac{1}{8} - \left(-\dfrac{1}{2}\right) = \dfrac{3}{8}$.

$$S_{20} = \frac{20}{2}\left[-1 + 19\left(\frac{3}{8}\right)\right]$$

$$= 61.25$$

**EXERCISE 6F**

1   The first term of a progression is 16 and the second term is 24. Find the sum of the first eight terms given that the progression is:

   **a**   arithmetic               **b**   geometric

2   The first term of a progression is 20 and the second term is 16.

   **a**   Given that the progression is geometric, find the sum to infinity.

   **b**   Given that the progression is arithmetic, find the number of terms in the progression if the sum of all the terms is −160.

3   The first, second and third terms of a geometric progression are the first, fourth and tenth terms, respectively, of an arithmetic progression. Given that the first term in each progression is 12 and the common ratio of the geometric progression is $r$, where $r \neq 1$, find:

   **a**   the value of $r$

   **b**   the sixth term of each progression.

4   A geometric progression has eight terms. The first term is 256 and the common ratio is $\frac{1}{2}$.

   An arithmetic progression has 51 terms and common difference $\frac{1}{2}$.

   The sum of all the terms in the geometric progression is equal to the sum of all the terms in the arithmetic progression. Find the first term and the last term in the arithmetic progression.

5   The first, second and third terms of a geometric progression are the first, sixth and ninth terms, respectively, of an arithmetic progression. Given that the first term in each progression is 100 and the common ratio of the geometric progression is $r$, where $r \neq 1$, find:

   **a**   the value of $r$

   **b**   the fifth term of each progression.

6   The first term of an arithmetic progression is 16 and the sum of the first 20 terms is 1080.

   **a**   Find the common difference of this progression.

   The first, third and $n$th terms of this arithmetic progression are the first, second and third terms, respectively, of a geometric progression.

   **b**   Find the common ratio of the geometric progression and the value of $n$.

7   The first term of a progression is $2x$ and the second term is $x^2$.

   **a**   For the case where the progression is arithmetic with a common difference of 15, find the two possible values of $x$ and corresponding values of the third term.

   **b**   For the case where the progression is geometric with a third term of $-\frac{1}{16}$, find the sum to infinity.

181

# Checklist of learning and understanding

**Binomial expansions**

Binomial coefficients, denoted by $^{n}C_{r}$ or $\begin{pmatrix} n \\ r \end{pmatrix}$, can be found using:

- Pascal's triangle

- the formulae $\begin{pmatrix} n \\ r \end{pmatrix} = \dfrac{n!}{r!(n-r)!}$ or $\begin{pmatrix} n \\ r \end{pmatrix} = \dfrac{n \times (n-1) \times (n-2) \times \cdots \times (n-r+1)}{r \times (r-1) \times (r-2) \times \cdots \times 3 \times 2 \times 1}$.

If $n$ is a positive integer, the Binomial theorem states that:

$$(1+x)^n = \begin{pmatrix} n \\ 0 \end{pmatrix} + \begin{pmatrix} n \\ 1 \end{pmatrix} x + \begin{pmatrix} n \\ 2 \end{pmatrix} x^2 + \cdots + \begin{pmatrix} n \\ n \end{pmatrix} x^n, \text{ where the } (r+1)\text{th term} = \begin{pmatrix} n \\ r \end{pmatrix} x^r.$$

We can extend this rule to give:

$$(a+b)^n = \begin{pmatrix} n \\ 0 \end{pmatrix} a^n + \begin{pmatrix} n \\ 1 \end{pmatrix} a^{n-1}b^1 + \begin{pmatrix} n \\ 2 \end{pmatrix} a^{n-2}b^2 + \ldots + \begin{pmatrix} n \\ n \end{pmatrix} b^n, \text{ where the } (r+1)\text{th term} = \begin{pmatrix} n \\ r \end{pmatrix} a^{n-r}b^r.$$

We can also write the expansion of $(1+x)^n$ as:

$$(1+x)^n = 1 + nx + \frac{n(n-1)}{2!} x^2 + \frac{n(n-1)(n-2)}{3!} x^3 + \ldots + x^n$$

**Arithmetic series**

For an arithmetic progression with first term $a$, common difference $d$ and $n$ terms:

- the $k$th term is $a + (k-1)d$
- the last term is $l = a + (n-1)d$
- the sum of the terms is $S_n = \dfrac{n}{2}(a+l) = \dfrac{n}{2}[2a + (n-1)d]$.

**Geometric series**

For a geometric progression with first term $a$, common ratio $r$ and $n$ terms:

- the $k$th term is $ar^{k-1}$
- the last term is $ar^{n-1}$
- sum of the terms is $S_n = \dfrac{a(1-r^n)}{1-r} = \dfrac{a(r^n-1)}{r-1}$.

The condition for an infinite geometric series to converge is $-1 < r < 1$.

When an infinite geometric series converges, $S_\infty = \dfrac{a}{1-r}$.

## END-OF-CHAPTER REVIEW EXERCISE 6

1 Find the coefficient of $x^2$ in the expansion of $\left( 2x + \dfrac{3}{x^2} \right)^5$. [3]

2 In the expansion of $(a + 2x)^6$, the coefficient of $x$ is equal to the coefficient of $x^2$. Find the value of the constant $a$. [3]

3 In the expansion of $\left( 1 - \dfrac{x}{a} \right)(5 + x)^6$, the coefficient of $x^2$ is zero. Find the value of $a$. [3]

4 Find the term independent of $x$ in the expansion of $\left( 3x - \dfrac{2}{5x} \right)^6$. [3]

5 In the expansion of $(2 + ax)^7$, where $a$ is a constant, the coefficient of $x$ is $-2240$. Find the coefficient of $x^2$. [4]

6 Find the coefficient of $x^5$ in the expansion of $\left( x^3 + \dfrac{2}{x^2} \right)^5$. [4]

7 Find the term independent of $x$ in the expansion of $\left( 3x^2 - \dfrac{1}{2x^3} \right)^5$. [4]

8 a Find the first three terms in the expansion of $(x - 3x^2)^8$, in descending powers of $x$. [3]

   b Find the coefficient of $x^{15}$ in the expansion of $(1 - x)(x - 3x^2)^8$. [2]

9 a Find the first three terms in the expansion of $(1 + px)^8$, in ascending powers of $x$. [3]

   b Given that the coefficient of $x^2$ in the expansion of $(1 - 2x)(1 + px)^8$ is 204, find the possible values of $p$. [4]

10 a Find the first three terms, in ascending powers of $x$, in the expansion of:

   i $(1 + 2x)^5$ [2]

   ii $(3 - x)^5$ [2]

   b Find the coefficient of $x^2$ in the expansion of $[(1 + 2x)(3 - x)]^5$. [3]

11 The first term of an arithmetic progression is 1.75 and the second term is 1.5. The sum of the first $n$ terms is $-n$. Find the value of $n$. [4]

12 The second term of a geometric progression is $-1458$ and the fifth term is 432. Find:

   a the common ratio [3]

   b the first term [1]

   c the sum to infinity. [2]

13 An arithmetic progression has first term $a$ and common difference $d$. The sum of the first 100 terms is 25 times the sum of the first 20 terms.

   a Find $d$ in terms of $a$. [3]

   b Write down an expression, in terms of $a$, for the 50th term. [2]

14 The tenth term of an arithmetic progression is 17 and the sum of the first five terms is 190.

   a Find the first term of the progression and the common difference. [4]

   b Given that the $n$th term of the progression is $-19$, find the value of $n$. [2]

**15 a** The fifth term of an arithmetic progression is 18 and the sum of the first eight terms is 186. Find the first term and the common difference. [4]

**b** The first term of a geometric progression is 32 and the fourth term is $\frac{1}{2}$. Find the sum to infinity of the progression. [3]

**16 a** The seventh term of an arithmetic progression is 19 and the sum of the first twelve terms is 224. Find the fourth term. [4]

**b** A geometric progression has first term 3 and common ratio $r$. A second geometric progression has first term 2 and common ratio $\frac{1}{5}r$. The two progressions have the same sum to infinity, $S$. Find the value of $r$ and the value of $S$. [3]

**17 a** A geometric progression has first term $a$, common ratio $r$ and sum to infinity $S$.

A second geometric progression has first term $5a$, common ratio $3r$ and sum to infinity $10S$. Find the value of $r$. [3]

**b** An arithmetic progression has first term –4. The $n$th term is 8 and the ($2n$)th term is 20.8. Find the value of $n$. [4]

**18** A television quiz show takes place every day. On day 1 the prize money is \$1000. If this is not won the prize money is increased for day 2. The prize money is increased in a similar way every day until it is won. The television company considered the following two different models for increasing the prize money.

Model 1: Increase the prize money by \$1000 each day.

Model 2: Increase the prize money by 10% each day.

On each day that the prize money is not won the television company makes a donation to charity. The amount donated is 5% of the value of the prize on that day. After 40 days the prize money has still not been won. Calculate the total amount donated to charity

**i** if Model 1 is used, [4]

**ii** if Model 2 is used. [3]

*Cambridge International AS & A Level Mathematics 9709 Paper 11 Q8 June 2011*

**19 a** The first two terms of an arithmetic progression are 1 and $\cos^2 x$ respectively. Show that the sum of the first ten terms can be expressed in the form $a - b \sin^2 x$, where $a$ and $b$ are constants to be found. [3]

**b** The first two terms of a geometric progression are 1 and $\frac{1}{3}\tan^2 \theta$ respectively, where $0 < \theta < \frac{1}{2}\pi$.

**i** Find the set of values of $\theta$ for which the progression is convergent. [2]

**ii** Find the exact value of the sum to infinity when $\theta = \frac{1}{6}\pi$. [2]

*Cambridge International AS & A Level Mathematics 9709 Paper 11 Q7 June 2012*

**20** The first term of a progression is $4x$ and the second term is $x^2$.

**i** For the case where the progression is arithmetic with a common difference of 12, find the possible values of $x$ and the corresponding values of the third term. [4]

**ii** For the case where the progression is geometric with a sum to infinity of 8, find the third term. [4]

*Cambridge International AS & A Level Mathematics 9709 Paper 11 Q8 November 2015*

**21** **a** The third and fourth terms of a geometric progression are $\frac{1}{3}$ and $\frac{2}{9}$ respectively. Find the sum to infinity of the progression. [4]

**b** A circle is divided into 5 sectors in such a way that the angles of the sectors are in arithmetic progression. Given that the angle of the largest sector is 4 times the angle of the smallest sector, find the angle of the largest sector. [4]

*Cambridge International AS & A Level Mathematics 9709 Paper 11 Q7 June 2015*

**22** **a** In an arithmetic progression the sum of the first ten terms is 400 and the sum of the next ten terms is 1000. Find the common difference and the first term. [5]

**b** A geometric progression has first term $a$, common ratio $r$ and sum to infinity 6. A second geometric progression has first term $2a$, common ratio $r^2$ and sum to infinity 7. Find the values of $a$ and $r$. [5]

*Cambridge International AS & A Level Mathematics 9709 Paper 11 Q9 November 2013*

1   Find the highest power of $x$ in the expansion of $\left[ (5x^4 + 3)^8 + (1 - 3x^3)^5(4x^2 - 5x^5)^6 \right]^4$.   [2]

2   Find the term independent of $x$ in the expansion of $\left( 4x - \dfrac{1}{x^2} \right)^6$.   [3]

3   **a**  Find the first three terms in the expansion of $\left( 3x - \dfrac{2}{x^2} \right)^6$, in descending powers of $x$.   [3]

    **b**  Hence, find the coefficient of $x^2$ in the expansion of $\left( 1 + \dfrac{2}{x} \right)\left( 3x - \dfrac{2}{x} \right)^6$.   [2]

4   **a**  Find the first three terms when $(1 - 2x)^5$ is expanded, in ascending powers of $x$.   [3]

    **b**  In the expansion of $(3 + ax)(1 - 2x)^5$, the coefficient of $x^2$ is zero.

       Find the value of $a$.   [2]

5   The first term of a geometric progression is 50 and the second term is $-40$.

    **a**  Find the fourth term.   [3]

    **b**  Find the sum to infinity.   [2]

6   The first three terms of a geometric progression are $3k + 14$, $k + 14$ and $k$, respectively.

    All the terms in the progression are positive.

    **a**  Find the value of $k$.   [3]

    **b**  Find the sum to infinity.   [2]

7   The sum of the 1st and 2nd terms of a geometric progression is 50 and the sum of the 2nd and 3rd terms is 30. Find the sum to infinity.   [6]

    *Cambridge International AS & A Level Mathematics 9709 Paper 11 Q5 November 2016*

8   **i**  Show that $\cos^4 x \equiv 1 - 2\sin^2 x + \sin^4 x$.   [1]

    **ii**  Hence, or otherwise, solve the equation $8\sin^4 x + \cos^4 x = 2\cos^2 x$ for $0° \leqslant x \leqslant 360°$.   [5]

    *Cambridge International AS & A Level Mathematics 9709 Paper 11 Q6 November 2016*

9   A sector of a circle, radius $r$ cm, has a perimeter of $60$ cm.

    **a**  Show that the area, $A$ cm$^2$, of the sector is given by $A = 30r - r^2$.   [2]

    **b**  Express $30r - r^2$ in the form $a - (r - b)^2$, where $a$ and $b$ are constants.   [2]

    Given that $r$ can vary:

    **c**  find the value of $r$ at which $A$ is a maximum   [1]

    **d**  find this stationary value of $A$.   [1]

**10**

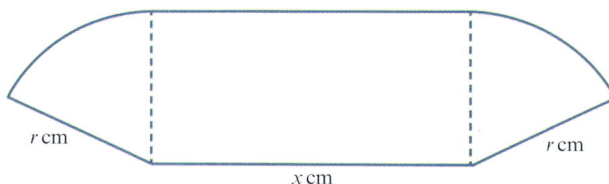

The diagram shows a metal plate consisting of a rectangle with sides $x$ cm and $r$ cm and two identical sectors of a circle of radius $r$ cm. The perimeter of the plate is 100 cm.

**a** Show that the area, $A$ cm$^2$, of the plate is given by $A = 50r - r^2$. [2]

**b** Express $50r - r^2$ in the form $a - (r - b)^2$, where $a$ and $b$ are constants. [2]

Given that $r$ can vary:

**c** find the value of $r$ at which $A$ is a maximum [1]

**d** find this stationary value of $A$. [1]

**11**

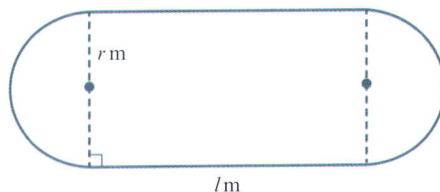

The diagram shows a running track. The track has a perimeter of 400 m and consists of two straight sections of length $l$ m and two semicircular sections of radius $r$ m.

**a** Show that the area, $A$ m$^2$, of the region enclosed by the track is given by $A = 400r - \pi r^2$. [2]

**b** Express $400r - \pi r^2$ in the form $\dfrac{a}{\pi} - \pi \left( r - \dfrac{b}{\pi} \right)^2$, where $a$ and $b$ are constants. [3]

Given that $l$ and $r$ can vary:

**c** show that $A$ has a maximum value when $l = 0$ [2]

**d** find this stationary value of $A$. [1]

187

**12**

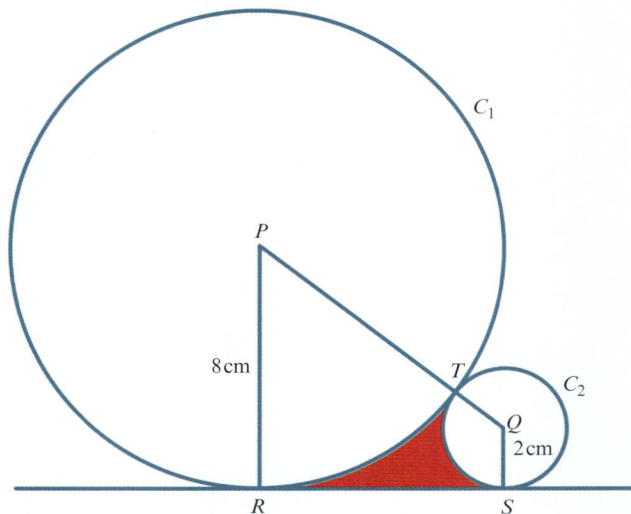

The diagram shows two circles, $C_1$ and $C_2$, touching at the point $T$. Circle $C_1$ has centre $P$ and radius 8 cm; circle $C_2$ has centre $Q$ and radius 2 cm. Points $R$ and $S$ lie on $C_1$ and $C_2$ respectively, and $RS$ is a tangent to both circles.

  i  Show that $RS = 8$ cm.   **[2]**

  ii  Find angle $RPQ$ in radians correct to 4 significant figures.   **[2]**

  iii  Find the area of the shaded region.   **[4]**

*Cambridge International AS & A Level Mathematics 9709 Paper 11 Q9 November 2010*

**13**

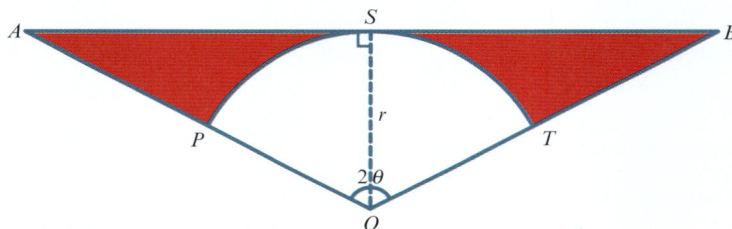

In the diagram, $OAB$ is an isosceles triangle with $OA = OB$ and angle $AOB = 2\theta$ radians. Arc $PST$ has centre $O$ and radius $r$, and the line $ASB$ is a tangent to the arc $PST$ at $S$.

  i  Find the total area of the shaded regions in terms of $r$ and $\theta$.   **[4]**

  ii  In the case where $\theta = \dfrac{1}{3}\pi$ and $r = 6$, find the total perimeter of the shaded regions, leaving your answer in terms of $\sqrt{3}$ and $\pi$.   **[5]**

*Cambridge International AS & A Level Mathematics 9709 Paper 11 Q9 June 2011*

**14**  The function f is such that $f(x) = 2\sin^2 x - 3\cos^2 x$ for $0 \leqslant x \leqslant \pi$.

  i  Express $f(x)$ in the form $a + b\cos^2 x$, stating the values of $a$ and $b$.   **[2]**

  ii  State the greatest and least values of $f(x)$.   **[2]**

  iii  Solve the equation $f(x) + 1 = 0$.   **[3]**

*Cambridge International AS & A Level Mathematics 9709 Paper 11 Q5 June 2010*

188

**15**  **i**  Prove the identity $\dfrac{\sin\theta}{1-\cos\theta} - \dfrac{1}{\sin\theta} \equiv \dfrac{1}{\tan\theta}$.  [4]

   **ii**  Hence solve the equation $\dfrac{\sin\theta}{1-\cos\theta} - \dfrac{1}{\sin\theta} = 4\tan\theta$ for $0° < \theta < 180°$.  [3]

   *Cambridge International AS & A Level Mathematics 9709 Paper 11 Q9 June 2014*

**16**  The function f is defined by $f : x \mapsto 4\sin x - 1$ for $-\dfrac{1}{2}\pi \leqslant x \leqslant \dfrac{1}{2}\pi$.

   **i**  State the range of f.  [2]

   **ii**  Find the coordinates of the points at which the curve $y = f(x)$ intersects the coordinate axes.  [3]

   **iii**  Sketch the graph of $y = f(x)$.  [2]

   **iv**  Obtain an expression for $f^{-1}(x)$, stating both the domain and range of $f^{-1}$.  [4]

   *Cambridge International AS & A Level Mathematics 9709 Paper 11 Q11 June 2016*

**17**  **a**  The first term of a geometric progression in which all the terms are positive is $50$. The third term is $32$. Find the sum to infinity of the progression.  [3]

   **b**  The first three terms of an arithmetic progression are $2\sin x$, $3\cos x$ and $(\sin x + 2\cos x)$ respectively, where $x$ is an acute angle.

   **i**  Show that $\tan x = \dfrac{4}{3}$.  [3]

   **ii**  Find the sum of the first twenty terms of the progression.  [3]

   *Cambridge International AS & A Level Mathematics 9709 Paper 11 Q9 June 2016*

189

# Chapter 7
# Differentiation

**In this chapter you will learn how to:**

■ understand that the gradient of a curve at a point is the limit of the gradients of a suitable sequence of chords

■ use the notations $f'(x)$, $f''(x)$, $\dfrac{dy}{dx}$ and $\dfrac{d^2 y}{dx^2}$ for the first and second derivatives

■ use the derivative of $x^n$ (for any rational $n$), together with constant multiples, sums, differences of functions, and of composite functions using the chain rule

■ apply differentiation to gradients, tangents and normals.

## PREREQUISITE KNOWLEDGE

| Where it comes from | What you should be able to do | Check your skills |
|---|---|---|
| IGCSE / O Level Mathematics | Use the rules of indices to simplify expressions to the form $ax^n$. | 1 Write in the form $ax^n$:<br>**a** $3x\sqrt{x}$<br>**b** $5\sqrt[3]{x^2}$<br>**c** $\dfrac{x}{2\sqrt{x}}$<br>**d** $\dfrac{1}{2x}$<br>**e** $\dfrac{3}{x^2}$<br>**f** $-\dfrac{2x^2}{5\sqrt[3]{x}}$ |
| IGCSE / O Level Mathematics | Write $\dfrac{k}{(ax+b)^n}$ in the form $k(ax+b)^{-n}$. | 2 Write in the form $k(ax+b)^{-n}$:<br>**a** $\dfrac{4}{(x-2)^3}$<br>**b** $\dfrac{2}{(3x+1)^5}$ |
| Chapter 3 | Find the gradient of a perpendicular line. | 3 The gradient of a line is $\dfrac{2}{3}$. Write down the gradient of a line that is perpendicular to it. |
| Chapter 3 | Find the equation of a line with a given gradient and a given point on the line. | 4 Find the equation of the line with gradient 2 that passes through the point (2, 5). |

## Why do we study differentiation?

Calculus is the mathematical study of change. Calculus has two basic tools, differentiation and integration, and it has widespread uses in science, medicine, engineering and economics. A few examples where calculus is used are:

- designing effective aircraft wings
- the study of radioactive decay
- the study of population change
- modelling the financial world.

In this chapter you will be studying the first of the two basic tools of calculus. You will learn the rules of differentiation and how to apply these to problems involving gradients, tangents and normals. In Chapter 8 you will then learn how to apply these rules of differentiation to more practical problems.

## 7.1 Derivatives and gradient functions

At IGCSE / O Level you learnt how to estimate the gradient of a curve at a point by drawing a suitable tangent and then calculating the gradient of the tangent. This method only gives an approximate answer (because of the inaccuracy of drawing the tangent) and it is also very time consuming.

In this chapter you will learn a method for finding the exact gradient of the graph of a function (which does not involve drawing the graph). This exact method is called **differentiation**.

**WEB LINK**

Try the *Calculus* resources on the Underground Mathematics website.

**EXPLORE 7.1**

Consider the quadratic function $y = x^2$ and a point $P(x, x^2)$ on the curve.

1 Let $P$ be the point $(2, 4)$.

The points $A(2.2, 4.84)$, $B(2.1, 4.41)$ and $C(2.01, 4.0401)$ also lie on the curve and are close to the point $P(2, 4)$.

   a Calculate the gradient of:

   i the chord $PA$

   ii the chord $PB$

   iii the chord $PC$.

   b Discuss your results with those of your classmates and make suggestions as to what is happening.

   c Suggest a value for the gradient of the curve $y = x^2$ at the point $(2, 4)$.

2 Let $P$ be the point $(3, 9)$.

The points $A(3.2, 10.24)$, $B(3.1, 9.61)$ and $C(3.01, 9.0601)$ also lie on the curve and are close to the point $P(3, 9)$.

   a Calculate the gradient of:

   i the chord $PA$

   ii the chord $PB$

   iii the chord $PC$.

   b Discuss your results with those of your classmates and make suggestions as to what is happening.

   c Suggest a value for the gradient of the curve $y = x^2$ at the point $(3, 9)$.

3 Use a spreadsheet to investigate the value of the gradient at other points on the curve $y = x^2$.

4 Can you suggest a general formula for the gradient of the curve $y = x^2$ at the point $(a, a^2)$? What would be the gradient at $(x, x^2)$?

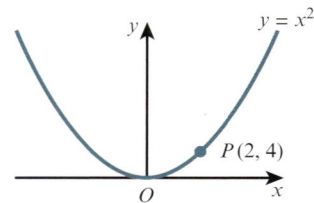

The general formula for the gradient of the curve $y = x^2$ at the point $(x, x^2)$ can be proved algebraically.

Take a point $P(x, y)$ on the curve $y = x^2$ and a point $A$ that is close to the point $P$.

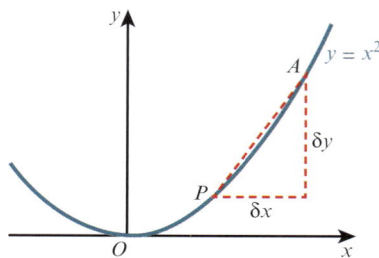

**WEB LINK**

There are other ways of thinking about the gradient of a curve. Try the following resources on the Underground Mathematics website *Zooming in* and *Mapping a derivative*.

The coordinates of $A$ are $(x + \delta x, y + \delta y)$, where $\delta x$ is a small increase in the value of $x$ and $\delta y$ is the corresponding small increase in the value of $y$.

We can also write the coordinates of $P$ and $A$ as $(x, x^2)$ and $\left(x + \delta x, (x + \delta x)^2\right)$.

Gradient of chord $PA = \dfrac{y_2 - y_1}{x_2 - x_1}$

$$= \frac{(x + \delta x)^2 - x^2}{(x + \delta x) - x}$$

$$= \frac{x^2 + 2x\,\delta x + (\delta x)^2 - x^2}{\delta x}$$

$$= \frac{2x\,\delta x + (\delta x)^2}{\delta x}$$

$$= 2x + \delta x$$

**TIP**

We use the Greek symbol delta, $\delta$, to denote a very small change in a quantity.

As $\delta x$ tends towards $0$, $A$ tends to $P$ and the gradient of the chord $PA$ tends to a value. We call this value the gradient of the curve at $P$.

In this case, therefore, the gradient of the curve at $P$ is $2x$.

This process of finding the gradient of a curve at any point is called differentiation.

Later in this chapter, you will learn some rules for differentiating functions without having to calculate the gradients of chords as we have done here. The process of calculating gradients using the limit of gradients of chords is sometimes called **differentiation from first principles**.

## Notation

There are three different notations that are used to describe the previous rule.

1. If $y = x^2$, then $\dfrac{dy}{dx} = 2x$.

2. If $f(x) = x^2$, then $f'(x) = 2x$.

3. $\dfrac{d}{dx}(x^2) = 2x$

If $y$ is a function of $x$, then $\dfrac{dy}{dx}$ is called the **derivative** of $y$ with respect to $x$.

Likewise $f'(x)$ is called the derivative of $f(x)$.

If $y = f(x)$ is the graph of a function, then $\dfrac{dy}{dx}$ or $f'(x)$ is sometimes also called the **gradient function** of this curve.

$\dfrac{d}{dx}(x^2) = 2x$ means 'if we differentiate $x^2$ with respect to $x$, the result is $2x$'.

You do not need to be able to differentiate from first principles but you are expected to understand that the gradient of a curve at a point is the limit of a suitable sequence of chords.

**DID YOU KNOW?**

193

Gottfried Wilhelm Leibniz and Isaac Newton are both credited with developing the modern calculus that we use today. Leibniz's notation for derivatives was $\dfrac{dy}{dx}$. Newton's notation for $\dfrac{dy}{dx}$ was $\dot{y}$. The notation $f'(x)$ is known as Lagrange's notation.

**EXPLORE 7.2**

1. Use a spreadsheet to investigate the gradient of the curve $y = x^3$.

2. Can you suggest a general formula for the gradient of the curve $y = x^3$ at the point $(x, x^3)$?

**E** 3. Differentiate $y = x^3$ from first principles to confirm your answer to **question 2**.

## Differentiation of power functions

We now know that $\dfrac{d}{dx}(x^2) = 2x$ and that $\dfrac{d}{dx}(x^3) = 3x^2$.

Investigating the gradient of the curves $y = x^4$, $y = x^5$ and $y = x^6$ would give the results:

$$\frac{d}{dx}(x^4) = 4x^3 \qquad \frac{d}{dx}(x^5) = 5x^4 \qquad \frac{d}{dx}(x^6) = 6x^5$$

This leads to the general rule for differentiating power functions:

---

### 🔍 KEY POINT 7.1

$$\frac{d}{dx}(x^n) = nx^{n-1}$$

This is true for any real power $n$, not only for positive integer values of $n$.

---

You may find it easier to remember this rule as:

'Multiply by the power $n$ and then subtract one from the power.'

So for the earlier example where $y = x^2$:

$$\frac{dy}{dx} = 2 \times x^{2-1}$$
$$= 2x^1$$
$$= 2x$$

### WORKED EXAMPLE 7.1

Find the derivative of each of the following.

**a** $x^7$      **b** $\dfrac{1}{x^2}$      **c** $f(x) = \sqrt{x}$      **d** $y = 2$

**Answer**

**a** $\dfrac{d}{dx}(x^7) = 7x^{7-1}$    Multiply by the power 7 and then subtract one from the power.

$\qquad\qquad = 7x^6$

**b** $\dfrac{d}{dx}\left(\dfrac{1}{x^2}\right) = \dfrac{d}{dx}(x^{-2})$    Write $\dfrac{1}{x^2}$ as $x^{-2}$.

$\qquad\qquad = -2x^{-2-1}$    Multiply by the power $-2$ and then subtract one from the power.

$\qquad\qquad = -2x^{-3}$

$\qquad\qquad = -\dfrac{2}{x^3}$

**c** $f(x) = \sqrt{x}$    Write $\sqrt{x}$ as $x^{\frac{1}{2}}$.

$\quad f(x) = x^{\frac{1}{2}}$    Multiply by the power $\dfrac{1}{2}$ and then subtract one from the power.

$\quad f'(x) = \dfrac{1}{2}x^{\frac{1}{2}-1}$

$\qquad\quad = \dfrac{1}{2}x^{-\frac{1}{2}}$

$\qquad\quad = \dfrac{1}{2\sqrt{x}}$

**d**     $y = 2$ ·············· Write 2 as $2x^0$.

$y = 2x^0$ ·············· Multiply by the power and then subtract one from the power.

$\dfrac{dy}{dx} = 0x^{0-1}$

$= 0$

**TIP**

It is worth remembering that when you differentiate a constant, the answer is always 0.

You need to know and be able to use the following two rules.

## Scalar multiple rule

If $k$ is a constant and $f(x)$ is a function then:

**KEY POINT 7.2**

$$\frac{d}{dx}[kf(x)] = k\frac{d}{dx}[f(x)]$$

## Addition/subtraction rule

If $f(x)$ and $g(x)$ are functions then

**KEY POINT 7.3**

$$\frac{d}{dx}[f(x) \pm g(x)] = \frac{d}{dx}[f(x)] \pm \frac{d}{dx}[g(x)]$$

**WORKED EXAMPLE 7.2**

Differentiate $3x^4 - \dfrac{1}{2x^2} + \dfrac{4}{\sqrt{x}} + 5$ with respect to $x$.

**Answer**

$$\frac{d}{dx}\left(3x^4 - \frac{1}{2x^2} + \frac{4}{\sqrt{x}} + 5\right) = \frac{d}{dx}\left(3x^4 - \frac{1}{2}x^{-2} + 4x^{-\frac{1}{2}} + 5x^0\right)$$

$$= 3\frac{d}{dx}(x^4) - \frac{1}{2}\frac{d}{dx}(x^{-2}) + 4\frac{d}{dx}\left(x^{-\frac{1}{2}}\right) + 5\frac{d}{dx}(x^0)$$

$$= 3(4x^3) - \frac{1}{2}(-2x^{-3}) + 4\left(-\frac{1}{2}x^{-\frac{3}{2}}\right) + 5(0x^{-1})$$

$$= 12x^3 + x^{-3} - 2x^{-\frac{3}{2}}$$

$$= 12x^3 + \frac{1}{x^3} - \frac{2}{\sqrt{x^3}}$$

**WORKED EXAMPLE 7.3**

Find the gradient of the tangent to the curve $y = x(2x - 1)(x + 3)$ at the point $(1, 4)$.

**Answer**

$$y = x(2x - 1)(x + 3)$$

Expand brackets and simplify.

$$y = 2x^3 + 5x^2 - 3x$$

$$\frac{dy}{dx} = 6x^2 + 10x - 3$$

When $x = 1$, $\dfrac{dy}{dx} = 6(1)^2 + 10(1) - 3$

$$= 13$$

Gradient of curve at $(1, 4)$ is $13$.

**WORKED EXAMPLE 7.4**

The curve $y = ax^4 + bx^2 + x$ has gradient $3$ when $x = 1$ and gradient $-51$ when $x = -2$.

Find the value of $a$ and the value of $b$.

**Answer**

$$y = ax^4 + bx^2 + x$$

$$\frac{dy}{dx} = 4ax^3 + 2bx + 1$$

Since $\dfrac{dy}{dx} = 3$ when $x = 1$:

$$4a(1)^3 + 2b(1) + 1 = 3$$

$$4a + 2b = 2$$

$$2a + b = 1 \quad \text{---------- (1)}$$

Since $\dfrac{dy}{dx} = -51$ when $x = -2$:

$$4a(-2)^3 + 2b(-2) + 1 = -51$$

$$-32a - 4b = -52$$

$$8a + b = 13 \quad \text{--------- (2)}$$

$(2) - (1)$ gives $6a = 12$

$$\therefore a = 2$$

Substitute $a = 2$ into $(1)$: $4 + b = 1$

$$\therefore b = -3$$

**EXERCISE 7A**

1 The points $A(0, 0)$, $B(0.5, 0.75)$, $C(0.8, 1.44)$, $D(0.95, 1.8525)$, $E(0.99, 1.9701)$ and $F(1, 2)$ lie on the curve $y = f(x)$.

a Copy and complete the table to show the gradients of the chords $CF$, $DF$ and $EF$.

| Chord | $AF$ | $BF$ | $CF$ | $DF$ | $EF$ |
|---|---|---|---|---|---|
| Gradient | 2 | 2.5 | | | |

b Use the values in the table to predict the value of $\dfrac{dy}{dx}$ when $x = 1$.

2 By considering the gradient of a suitable sequence of chords, find a value for the gradient of the curve at the given point.

a $y = x^4$ at $(1, 1)$

b $y = x^2 - 2x + 3$ at $(0, 3)$

c $y = 2\sqrt{x}$ at $(4, 4)$

d $y = \dfrac{12}{x}$ at $(2, 6)$

3 Differentiate with respect to $x$:

a $x^5$

b $x^9$

c $x^{-4}$

d $\dfrac{1}{x}$

e $8$

f $\sqrt[3]{x^2}$

g $x^3 \times x^2$

h $\dfrac{x^5}{x^2}$

4 Find $f'(x)$ for each of the following.

a $f(x) = 2x^4$

b $f(x) = 3x^5$

c $f(x) = \dfrac{x^6}{2}$

d $f(x) = \dfrac{3}{x}$

e $f(x) = \dfrac{5}{3x^2}$

f $f(x) = -2$

g $f(x) = \dfrac{4x}{\sqrt{x}}$

h $f(x) = \dfrac{2x\sqrt{x}}{3x^3}$

5 Find $\dfrac{dy}{dx}$ for each of the following.

a $y = 5x^2 - x + 1$

b $y = 2x^3 + 8x - 4$

c $y = 7 - 3x + 5x^2$

d $y = (x + 5)(x - 4)$

e $y = \left(2x^2 - 3\right)^2$

f $y = \dfrac{2x - 5}{x^2}$

g $y = 7x^2 - \dfrac{3}{x} + \dfrac{2}{x^2}$

h $y = 3x + \dfrac{5}{x} - \dfrac{1}{2\sqrt{x}}$

i $y = \dfrac{4x^2 + 3x - 2}{\sqrt{x}}$

6 Find the value of $\dfrac{dy}{dx}$ for each curve at the given point.

a $y = x^2 + x - 4$ at the point $(1, -2)$

b $y = 5 - \dfrac{2}{x}$ at the point $(2, 4)$

c $y = \dfrac{3x - 2}{x^2}$ at the point $(-2, -2)$

7 Find the gradient of the curve $y = (2x - 5)(x + 4)$ at the point $(3, 7)$.

8 Given that $xy = 12$, find the value of $\dfrac{dy}{dx}$ when $x = 2$.

9 Find the gradient of the curve $y = 5x^2 - 8x + 3$ at the point where the curve crosses the $y$-axis.

197

10 Find the coordinates of the points on the curve $y = x^3 - 3x - 8$ where the gradient is 9.

11 Find the gradient of the curve $y = \dfrac{5x - 10}{x^2}$ at the point where the curve crosses the $x$-axis.

12 The curve $y = x^2 - 4x - 5$ and the line $y = 1 - 3x$ meet at the points $A$ and $B$.

    a   Find the coordinates of the points $A$ and $B$.

    b   Find the gradient of the curve at each of the points $A$ and $B$.

13 The gradient of the curve $y = ax^2 + bx$ at the point $(3, -3)$ is 5. Find the value of $a$ and the value of $b$.

14 The gradient of the curve $y = x^3 + ax^2 + bx + 7$ at the point $(1, 5)$ is $-5$. Find the value of $a$ and the value of $b$.

15 The curve $y = ax + \dfrac{b}{x^2}$ has gradient 16 when $x = 1$ and gradient $-8$ when $x = -1$. Find the value of $a$ and the value of $b$.

16 Given that the gradient of the curve $y = x^3 + ax^2 + bx + 3$ is zero when $x = 1$ and when $x = 6$, find the value of $a$ and the value of $b$.

17 Given that $y = 2x^3 - 3x^2 - 36x + 5$, find the range of values of $x$ for which $\dfrac{dy}{dx} < 0$.

18 Given that $y = 4x^3 + 3x^2 - 6x - 9$, find the range of values of $x$ for which $\dfrac{dy}{dx} \geqslant 0$.

19 A curve has equation $y = 3x^3 + 6x^2 + 4x - 5$. Show that the gradient of the curve is never negative.

**WEB LINK**

Try the following resources on the Underground Mathematics website:

• *Slippery slopes*

• *Gradient match.*

198

## 7.2 The chain rule

To differentiate $y = (3x - 2)^7$, we could expand the brackets and then differentiate each term separately. This would take a long time to do. There is a more efficient method available that allows us to find the derivative without expanding.

Let $u = 3x - 2$, then $y = (3x - 2)^7$ becomes $y = u^7$.

This means that $y$ has changed from a function in terms of $x$ to a function in terms of $u$.

We can find the derivative of the composite function $y = (3x - 2)^7$ using the **chain rule**:

**KEY POINT 7.4**

$$\frac{dy}{dx} = \frac{dy}{du} \times \frac{du}{dx}$$

**WEB LINK**

Try the *Chain mapping* resource on the Underground Mathematics website.

**WORKED EXAMPLE 7.5**

Find the derivative of $y = (3x - 2)^7$.

**Answer**

$$y = (3x - 2)^7$$

Let $u = 3x - 2$      so      $y = u^7$

$$\frac{du}{dx} = 3 \quad \text{and} \quad \frac{dy}{du} = 7u^6$$

$$\frac{dy}{dx} = \frac{dy}{du} \times \frac{du}{dx} \qquad\qquad\qquad\qquad\qquad \text{Use the chain rule.}$$

$$= 7u^6 \times 3$$

$$= 7(3x - 2)^6 \times 3$$

$$= 21(3x - 2)^6$$

With practice you will be able to do this mentally.

Consider the 'inside' of $(3x - 2)^7$ to be $3x - 2$.

To differentiate $(3x - 2)^7$:

**Step 1**: Differentiate the 'outside':      $7(3x - 2)^6$

**Step 2**: Differentiate the 'inside':      $3$

**Step 3**: Multiply these two expressions:    $21(3x - 2)^6$

199

**WORKED EXAMPLE 7.6**

Find the derivative of $y = \dfrac{2}{(3x^2 + 1)^5}$.

**Answer**

$$y = \frac{2}{(3x^2 + 1)^5}$$

Let $u = 3x^2 + 1$      so      $y = 2u^{-5}$

$$\frac{du}{dx} = 6x \quad \text{and} \quad \frac{dy}{du} = -10u^{-6}$$

$$\frac{dy}{dx} = \frac{dy}{du} \times \frac{du}{dx} \qquad\qquad\qquad\qquad\qquad \text{Use the chain rule.}$$

$$= -10u^{-6} \times 6x$$

$$= -10(3x^2 + 1)^{-6} \times 6x$$

$$= -\frac{60x}{(3x^2 + 1)^6}$$

Alternatively, to differentiate the expression mentally:

Write $\dfrac{2}{(3x^2 + 1)^5}$ as $2(3x^2 + 1)^{-5}$.

**Step 1**: Differentiate the 'outside': $-10(3x^2 + 1)^{-6}$

**Step 2**: Differentiate the 'inside': $6x$

**Step 3**: Multiply the two expressions: $-60x(3x^2 + 1)^{-6} = -\dfrac{60x}{(3x^2 + 1)^6}$

---

**WORKED EXAMPLE 7.7**

The curve $y = \sqrt{ax + b}$ passes through the point $(12, 4)$ and has gradient $\dfrac{1}{4}$ at this point.

Find the value of $a$ and the value of $b$.

**Answer**

$y = \sqrt{ax + b}$ .................................................. Substitute $x = 12$ and $y = 4$.

$4 = \sqrt{12a + b}$ -------(1)

$y = (ax + b)^{\frac{1}{2}}$ .................................................. Write $\sqrt{ax + b}$ in the form $(ax + b)^{\frac{1}{2}}$.

Let $u = ax + b$ so $y = u^{\frac{1}{2}}$

$\dfrac{du}{dx} = a$ and $\dfrac{dy}{du} = \dfrac{1}{2} u^{-\frac{1}{2}}$

$\dfrac{dy}{dx} = \dfrac{dy}{du} \times \dfrac{du}{dx}$ .................................................. Use the chain rule.

$\quad = \dfrac{1}{2} u^{-\frac{1}{2}} \times a$

$\dfrac{dy}{dx} = \dfrac{a}{2\sqrt{ax + b}}$ .................................................. Substitute $x = 12$ and $\dfrac{dy}{dx} = \dfrac{1}{4}$.

$\dfrac{1}{4} = \dfrac{a}{2\sqrt{12a + b}}$

$2a = \sqrt{12a + b}$ -------(2)

(1) and (2) give $2a = 4$

$\qquad\qquad a = 2$

Substituting $a = 2$ into (1) gives:

$4 = \sqrt{24 + b}$

$16 = 24 + b$

$b = -8$

$\therefore a = 2, \ b = -8$

200

**1** Differentiate with respect to $x$:

a $(x + 4)^6$

b $(2x + 3)^8$

c $(3 - 4x)^5$

d $\left(\dfrac{1}{2}x + 1\right)^9$

e $\dfrac{(5x - 2)^8}{4}$

f $5(2x - 1)^5$

g $2(4 - 7x)^4$

h $\dfrac{1}{5}(3x - 1)^7$

i $(x^2 + 3)^5$

j $(2 - x^2)^8$

k $(x^2 + 4x)^3$

l $\left(x^2 - \dfrac{5}{x}\right)^5$

**2** Differentiate with respect to $x$:

a $\dfrac{1}{x + 2}$

b $\dfrac{3}{x - 5}$

c $\dfrac{8}{3 - 2x}$

d $\dfrac{16}{x^2 + 2}$

e $\dfrac{4}{(3x + 1)^6}$

f $\dfrac{3}{2(3x + 1)^5}$

g $\dfrac{8}{x^2 + 2x}$

h $\dfrac{7}{(2x^2 - 5x)^7}$

**3** Differentiate with respect to $x$:

a $\sqrt{x - 5}$

b $\sqrt{2x + 3}$

c $\sqrt{2x^2 - 1}$

d $\sqrt{x^3 - 5x}$

e $\sqrt[3]{5 - 2x}$

f $2\sqrt{3x + 1}$

g $\dfrac{1}{\sqrt{2x - 5}}$

h $\dfrac{6}{\sqrt[3]{2 - 3x}}$

**4** Find the gradient of the curve $y = (2x - 3)^5$ at the point $(2, 1)$.

**5** Find the gradient of the curve $y = \dfrac{6}{(x - 1)^2}$ at the point where the curve crosses the $y$-axis.

**6** Find the gradient of the curve $y = x - \dfrac{3}{x + 2}$ at the points where the curve crosses the $x$-axis.

**7** Find the coordinates of the point on the curve $y = \sqrt{(x^2 - 10x + 26)}$ where the gradient is $0$.

**8** The curve $y = \dfrac{a}{bx - 1}$ passes through the point $(2, 1)$ and has gradient $-\dfrac{3}{5}$ at this point.
Find the value of $a$ and the value of $b$.

## 7.3 Tangents and normals

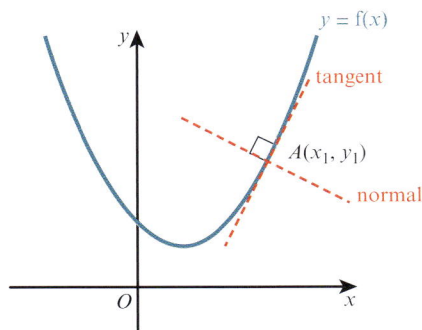

The line perpendicular to the tangent at the point $A$ is called the **normal** at $A$.

If the value of $\dfrac{dy}{dx}$ at the point $A(x_1, y_1)$ is $m$, then the equation of the tangent at $A$ is given by:

> 🔍 **KEY POINT 7.5**
>
> $y - y_1 = m(x - x_1)$

> 💡 **TIP**
>
> We use the numerical form for $m$ in this formula (not the derivative formula).

The normal at the point $(x_1, y_1)$ is perpendicular to the tangent, so the gradient of the normal is $-\dfrac{1}{m}$ and the equation of the normal is given by:

---

### KEY POINT 7.6

$$y - y_1 = -\frac{1}{m}(x - x_1)$$

---

This formula only makes sense when $m \neq 0$. If $m = 0$, it means that the tangent is horizontal and the normal is vertical, so it has equation $x = x_1$ instead.

### WORKED EXAMPLE 7.8

Find the equation of the tangent and the normal to the curve $y = 2x^2 + \dfrac{8}{x^2} - 9$ at the point where $x = 2$.

**Answer**

$y = 2x^2 + 8x^{-2} - 9$

$\dfrac{dy}{dx} = 4x - 16x^{-3}$

When $x = 2$, $y = 2(2)^2 + 8(2)^{-2} - 9 = 1$

and $\dfrac{dy}{dx} = 4(2) - 16(2)^{-3} = 6$

Tangent: passes through the point $(2, 1)$ and gradient $= 6$

$\quad y - 1 = 6(x - 2)$

$\quad y = 6x - 11$

Normal: passes through the point $(2, 1)$ and gradient $= -\dfrac{1}{6}$

$\quad y - 1 = -\dfrac{1}{6}(x - 2)$

$x + 6y = 8$

### WORKED EXAMPLE 7.9

A curve has equation $y = \left(4 - \sqrt{x}\right)^3$.

The normal at the point $P(4, 8)$ and the normal at the point $Q(9, 1)$ intersect at the point $R$.

    **a** Find the coordinates of $R$.

    **b** Find the area of triangle $PQR$.

**Answer**

**a** $\qquad\qquad y = \left(4 - \sqrt{x}\right)^3$

$\qquad\qquad \dfrac{dy}{dx} = 3\left(4 - \sqrt{x}\right)^2\left(-\frac{1}{2}x^{-\frac{1}{2}}\right) = -\dfrac{3\left(4 - \sqrt{x}\right)^2}{2\sqrt{x}}$

When $x = 4$, $\dfrac{dy}{dx} = -\dfrac{3\left(4 - \sqrt{4}\right)^2}{2\sqrt{4}} = -3$

When $x = 9$, $\dfrac{dy}{dx} = -\dfrac{3\left(4 - \sqrt{9}\right)^2}{2\sqrt{9}} = -\dfrac{1}{2}$

Normal at $P$: passes through the point $(4, 8)$ and gradient $= \dfrac{1}{3}$

$$y - 8 = \frac{1}{3}(x - 4)$$
$$3y = x + 20 \ \text{-----}(1)$$

Normal at $Q$: passes through the point $(9, 1)$ and gradient $= 2$

$$y - 1 = 2(x - 9)$$
$$y = 2x - 17 \ \text{----}(2)$$

Solving equations (1) and (2) gives:

$$3(2x - 17) = x + 20$$
$$x = 14.2$$

When $x = 14.2, \ y = 2(14.2) - 17 = 11.4$

Hence, $R$ is the point $(14.2, 11.4)$.

b

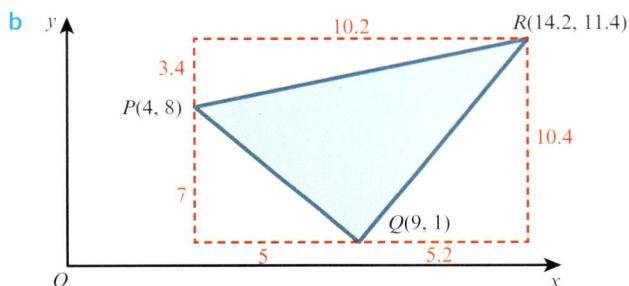

Area of triangle $PQR$ = area of rectangle − sum of areas of outside triangles

$$= 10.2 \times 10.4 - \left[ \left( \frac{1}{2} \times 5 \times 7 \right) + \left( \frac{1}{2} \times 5.2 \times 10.4 \right) + \left( \frac{1}{2} \times 10.2 \times 3.4 \right) \right]$$
$$= 106.08 - [17.5 + 27.04 + 17.34]$$
$$= 44.2 \ \text{units}^2$$

## EXERCISE 7C

**1** Find the equation of the tangent to each curve at the given point.

**a** $y = x^2 - 3x + 2$ at the point $(3, 2)$

**b** $y = (2x - 5)^4$ at the point $(2, 1)$

**c** $y = \dfrac{x^3 - 5}{x}$ at the point $(-1, 6)$

**d** $y = 2\sqrt{x - 5}$ at the point $(9, 4)$

**2** Find the equation of the normal to each curve at the given point.

**a** $y = 3x^3 + x^2 - 4x + 1$ at the point $(0, 1)$

**b** $y = \dfrac{3}{\sqrt[3]{x + 1}}$ at the point $(-2, -3)$

**c** $y = (5 - 2x)^3$ at the point $(3, -1)$

**d** $y = \dfrac{20}{x^2 + 1}$ at the point $(3, 2)$

203

**3** A curve passes through the point $A\left(2, \frac{1}{2}\right)$ and has equation $y = \dfrac{8}{(x+2)^2}$.

    **a** Find the equation of the tangent to the curve at the point $A$.

    **b** Find the equation of the normal to the curve at the point $A$.

**4** The equation of a curve is $y = 5 - 3x - 2x^2$.

    **a** Show that the equation of the normal to the curve at the point $(-2, 3)$ is
$x + 5y = 13$.

    **b** Find the coordinates of the point at which the normal meets the curve again.

**5** The normal to the curve $y = x^3 - 5x + 3$ at the point $(-1, 7)$ intersects the
$y$-axis at the point $P$.

    Find the coordinates of $P$.

**6** The tangents to the curve $y = 5 - 3x - x^2$ at the points $(-1, 7)$ and $(-4, 1)$ meet at
the point $Q$.

    Find the coordinates of $Q$.

**7** The normal to the curve $y = 4 - 2\sqrt{x}$ at the point $P(16, -4)$ meets the $x$-axis at
the point $Q$.

    **a** Find the equation of the normal $PQ$.

    **b** Find the coordinates of $Q$.

**8** The equation of a curve is $y = 2x - \dfrac{10}{x^2} + 8$.

    **a** Find $\dfrac{dy}{dx}$.

    **b** Show that the normal to the curve at the point $\left(-4, -\dfrac{5}{8}\right)$ meets the $y$-axis at
the point $(0, -3)$.

**9** The normal to the curve $y = \dfrac{6}{\sqrt{x-2}}$ at the point $(3, 6)$ meets the $x$-axis at $P$ and
the $y$-axis at $Q$.

    Find the midpoint of $PQ$.

**10** A curve has equation $y = x^5 - 8x^3 + 16x$. The normal at the point $P(1, 9)$ and
the tangent at the point $Q(-1, -9)$ intersect at the point $R$.

    Find the coordinates of $R$.

**11** A curve has equation $y = 2\left(\sqrt{x} - 1\right)^3 + 2$. The normal at the point $P(4, 4)$ and
the normal at the point $Q(9, 18)$ intersect at the point $R$.

    **a** Find the coordinates of $R$.

    **b** Find the area of triangle $PQR$.

**12** A curve has equation $y = 3x + \dfrac{12}{x}$ and passes through the points $A(2, 12)$ and

    $B(6, 20)$. At each of the points $C$ and $D$ on the curve, the tangent is parallel to $AB$.

    **a** Find the coordinates of the points $C$ and $D$. Give your answer in exact form.

    **b** Find the equation of the perpendicular bisector of $CD$.

**13** The curve $y = x(x - 1)(x + 2)$ crosses the $x$-axis at the points $O(0, 0)$, $A(1, 0)$ and $B(-2, 0)$. The normals to the curve at the points $A$ and $B$ meet at the point $C$. Find the coordinates of the point $C$.

**14** A curve has equation $y = \dfrac{5}{2 - 3x}$ and passes through the points $P(-1, 1)$. Find the equation of the tangent to the curve at $P$ and find the angle that this tangent makes with the $x$-axis.

**15** The curve $y = \dfrac{12}{2x - 3} - 4$ intersects the $x$-axis at $P$. The tangent to the curve at $P$ intersects the $y$-axis at $Q$. Find the distance $PQ$.

**16** The normal to the curve $y = 2x^2 + kx - 3$ at the point $(3, -6)$ is parallel to the line $x + 5y = 10$.

   **a** Find the value of $k$.

   **b** Find the coordinates of the point where the normal meets the curve again.

**WEB LINK**

Try the *Tangent or normal* resource on the Underground Mathematics website.

## 7.4 Second derivatives

If we differentiate $y$ with respect to $x$ we obtain $\dfrac{\mathrm{d}y}{\mathrm{d}x}$.

$\dfrac{\mathrm{d}y}{\mathrm{d}x}$ is called the **first derivative** of $y$ with respect to $x$.

If we then differentiate $\dfrac{\mathrm{d}y}{\mathrm{d}x}$ with respect to $x$ we obtain $\dfrac{\mathrm{d}}{\mathrm{d}x}\left(\dfrac{\mathrm{d}y}{\mathrm{d}x}\right)$, which is usually written as $\dfrac{\mathrm{d}^2 y}{\mathrm{d}x^2}$.

$\dfrac{\mathrm{d}^2 y}{\mathrm{d}x^2}$ is called the **second derivative** of $y$ with respect to $x$.

So for  $y = x^3 + 5x^2 - 3x + 2$      or      $f(x) = x^3 + 5x^2 - 3x + 2$

$\qquad \dfrac{\mathrm{d}y}{\mathrm{d}x} = 3x^2 + 10x - 3$      or      $f'(x) = 3x^2 + 10x - 3$

$\qquad \dfrac{\mathrm{d}^2 y}{\mathrm{d}x^2} = 6x + 10$      or      $f''(x) = 6x + 10$

**WORKED EXAMPLE 7.10**

Given that $y = \dfrac{5}{(2x - 3)^3}$, find $\dfrac{\mathrm{d}^2 y}{\mathrm{d}x^2}$.

**Answer**

$y = 5(2x - 3)^{-3}$

$\dfrac{\mathrm{d}y}{\mathrm{d}x} = -15(2x - 3)^{-4} \times 2$  ......... Use the chain rule.

$\quad = -30(2x - 3)^{-4}$

$\dfrac{\mathrm{d}^2 y}{\mathrm{d}x^2} = 120(2x - 3)^{-5} \times 2$  ......... Use the chain rule.

$\quad = \dfrac{240}{(2x - 3)^5}$

**WORKED EXAMPLE 7.11**

A curve has equation $y = x^3 + 3x^2 - 9x + 2$.

**a** Find the range of values of $x$ for which $\dfrac{dy}{dx}$ and $\dfrac{d^2y}{dx^2}$ are negative.

**b** Show that $\dfrac{d^2y}{dx^2} \neq \left(\dfrac{dy}{dx}\right)^2$.

**Answer**

**a**   $y = x^3 + 3x^2 - 9x + 2$

$\dfrac{dy}{dx} = 3x^2 + 6x - 9$

$\dfrac{dy}{dx} < 0$ when:   $3x^2 + 6x - 9 < 0$

$x^2 + 2x - 3 < 0$

$(x+3)(x-1) < 0$

$-3 < x < 1$ - - - - - - - (1)

$\dfrac{d^2y}{dx^2} = 6x + 6$

$\dfrac{d^2y}{dx^2} < 0$ when:   $6x + 6 < 0$

$x < -1$ - - - - - - - - - (2)

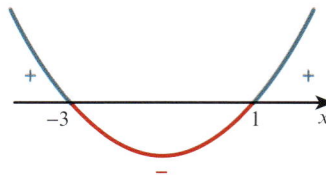

Combining (1) and (2) on a number line:

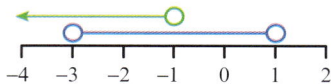

$\therefore -3 < x < -1$

**b**   $\dfrac{d^2y}{dx^2} = 6x + 6$   and   $\left(\dfrac{dy}{dx}\right)^2 = (3x^2 + 6x - 9)^2$.

Hence, $\dfrac{d^2y}{dx^2} \neq \left(\dfrac{dy}{dx}\right)^2$.

**EXERCISE 7D**

**1** Find $\dfrac{d^2y}{dx^2}$ for each of the following functions.

**a** $y = x^2 + 8x - 4$   **b** $y = 5x^3 - 7x^2 + 5$   **c** $y = 2 - \dfrac{6}{x^2}$

**d** $y = (2x - 3)^4$   **e** $y = \sqrt{4x - 9}$   **f** $y = \dfrac{2}{\sqrt{3x+1}}$

**g** $y = \dfrac{2x - 5}{x^2}$   **h** $y = 2x^2(5 - 3x + x^2)$   **i** $y = \dfrac{5x - 4}{\sqrt{x}}$

**2** Find $f''(x)$ for each of the following functions.

**a** $f(x) = \dfrac{5}{x^2} - \dfrac{3}{2x^5}$   **b** $f(x) = \dfrac{4x^2 - 3}{2x}$   **c** $f(x) = \dfrac{2x - 3\sqrt{x}}{x^2}$

**d** $f(x) = \sqrt{1 - 3x}$   **e** $f(x) = x^2(\sqrt{x} - 3)$   **f** $f(x) = \dfrac{15}{\sqrt[3]{2x+1}}$

**3** Given that $y = 4x - (2x - 1)^4$, find $\dfrac{dy}{dx}$ and $\dfrac{d^2y}{dx^2}$ .

**4** Given that $f(x) = x^3 + 2x^2 - 3x - 1$, find:

   **a**    $f(1)$                      **b**    $f'(1)$                      **c**    $f''(1)$

**5** Given that $f'(x) = \dfrac{3}{(2x - 1)^8}$, find $f''(x)$ .

**6** Given that $f(x) = \dfrac{2}{\sqrt{1 - 2x}}$ , find the value of $f''(-4)$.

**7** A curve has equation $y = 2x^3 - 21x^2 + 60x + 5$. Copy and complete the table to show whether $\dfrac{dy}{dx}$ and $\dfrac{d^2y}{dx^2}$ are positive (+), negative (−) or zero (0) for the given values of $x$.

| $x$ | 0 | 1 | 2 | 3 | 4 | 5 | 6 | 7 |
|---|---|---|---|---|---|---|---|---|
| $\dfrac{dy}{dx}$ | | | | | | | | |
| $\dfrac{d^2y}{dx^2}$ | | | | | | | | |

**8** A curve has equation $y = x^3 - 6x^2 - 15x - 7$. Find the range of values of $x$ for which both $\dfrac{dy}{dx}$ and $\dfrac{d^2y}{dx^2}$ are positive.

**9** Given that $y = x^2 - 2x + 5$, show that $4\dfrac{d^2y}{dx^2} + (x - 1)\dfrac{dy}{dx} = 2y$.

**10** Given that $y = 4\sqrt{x}$, show that $4x^2\dfrac{d^2y}{dx^2} + 4x\dfrac{dy}{dx} = y$.

**11** A curve has equation $y = x^3 + 2x^2 - 4x + 6$.

   **a**   Show that $\dfrac{dy}{dx} = 0$ when $x = -2$ and when $x = \dfrac{2}{3}$.

   **b**   Find the value of $\dfrac{d^2y}{dx^2}$ when $x = -2$ and when $x = \dfrac{2}{3}$.

**12** A curve has equation $y = \dfrac{ax + b}{x^2}$ . Given that $\dfrac{dy}{dx} = 0$ and $\dfrac{d^2y}{dx^2} = \dfrac{1}{2}$ when $x = 2$, find the value of $a$ and the value of $b$.

207

**WEB LINK**

Try the *Gradients of gradients* resource on the Underground Mathematics website.

# Checklist of learning and understanding

**Gradient of a curve**

- $\dfrac{\mathrm{d}y}{\mathrm{d}x}$ represents the gradient of the curve $y = \mathrm{f}(x)$.

**The four rules of differentiation**

- Power rule: $\dfrac{\mathrm{d}}{\mathrm{d}x}(x^n) = nx^{n-1}$

- Scalar multiple rule: $\dfrac{\mathrm{d}}{\mathrm{d}x}[k\mathrm{f}(x)] = k\,\dfrac{\mathrm{d}}{\mathrm{d}x}[\mathrm{f}(x)]$

- Addition/subtraction rule: $\dfrac{\mathrm{d}}{\mathrm{d}x}[\mathrm{f}(x) \pm \mathrm{g}(x)] = \dfrac{\mathrm{d}}{\mathrm{d}x}[\mathrm{f}(x)] \pm \dfrac{\mathrm{d}}{\mathrm{d}x}[\mathrm{g}(x)]$

- Chain rule: $\dfrac{\mathrm{d}y}{\mathrm{d}x} = \dfrac{\mathrm{d}y}{\mathrm{d}u} \times \dfrac{\mathrm{d}u}{\mathrm{d}x}$

**Tangents and normals**

If the value of $\dfrac{\mathrm{d}y}{\mathrm{d}x}$ at the point $(x_1, y_1)$ is $m$, then:

- the equation of the tangent at that point is given by $y - y_1 = m(x - x_1)$
- the equation of the normal at that point is given by $y - y_1 = -\dfrac{1}{m}(x - x_1)$.

**Second derivatives**

- $\dfrac{\mathrm{d}}{\mathrm{d}x}\left(\dfrac{\mathrm{d}y}{\mathrm{d}x}\right) = \dfrac{\mathrm{d}^2 y}{\mathrm{d}x^2}$

**END-OF-CHAPTER REVIEW EXERCISE 7**

1   Differentiate $\dfrac{3x^5 - 7}{4x}$ with respect to $x$. [3]

2   Find the gradient of the curve $y = \dfrac{8}{4x - 5}$ at the point where $x = 2$. [3]

3   A curve has equation $y = 3x^3 - 3x^2 + x - 7$. Show that the gradient of the curve is never negative. [3]

4   The equation of a curve is $y = (3 - 5x)^3 - 2x$. Find $\dfrac{dy}{dx}$ and $\dfrac{d^2y}{dx^2}$. [3]

5   Find the gradient of the curve $y = \dfrac{15}{x^2 - 2x}$ at the point where $x = 5$. [4]

6   The normal to the curve $y = 5\sqrt{x}$ at the point $P(4, 10)$ meets the $x$-axis at the point $Q$.

   a   Find the equation of the normal $PQ$. [4]

   b   Find the coordinates of $Q$. [1]

7   The equation of a curve is $y = 5x + \dfrac{12}{x^2}$.

   a   Find $\dfrac{dy}{dx}$. [2]

   b   Show that the normal to the curve at the point $(2, 13)$ meets the $x$-axis at the point $(28, 0)$. [3]

8   The normal to the curve $y = \dfrac{12}{\sqrt{x}}$ at the point $(9, 4)$ meets the $x$-axis at $P$ and the $y$-axis at $Q$.

Find the length of $PQ$, correct to 3 significant figures. [6]

9   The curve $y = x(x - 3)(x - 5)$ crosses the $x$-axis at the points $O(0, 0)$, $A(3, 0)$ and $B(5, 0)$.
The tangents to the curve at the points $A$ and $B$ meet at the point $C$.
Find the coordinates of the point $C$. [6]

10  A curve passes through the point $A(4, 2)$ and has equation $y = \dfrac{2}{(x - 3)^2}$.

   a   Find the equation of the tangent to the curve at the point $A$. [5]

   b   Find the equation of the normal to the curve at the point $A$. [2]

11  A curve passes through the point $P(5, 1)$ and has equation $y = 3 - \dfrac{10}{x}$.

   a   Show that the equation of the normal to the curve at the point $P$ is $5x + 2y = 27$. [4]

The normal meets the curve again at the point $Q$.

   b  i   Find the coordinates of $Q$. [3]

      ii   Find the midpoint of $PQ$. [1]

12  A curve has equation $y = 3x - \dfrac{4}{x}$ and passes through the points $A(1, -1)$ and $B(4, 11)$.

At each of the points $C$ and $D$ on the curve, the tangent is parallel to $AB$. Find the equation of the perpendicular bisector of $CD$. [7]

*Cambridge International AS & A Level Mathematics 9709 Paper 11 Q8 June 2016*

13

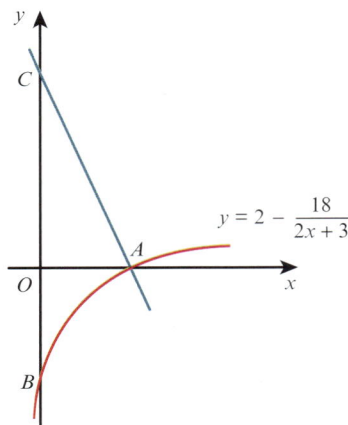

The diagram shows part of the curve $y = 2 - \dfrac{18}{2x + 3}$, which crosses the $x$-axis at $A$ and the $y$-axis at $B$.

The normal to the curve at $A$ crosses the $y$-axis at $C$.

i    Show that the equation of the line $AC$ is $9x + 4y = 27$.                    [6]

ii   Find the length of $BC$.                    [2]

*Cambridge International AS & A Level Mathematics 9709 Paper 11 Q7 June 2010*

14   The equation of a curve is $y = 3 + 4x - x^2$.

i    Show that the equation of the normal to the curve at the point $(3, 6)$ is $2y = x + 9$.    [4]

ii   Given that the normal meets the coordinate axes at points $A$ and $B$, find the coordinates of the mid-point of $AB$.                    [2]

iii  Find the coordinates of the point at which the normal meets the curve again.    [4]

*Cambridge International AS & A Level Mathematics 9709 Paper 11 Q10 November 2010*

15

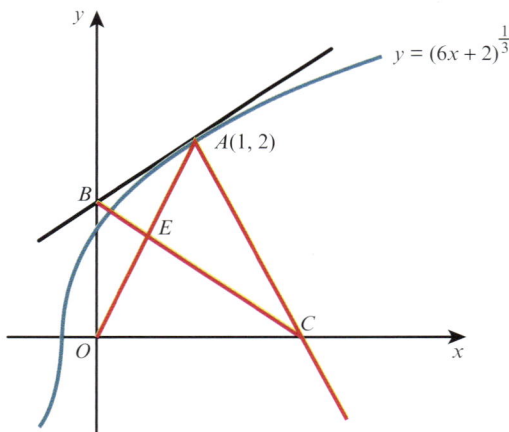

The diagram shows the curve $y = (6x + 2)^{\frac{1}{3}}$ and the point $A(1, 2)$ which lies on the curve. The tangent to the curve at $A$ cuts the $y$-axis at $B$ and the normal to the curve at $A$ cuts the $x$-axis at $C$.

i    Find the equation of the tangent $AB$ and the equation of the normal $AC$.    [5]

ii   Find the distance $BC$.                    [3]

iii  Find the coordinates of the point of intersection, $E$, of $OA$ and $BC$, and determine whether $E$ is the mid-point of $OA$.                    [4]

*Cambridge International AS & A Level Mathematics 9709 Paper 11 Q11 November 2012*

# Chapter 8
# Further differentiation

**In this chapter you will learn how to:**

- apply differentiation to increasing and decreasing functions and rates of change
- locate stationary points and determine their nature, and use information about stationary points when sketching graphs.

## PREREQUISITE KNOWLEDGE

| Where it comes from | What you should be able to do | Check your skills |
|---|---|---|
| Chapter 1 | Solve quadratic inequalities. | 1  Solve:<br><br>   **a**  $x^2 - 2x - 3 > 0$<br><br>   **b**  $6 + x - x^2 > 0$ |
| Chapter 7 | Find the first and second derivatives of $x^n$. | 2  Find $\dfrac{\mathrm{d}y}{\mathrm{d}x}$ and $\dfrac{\mathrm{d}^2 y}{\mathrm{d}x^2}$ for the following.<br><br>   **a**  $y = 3x^2 - x + 2$<br><br>   **b**  $y = \dfrac{3}{2x^2}$<br><br>   **c**  $y = 3x\sqrt{x}$ |
| Chapter 7 | Differentiate composite functions. | 3  Find $\dfrac{\mathrm{d}y}{\mathrm{d}x}$ for the following.<br><br>   **a**  $y = (2x - 1)^5$<br><br>   **b**  $y = \dfrac{3}{(1 - 3x)^2}$ |

212

## Why do we study differentiation?

In Chapter 7, you learnt how to differentiate functions and how to use differentiation to find gradients, tangents and normals.

In this chapter you will build on this knowledge and learn how to apply differentiation to problems that involve finding when a function is increasing (or decreasing) or when a function is at a maximum (or minimum) value. You will also learn how to solve practical problems involving rates of change.

There are many situations in real life where these skills are needed. Some examples are:

- manufacturers of canned food and drinks needing to minimise the cost of their manufacturing by minimising the amount of metal required to make a can for a given volume
- doctors calculating the time interval when the concentration of a drug in the bloodstream is increasing
- economists might use these tools to advise a company on its pricing strategy
- scientists calculating the rate at which the area of an oil slick is increasing.

▶▶| **FAST FORWARD**

In the Mechanics Coursebook, Chapter 6 you will learn to apply these skills to problems concerning displacement, velocity and time.

🌐 **WEB LINK**

Explore the *Calculus meets functions* station on the Underground Mathematics website.

**EXPLORE 8.1**

### Section A: Increasing functions

Consider the graph of $y = f(x)$.

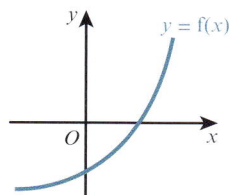

1   Complete the following two statements about $y = f(x)$.

'As the value of $x$ increases the value of $y$…'

'The sign of the gradient at any point is always …'

2   Sketch other graphs that satisfy these statements.

These types of functions are called **increasing functions**.

### Section B: Decreasing functions

Consider the graph of $y = g(x)$.

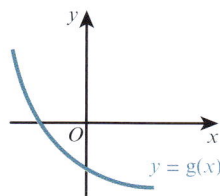

1   Complete the following two statements about $y = g(x)$.

'As the value of $x$ increases the value of $y$ …'

'The sign of the gradient at any point is always …'

2   Sketch other graphs that satisfy these statements.

These types of functions are called **decreasing functions**.

**WEB LINK**

Try the *Choose your families* resource on the Underground Mathematics website.

## 8.1 Increasing and decreasing functions

As you probably worked out from Explore 8.1, an **increasing function** $f(x)$ is one where the $f(x)$ values increase whenever the $x$ value increases. More precisely, this means that $f(a) < f(b)$ whenever $a < b$.

Likewise, a **decreasing function** $f(x)$ is one where the $f(x)$ values decrease whenever the $x$ value increases, or $f(a) > f(b)$ whenever $a < b$.

Sometimes we talk about a function **increasing** at a point, meaning that the function values are increasing around that point. If the gradient of the function is positive at a point, then the function is increasing there.

In the same way, we can talk about a function **decreasing** at a point. If the gradient of the function is negative at a point, then the function is decreasing there.

Now consider the function $y = h(x)$, shown on the graph.

We can divide the graph into two distinct sections:

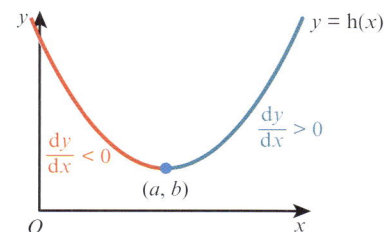

● $h(x)$ is increasing when $x > a$, i.e. $\dfrac{dy}{dx} > 0$ for $x > a$.

● $h(x)$ is decreasing when $x < a$, i.e. $\dfrac{dy}{dx} < 0$ for $x < a$.

Find the set of values of $x$ for which $y = 8 - 3x - x^2$ is decreasing.

**Answer**

$y = 8 - 3x - x^2$

$\dfrac{dy}{dx} = -3 - 2x$

When $\dfrac{dy}{dx} < 0$, $y$ is decreasing.

$-3 - 2x < 0$

$\quad\quad 2x > -3$

$\quad\quad\quad x > -\dfrac{3}{2}$

For the function $f(x) = 4x^3 - 15x^2 - 72x - 8$:

  **a** Find $f'(x)$.

  **b** Find the range of values of $x$ for which $f(x) = 4x^3 - 15x^2 - 72x - 8$ is increasing.

  **c** Find the range of values of $x$ for which $f(x) = 4x^3 - 15x^2 - 72x - 8$ is decreasing.

**Answer**

  **a**   $f(x) = 4x^3 - 15x^2 - 72x - 8$

       $f'(x) = 12x^2 - 30x - 72$

  **b**   When $f'(x) > 0$, $f(x)$ is increasing.

       $12x^2 - 30x - 72 > 0$

         $2x^2 - 5x - 12 > 0$

       $(2x + 3)(x - 4) > 0$

       Critical values are $-\dfrac{3}{2}$ and $4$.

       $\therefore x < -\dfrac{3}{2}$ and $x > 4$.

  **c**   When $f'(x) < 0$, $f(x)$ is decreasing.

       $\therefore -\dfrac{3}{2} < x < 4$

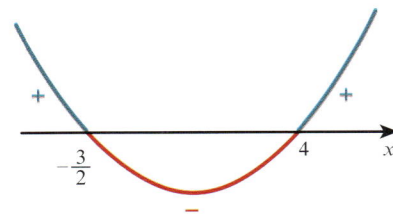

**WORKED EXAMPLE 8.3**

A function f is defined as $f(x) = \dfrac{5}{2x - 3}$ for $x > \dfrac{3}{2}$. Find an expression for $f'(x)$ and determine whether f is an increasing function, a decreasing function or neither.

**Answer**

$f(x) = \dfrac{5}{2x - 3}$                 Write in a form ready for differentiating.

$\phantom{f(x)} = 5(2x - 3)^{-1}$         Differentiate using the chain rule.

$f'(x) = -5(2x - 3)^{-2}(2)$

$\phantom{f'(x)} = -\dfrac{10}{(2x - 3)^2}$

If $x > \dfrac{3}{2}$, then $(2x - 3)^2 > 0$ for all values of $x$.

Hence, $f'(x) < 0$ for all values of $x$ in the domain of f.

$\therefore$ f is a decreasing function.

**EXERCISE 8A**

1   Find the set of values of $x$ for which each of the following is increasing.

    **a**   $f(x) = x^2 - 8x + 2$                       **b**   $f(x) = 2x^2 - 4x + 7$

    **c**   $f(x) = 5 - 7x - 2x^2$                 **d**   $f(x) = x^3 - 12x^2 + 2$

    **e**   $f(x) = 2x^3 - 15x^2 + 24x + 6$      **f**   $f(x) = 16 + 16x - x^2 - x^3$

2   Find the set of values of $x$ for which each of the following is decreasing.

    **a**   $f(x) = 3x^2 - 8x + 2$                    **b**   $f(x) = 10 + 9x - x^2$

    **c**   $f(x) = 2x^3 - 21x^2 + 60x - 5$     **d**   $f(x) = x^3 - 3x^2 - 9x + 5$

    **e**   $f(x) = -40x + 13x^2 - x^3$        **f**   $f(x) = 11 + 24x - 3x^2 - x^3$

3   Find the set of values of $x$ for which $f(x) = \dfrac{1}{6}(5 - 2x)^3 + 4x$ is increasing.

4   A function f is defined as $f(x) = \dfrac{4}{1 - 2x}$ for $x \geqslant 1$. Find an expression for $f'(x)$ and determine whether f is an increasing function, a decreasing function or neither.

5   A function f is defined as $f(x) = \dfrac{5}{(x + 2)^2} - \dfrac{2}{x + 2}$ for $x \geqslant 0$. Find an expression for $f'(x)$ and determine whether f is an increasing function, a decreasing function or neither.

6   Show that $f(x) = \dfrac{x^2 - 4}{x}$ is an increasing function.

7   A function f is defined as $f(x) = (2x + 5)^2 - 3$ for $x \geqslant 0$. Find an expression for $f'(x)$ and explain why f is an increasing function.

8   It is given that $f(x) = \dfrac{2}{x^4} - x^2$ for $x > 0$. Show that f is a decreasing function.

9   A manufacturing company produces $x$ articles per day. The profit function, $P(x)$, can be modeled by the function $P(x) = 2x^3 - 81x^2 + 840x$. Find the range of values of $x$ for which the profit is decreasing.

## 8.2 Stationary points

Consider the following graph of the function $y = f(x)$.

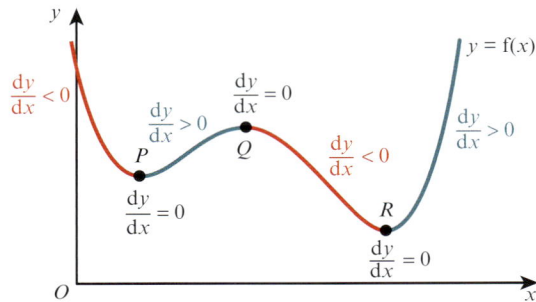

The red sections of the curve show where the gradient is negative (where $f(x)$ is a decreasing function) and the blue sections show where the gradient is positive (where $f(x)$ is an increasing function). The gradient of the curve is zero at the points $P$, $Q$ and $R$.

A point where the gradient is zero is called a stationary point or a turning point.

### Maximum points

The stationary point $Q$ is called a maximum point because the value of $y$ at this point is greater than the value of $y$ at other points close to $Q$.

At a maximum point:

- $\dfrac{dy}{dx} = 0$

- the gradient is positive to the left of the maximum and negative to the right.

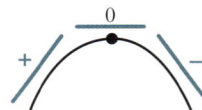

### Minimum points

The stationary points $P$ and $R$ are called minimum points.

At a minimum point:

- $\dfrac{dy}{dx} = 0$

- the gradient is negative to the left of the minimum and positive to the right.

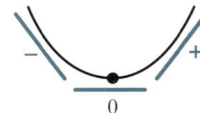

### Stationary points of inflexion

There is a third type of stationary point (turning point) called a **point of inflexion**.

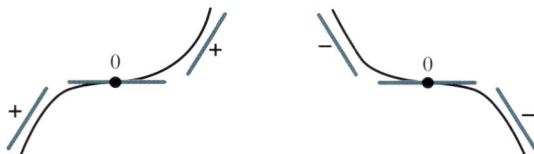

At a stationary point of inflexion:

- $\dfrac{dy}{dx} = 0$

- the gradient changes $\begin{cases} \text{from positive to zero and then to positive again} \\ \text{or} \\ \text{from negative to zero and then to negative again.} \end{cases}$

**WORKED EXAMPLE 8.4**

Find the coordinates of the stationary points on the curve $y = x^3 - 12x + 5$ and determine the nature of these points. Sketch the graph of $y = x^3 - 12x + 5$.

**Answer**

$$y = x^3 - 12x + 5$$
$$\frac{dy}{dx} = 3x^2 - 12$$

For stationary points:
$$\frac{dy}{dx} = 0$$
$$3x^2 - 12 = 0$$
$$x^2 - 4 = 0$$
$$(x + 2)(x - 2) = 0$$
$$x = -2 \text{ or } x = 2$$

When $x = -2$, $y = (-2)^3 - 12(-2) + 5 = 21$
When $x = 2$, $y = (2)^3 - 12(2) + 5 = -11$

The stationary points are $(-2, 21)$ and $(2, -11)$.

Now consider the gradient on either side of the points $(-2, 21)$ and $(2, -11)$:

| $x$ | $-2.1$ | $-2$ | $-1.9$ |
|---|---|---|---|
| $\frac{dy}{dx}$ | $3(-2.1)^2 - 12 =$ positive | 0 | $3(-1.9)^2 - 12 =$ negative |
| direction of tangent | / | — | \ |
| shape of curve | | ⌒ | |

| $x$ | $1.9$ | $2$ | $2.1$ |
|---|---|---|---|
| $\frac{dy}{dx}$ | $3(1.9)^2 - 12 =$ negative | 0 | $3(2.1)^2 - 12 =$ positive |
| direction of tangent | \ | — | / |
| shape of curve | | ⌣ | |

So $(-2, 21)$ is a maximum point and $(2, -11)$ is a minimum point.

The sketch graph of $y = x^3 - 12x + 5$ is:

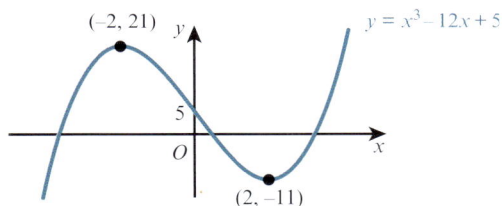

> **TIP**
> This is called the First Derivative Test.

## Second derivatives and stationary points

Consider moving from left to right along a curve, passing through a maximum point.

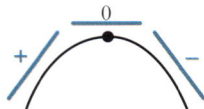

The gradient, $\dfrac{dy}{dx}$, starts as a positive value, decreases to zero at the maximum point and then decreases to a negative value.

Since $\dfrac{dy}{dx}$ decreases as $x$ increases, then the rate of change of $\dfrac{dy}{dx}$ is negative.

The rate of change of $\dfrac{dy}{dx}$ is written as $\dfrac{d}{dx}\left(\dfrac{dy}{dx}\right) = \dfrac{d^2y}{dx^2}$.

$\dfrac{d^2y}{dx^2}$ is called the second derivative of $y$ with respect $x$.

This leads to the rule:

> **KEY POINT 8.1**
>
> If $\dfrac{dy}{dx} = 0$ and $\dfrac{d^2y}{dx^2} < 0$, then the point is a maximum point.

Now, consider moving from left to right along a curve, passing through a minimum point.

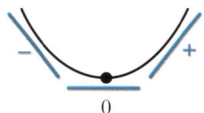

The gradient, $\dfrac{dy}{dx}$, starts as a negative value, increases to zero at the minimum point and then increases to a positive value.

Since $\dfrac{dy}{dx}$ increases as $x$ increases, then the rate of change of $\dfrac{dy}{dx}$ is positive.

This leads to the rule:

> **KEY POINT 8.2**
>
> If $\dfrac{dy}{dx} = 0$ and $\dfrac{d^2y}{dx^2} > 0$, then the point is a minimum point.

If $\dfrac{dy}{dx} = 0$ and $\dfrac{d^2y}{dx^2} = 0$, then the nature of the stationary point can be found using the first derivative test.

**WORKED EXAMPLE 8.5**

Find the coordinates of the stationary points on the curve $y = \dfrac{x^2 + 9}{x}$ and use the second derivative to determine the nature of these points.

**Answer**

$$y = \frac{x^2 + 9}{x} = x + 9x^{-1}$$

$$\frac{dy}{dx} = 1 - 9x^{-2} = 1 - \frac{9}{x^2}$$

For stationary points: $\qquad \dfrac{dy}{dx} = 0$

$$1 - \frac{9}{x^2} = 0$$

$$x^2 - 9 = 0$$

$$(x + 3)(x - 3) = 0$$

$$x = -3 \text{ or } x = 3$$

When $x = -3$, $\qquad y = \dfrac{(-3)^2 + 9}{-3} = -6$

When $x = 3$, $\qquad y = \dfrac{3^2 + 9}{3} = 6$

The stationary points are $(-3, -6)$ and $(3, 6)$.

$$\frac{d^2y}{dx^2} = 18x^{-3} = \frac{18}{x^3}$$

When $x = -3$, $\dfrac{d^2y}{dx^2} = \dfrac{18}{(-3)^3} < 0$

When $x = 3$, $\dfrac{d^2y}{dx^2} = \dfrac{18}{3^3} > 0$

$\therefore$ $(-3, -6)$ is a maximum point and $(3, 6)$ is a minimum point.

**EXPLORE 8.2**

The graph of the function $y = f(x)$ passes through the point $(1, -35)$ and the point $(6, 90)$.

The following graphs show $y = f'(x)$ and $y = f''(x)$.

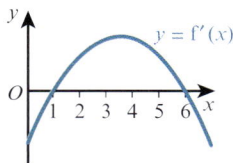

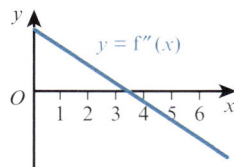

1 Discuss the properties of these two graphs and the information that can be obtained from them.

2 Without finding the equation of the function $y = f(x)$, determine, giving reasons:

   a the coordinates of the maximum point on the curve

   b the coordinates of the minimum point on the curve.

3 Sketch the graph of the function $y = f(x)$.

**EXERCISE 8B**

1 Find the coordinates of the stationary points on each of the following curves and determine the nature of each stationary point. Sketch the graph of each function and use graphing software to check your graphs.

a $y = x^2 - 4x + 8$          b $y = (3 + x)(2 - x)$

c $y = x^3 - 12x + 6$          d $y = 10 + 9x - 3x^2 - x^3$

e $y = x^4 + 4x - 1$          f $y = (2x - 3)^3 - 6x$

2 Find the coordinates of the stationary points on each of the following curves and determine the nature of each stationary point.

a $y = \sqrt{x} + \dfrac{9}{\sqrt{x}}$          b $y = 4x^2 + \dfrac{8}{x}$

c $y = \dfrac{(x - 3)^2}{x}$          d $y = x^3 + \dfrac{48}{x} + 4$

e $y = 4\sqrt{x} - x$          f $y = 2x + \dfrac{8}{x^2}$

3 The equation of a curve is $y = \dfrac{x^2 - 9}{x^2}$.

Find $\dfrac{dy}{dx}$ and, hence, explain why the curve does not have a stationary point.

4 A curve has equation $y = 2x^3 - 3x^2 - 36x + k$.

a Find the $x$-coordinates of the two stationary points on the curve.

b Hence, find the two values of $k$ for which the curve has a stationary point on the $x$-axis.

5 The curve $y = x^3 + ax^2 - 9x + 2$ has a maximum point at $x = -3$.

a Find the value of $a$.

b Find the range of values of $x$ for which the curve is a decreasing function.

6 The curve $y = 2x^3 + ax^2 + bx - 30$ has a stationary point when $x = 3$.

The curve passes through the point $(4, 2)$.

a Find the value of $a$ and the value of $b$.

b Find the coordinates of the other stationary point on the curve and determine the nature of this point.

7 The curve $y = 2x^3 + ax^2 + bx - 30$ has no stationary points.

Show that $a^2 < 6b$.

8 A curve has equation $y = 1 + 2x + \dfrac{k^2}{2x - 3}$, where $k$ is a positive constant. Find, in terms of $k$, the values of $x$ for which the curve has stationary points and determine the nature of each stationary point.

9 Find the coordinates of the stationary points on the curve $y = x^4 - 4x^3 + 4x^2 + 1$ and determine the nature of each of these points. Sketch the graph of the curve.

10 The curve $y = x^3 + ax^2 + b$ has a stationary point at $(4, -27)$.

a Find the value of $a$ and the value of $b$.

b Determine the nature of the stationary point $(4, -27)$.

c Find the coordinates of the other stationary point on the curve and determine the nature of this stationary point.

d Find the coordinates of the point on the curve where the gradient is minimum and state the value of the minimum gradient.

11 The curve $y = ax + \dfrac{b}{x^2}$ has a stationary point at $(2, 12)$.

   a Find the value of $a$ and the value of $b$.

   b Determine the nature of the stationary point $(2, 12)$.

   c Find the range of values of $x$ for which $ax + \dfrac{b}{x^2}$ is increasing.

12 The curve $y = x^2 + \dfrac{a}{x} + b$ has a stationary point at $(3, 5)$.

   a Find the value of $a$ and the value of $b$.

   b Determine the nature of the stationary point $(3, 5)$.

   c Find the range of values of $x$ for which $x^2 + \dfrac{a}{x} + b$ is decreasing.

13 The curve $y = 2x^3 + ax^2 + bx + 7$ has a stationary point at the point $(2, -13)$.

   a Find the value of $a$ and the value of $b$.

   b Find the coordinates of the second stationary point on the curve.

   c Determine the nature of the two stationary points.

   d Find the coordinates of the point on the curve where the gradient is minimum and state the value of the minimum gradient.

**WEB LINK**

Try the following resources on the Underground Mathematics website:
- *Floppy hair*
- *Two-way calculus*
- *Curvy cubics*
- *Can you find… curvy cubics edition.*

## 8.3 Practical maximum and minimum problems

There are many problems for which we need to find the maximum or minimum value of an expression. For example, the manufacturers of canned food and drinks often need to minimise the cost of their manufacturing. To do this they need to find the minimum amount of metal required to make a container for a given volume. Other situations might involve finding the maximum area that can be enclosed within a shape.

**WORKED EXAMPLE 8.6**

The surface area of the solid cuboid is $100\,\text{cm}^2$ and the volume is $V\,\text{cm}^3$.

   a Express $h$ in terms of $x$.

   b Show that $V = 25x - \dfrac{1}{2}x^3$.

   c Given that $x$ can vary, find the stationary value of $V$ and determine whether this stationary value is a maximum or a minimum.

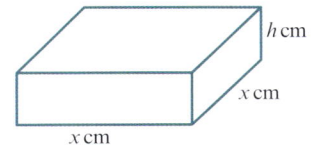

**Answer**

a Surface area $= 2x^2 + 4xh$

$2x^2 + 4xh = 100$

$h = \dfrac{100 - 2x^2}{4x}$

$h = \dfrac{25}{x} - \dfrac{1}{2}x$

**b** $V = x^2 h$ ............................................................. Substitute for $h$.

$$= x^2 \left( \frac{25}{x} - \frac{1}{2} x \right)$$

$$= 25x - \frac{1}{2} x^3$$

**c** $\dfrac{dV}{dx} = 25 - \dfrac{3}{2} x^2$

Stationary values occur when $\dfrac{dV}{dx} = 0$:

$$25 - \frac{3}{2} x^2 = 0$$

$$x = \frac{5\sqrt{6}}{3}$$

When $x = \dfrac{5\sqrt{6}}{3}$, $V = 25 \left( \dfrac{5\sqrt{6}}{3} \right) - \dfrac{1}{2} \left( \dfrac{5\sqrt{6}}{3} \right)^3 = 68.04$ (to 2 decimal places)

$$\frac{d^2V}{dx^2} = -3x$$

When $x = \dfrac{5\sqrt{6}}{3}$, $\dfrac{d^2V}{dx^2} = -5\sqrt{6}$, which is $< 0$.

The stationary value of $V$ is 68.04 and it is a maximum value.

222

### WORKED EXAMPLE 8.7

The diagram shows a solid cylinder of radius $r$ cm and height $2h$ cm cut from a solid sphere of radius $5$ cm. The volume of the cylinder is $C$ cm$^3$.

**a** Express $r$ in terms of $h$.

**b** Show that $V = 50\pi h - 2\pi h^3$.

**c** Find the value for $h$ for which there is a stationary value of $V$.

**d** Determine the nature of this stationary value.

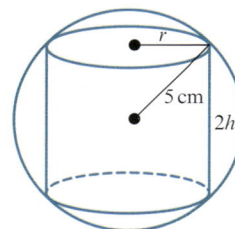

**Answer**

**a** $r^2 + h^2 = 5^2$ ............................................................. Use Pythagoras' theorem.

$$r = \sqrt{25 - h^2}$$

**b** $V = \pi r^2 (2h)$ ............................................................. Substitute for $r$.

$$= \pi \left( 25 - h^2 \right) (2h)$$

$$= 50\pi h - 2\pi h^3$$

**c** $\dfrac{dV}{dh} = 50\pi - 6\pi h^2$

Stationary values occur when $\dfrac{dV}{dh} = 0$:

$$50\pi - 6\pi h^2 = 0$$

$$h^2 = \frac{50\pi}{6\pi}$$

$$h = \frac{5\sqrt{3}}{3}$$

d $\quad \dfrac{d^2V}{dh^2} = -12\pi h$

When $h = \dfrac{5\sqrt{3}}{3}$, $\dfrac{d^2V}{dx^2} = -12\pi\left(\dfrac{5\sqrt{3}}{3}\right)$, which is $< 0$.

The stationary value is a maximum value.

### WORKED EXAMPLE 8.8

The diagram shows a hollow cone with base radius 12 cm and height 24 cm.

A solid cylinder stands on the base of the cone and the upper edge touches the inside of the cone.

The cylinder has base radius $r$ cm, height $h$ cm and volume $V$ cm$^3$.

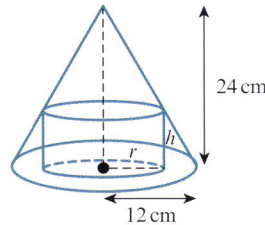

24 cm

12 cm

a Express $h$ in terms of $r$.

b Show that $V = 24\pi r^2 - 2\pi r^3$.

c Find the volume of the largest cylinder that can stand inside the cone.

**Answer**

a $\quad \dfrac{r}{12} = \dfrac{24 - h}{24}$      Use similar triangles.

$\qquad 2r = 24 - h$

$\qquad h = 24 - 2r$

b $\quad V = \pi r^2 h$      Substitute for $h$.

$\qquad = \pi r^2(24 - 2r)$

$\qquad = 24\pi r^2 - 2\pi r^3$

c $\quad \dfrac{dV}{dr} = 48\pi r - 6\pi r^2$

Stationary values occur when $\dfrac{dV}{dr} = 0$:

$$48\pi r - 6\pi r^2 = 0$$

$$6\pi r(8 - r) = 0$$

$$r = 8$$

When $r = 8$, $V = 24\pi(8)^2 - 2\pi(8)^3 = 512\pi$

$\dfrac{d^2V}{dr^2} = 48\pi - 12\pi r$

When $r = 8$, $\dfrac{d^2V}{dx^2} = 48\pi - 12\pi(8)$, which is $< 0$.

The stationary value is a maximum value.

Volume of the largest cylinder is $512\pi$ cm$^3$.

**i DID YOU KNOW?**

Differentiation can be used in business to find how to maximise company profits and to find how to minimise production costs.

223

1   The sum of two real numbers, $x$ and $y$, is 9.

   a   Express $y$ in terms of $x$.

   b   i   Given that $P = x^2 y$, write down an expression for $P$, in terms of $x$.

      ii   Find the maximum value of $P$.

   c   i   Given that $Q = 3x^2 + 2y^2$, write down an expression for $Q$, in terms of $x$.

      ii   Find the minimum value of $Q$.

2   A piece of wire, of length 40 cm, is bent to form a sector of a circle with radius $r$ cm and sector angle $\theta$ radians, as shown in the diagram. The total area enclosed by the shape is $A$ cm$^2$.

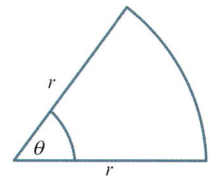

   a   Express $\theta$ in terms of $r$.

   b   Show that $A = 20r - r^2$.

   c   Find the value of $r$ for which there is a stationary value of $A$.

   d   Determine the magnitude and nature of this stationary value.

3   The diagram shows a rectangular enclosure for keeping animals.

   There is a fence on three sides of the enclosure and a wall on its fourth side.

   The total length of the fence is 50 m and the area enclosed is $A$ m$^2$.

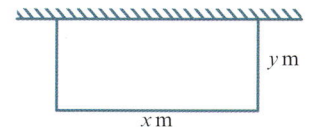

   a   Express $y$ in terms of $x$.

   b   Show that $A = \dfrac{1}{2}x(50 - x)$.

   c   Find the maximum possible area enclosed and the value of $x$ for which this occurs.

4   The diagram shows a rectangle, $ABCD$, where $AB = 20$ cm and $BC = 16$ cm.

   $PQRS$ is a quadrilateral where $PB = AS = 2x$ cm, $BQ = x$ cm and $DR = 4x$ cm.

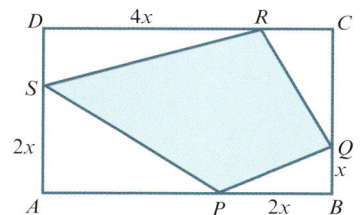

   a   Express the area of $PQRS$ in terms of $x$.

   b   Given that $x$ can vary, find the value of $x$ for which the area of $PQRS$ is a minimum and find the magnitude of this minimum area.

5   The diagram shows the graph of $3x + 2y = 30$. $OPQR$ is a rectangle with area $A$ cm$^2$. The point $O$ is the origin, $P$ lies on the $x$-axis, $R$ lies on the $y$-axis and $Q$ has coordinates $(x, y)$ and lies on the line $3x + 2y = 30$.

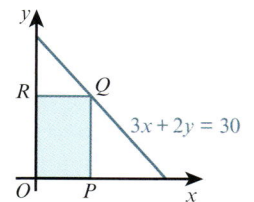

   a   Show that $A = 15x - \dfrac{3}{2}x^2$.

   b   Given that $x$ can vary, find the stationary value of $A$ and determine its nature.

6   $PQRS$ is a rectangle with base length $2p$ units and area $A$ units$^2$.

   The points $P$ and $Q$ lie on the $x$-axis and the points $R$ and $S$ lie on the curve $y = 9 - x^2$.

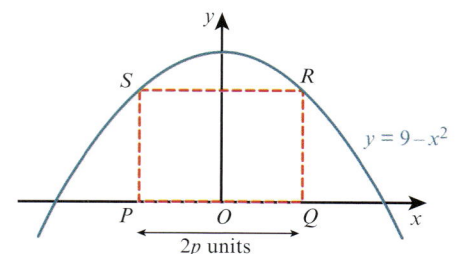

   a   Express $QR$ in terms of $p$.

   b   Show that $A = 2p\left(9 - p^2\right)$.

   c   Find the value of $p$ for which $A$ has a stationary value.

   d   Find this stationary value and determine its nature.

**7**

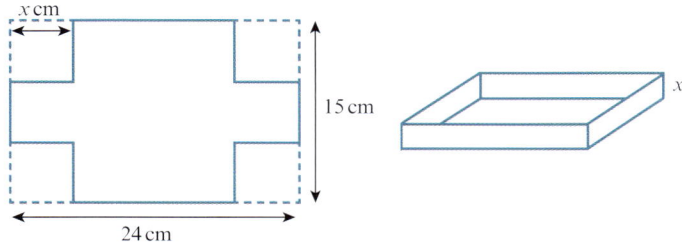

The diagram shows a 24 cm by 15 cm sheet of metal with a square of side $x$ cm removed from each corner. The metal is then folded to make an open rectangular box of depth $x$ cm and volume $V$ cm$^3$.

**a** Show that $V = 4x^3 - 78x^2 + 360x$.

**b** Find the stationary value of $V$ and the value of $x$ for which this occurs.

**c** Determine the nature of this stationary value.

**8** The volume of the solid cuboid shown in the diagram is 576 cm$^3$ and the surface area is $A$ cm$^2$.

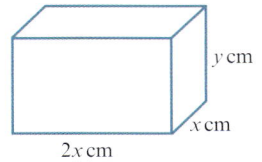

**a** Express $y$ in terms of $x$.

**b** Show that $A = 4x^2 + \dfrac{1728}{x}$.

**c** Find the maximum value of $A$ and state the dimensions of the cuboid for which this occurs.

**9** The diagram shows a piece of wire, of length 2 m, is bent to form the shape $PQRST$.

$PQST$ is a rectangle and $QRS$ is a semicircle with diameter $SQ$.

$PT = x$ m and $PQ = ST = y$ m.

The total area enclosed by the shape is $A$ m$^2$.

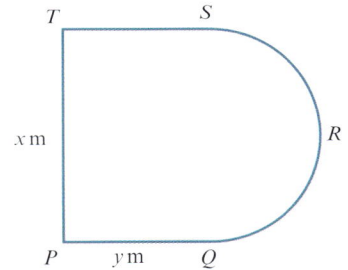

**a** Express $y$ in terms of $x$.

**b** Show that $A = x - \dfrac{1}{2}x^2 - \dfrac{1}{8}\pi x^2$.

**c** Find $\dfrac{\mathrm{d}A}{\mathrm{d}x}$ and $\dfrac{\mathrm{d}^2A}{\mathrm{d}x^2}$.

**d** Find the value for $x$ for which there is a stationary value of $A$.

**e** Determine the magnitude and nature of this stationary value.

**10** The diagram shows a window made from a rectangle with base $2r$ m and height $h$ m and a semicircle of radius $r$ m. The perimeter of the window is 5 m and the area is $A$ m$^2$.

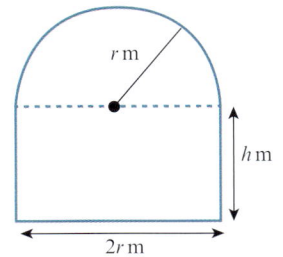

**a** Express $h$ in terms of $r$.

**b** Show that $A = 5r - 2r^2 - \dfrac{1}{2}\pi r^2$.

**c** Find $\dfrac{\mathrm{d}A}{\mathrm{d}r}$ and $\dfrac{\mathrm{d}^2A}{\mathrm{d}r^2}$.

**d** Find the value for $r$ for which there is a stationary value of $A$.

**e** Determine the magnitude and nature of this stationary value.

**11** A piece of wire, of length $100\,\text{cm}$, is cut into two pieces.

One piece is bent to make a square of side $x\,\text{cm}$ and the other is bent to make a circle of radius $r\,\text{cm}$. The total area enclosed by the two shapes is $A\,\text{cm}^2$.

**a** Express $r$ in terms of $x$.

**b** Show that $A = \dfrac{(\pi + 4)x^2 - 200x + 2500}{\pi}$.

**c** Find the value of $x$ for which $A$ has a stationary value and determine the nature and magnitude of this stationary value.

**12** A solid cylinder has radius $r\,\text{cm}$ and height $h\,\text{cm}$.

The volume of this cylinder is $432\pi\,\text{cm}^3$ and the surface area is $A\,\text{cm}^2$.

**a** Express $h$ in terms of $r$.

**b** Show that $A = 2\pi r^2 + \dfrac{864\pi}{r}$.

**c** Find the value for $r$ for which there is a stationary value of $A$.

**d** Determine the magnitude and nature of this stationary value.

**13**

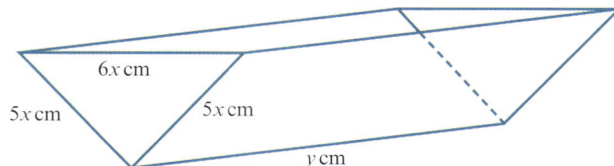

The diagram shows an open water container in the shape of a triangular prism of length $y\,\text{cm}$.

The vertical cross-section is an isosceles triangle with sides $5x\,\text{cm}$, $5x\,\text{cm}$ and $6x\,\text{cm}$.

The water container is made from $500\,\text{cm}^2$ of sheet metal and has a volume of $V\,\text{cm}^3$.

**a** Express $y$ in terms of $x$.

**b** Show that $V = 600x - \dfrac{144}{5}x^3$.

**c** Find the value of $x$ for which $V$ has a stationary value.

**d** Show that the value in part **c** is a maximum value.

**PS** **14** The diagram shows a solid formed by joining a hemisphere of radius $r$ cm to a cylinder of radius $r\,\text{cm}$ and height $h\,\text{cm}$. The surface area of the solid is $320\pi\,\text{cm}^2$ and the volume is $V\,\text{cm}^3$.

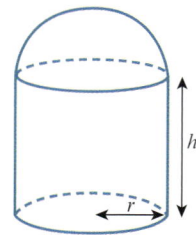

**a** Express $h$ in terms of $r$.

**b** Show that $V = 160\pi r - \dfrac{5}{6}\pi r^3$.

**c** Find the exact value of $r$ such that $V$ is a maximum.

**PS** **15** The diagram shows a right circular cone of base radius $r\,\text{cm}$ and height $h\,\text{cm}$ cut from a solid sphere of radius $10\,\text{cm}$. The volume of the cone is $V\,\text{cm}^3$.

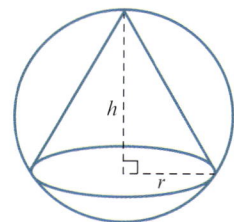

**a** Express $r$ in terms of $h$.

**b** Show that $V = \dfrac{1}{3}\pi h^2(20 - h)$.

**c** Find the value for $h$ for which there is a stationary value of $V$.

**d** Determine the magnitude and nature of this stationary value.

## 8.4 Rates of change

A                          B                          C

Consider pouring water at a constant rate of $10\,\text{cm}^3\,\text{s}^{-1}$ into each of these three large containers.

1   Discuss how the height of water in container A changes with time.

2   Discuss how the height of water in container B changes with time.

3   Discuss how the height of water in container C changes with time.

4   On copies of the following axes, sketch graphs to show how the height of water in a container ($h\,\text{cm}$) varies with time ($t$ seconds) for each container.

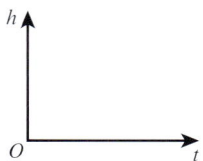 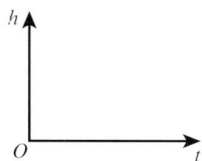 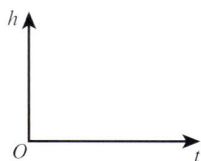

5   What can you say about the gradients?

You should have come to the conclusion that:

- the height of water in container A increases at a constant rate

- the height of water in containers B and C does not increase at a constant rate.

The (constant) rate of change of the height of the water in container A can be found by finding the gradient of the straight-line graph.

The rate of change of the height of the water in containers B and C at a particular time, $t$ seconds, can be estimated by drawing a tangent to the curve and then finding the gradient of the tangent. A more accurate method is to use differentiation if we know the equation of the graph.

**WORKED EXAMPLE 8.9**

Given that $h = \dfrac{1}{5}t^2$, find the rate of change of $h$ with respect to $t$ when $t = 2$.

**Answer**

$h = \dfrac{1}{5}t^2$  · · · · · · · · · · · · · · · · · · · · · · · · ·   Differentiate to obtain $\dfrac{\text{d}h}{\text{d}t}$ (the rate of change of $h$ with respect to $t$).

$\dfrac{\text{d}h}{\text{d}t} = \dfrac{2}{5}t$

When $t = 2$, $\dfrac{\text{d}h}{\text{d}t} = \dfrac{2}{5}(2) = \dfrac{4}{5}$

227

**WORKED EXAMPLE 8.10**

Variables $V$ and $t$ are connected by the equation $V = 2t^2 - 3t + 8$.

Find the rate of change of $V$ with respect to $t$ when $t = 4$.

**Answer**

$V = 2t^2 - 3t + 8$ .................................................... Differentiate to obtain $\dfrac{dV}{dt}$ (the rate of change of $V$ with respect to $t$).

$\dfrac{dV}{dt} = 4t - 3$

When $t = 4$, $\dfrac{dV}{dt} = 4(4) - 3 = 13$

## Connected rates of change

When two variables, $x$ and $y$, both vary with a third variable, $t$, we can connect the three variables using the chain rule.

**KEY POINT 8.3**

The chain rule states: $\dfrac{dy}{dt} = \dfrac{dy}{dx} \times \dfrac{dx}{dt}$

**REWIND**

In Chapter 7, Section 7.2 we learnt how to differentiate using the chain rule. Here we will look at how the chain rule can be used for problems involving connected rates of change.

We may also need to use the rule:

**KEY POINT 8.4**

$\dfrac{dx}{dy} = \dfrac{1}{\dfrac{dy}{dx}}$

We can deduce this as: if we set $t = y$ in the chain rule, we get

$$\dfrac{dy}{dy} = \dfrac{dy}{dx} \times \dfrac{dx}{dy}$$

Since $\dfrac{dy}{dy} = 1$, the rule follows.

**WORKED EXAMPLE 8.11**

A point with coordinates $(x, y)$ moves along the curve $y = x + \sqrt{2x + 3}$ in such a way that the rate of increase of $x$ has the constant value 0.06 units per second. Find the rate of increase of $y$ at the instant when $x = 3$. State whether the $y$-coordinate is increasing or decreasing.

**Answer**

$y = x + (2x + 3)^{\frac{1}{2}}$ and $\dfrac{dx}{dt} = 0.06$

$\dfrac{dy}{dx} = 1 + \dfrac{1}{2}(2x + 3)^{-\frac{1}{2}}(2)$ .................................................... Differentiate to find $\dfrac{dy}{dx}$.

$= 1 + \dfrac{1}{\sqrt{2x + 3}}$

When $x = 3$, $\dfrac{dy}{dx} = 1 + \dfrac{1}{\sqrt{2\,(3) + 3}} = \dfrac{4}{3}$

Using the chain rule: $\dfrac{dy}{dt} = \dfrac{dy}{dx} \times \dfrac{dx}{dt}$

$$= \dfrac{4}{3} \times 0.06$$

$$= 0.08$$

Rate of change of $y$ is 0.08 units per second.

The $y$-coordinate is increasing (since $\dfrac{dy}{dt}$ is a positive quantity).

### EXERCISE 8D

1 A point is moving along the curve $y = 3x - 2x^3$ in such a way that the $x$-coordinate is increasing at 0.015 units per second. Find the rate at which the $y$-coordinate is changing when $x = 2$, stating whether the $y$-coordinate is increasing or decreasing.

2 A point with coordinates $(x, y)$ moves along the curve $y = \sqrt{1 + 2x}$ in such a way that the rate of increase of $x$ has the constant value 0.01 units per second. Find the rate of increase of $y$ at the instant when $x = 4$.

3 A point is moving along the curve $y = \dfrac{8}{x^2 - 2}$ in such a way that the $x$-coordinate is increasing at a constant rate of 0.005 units per second. Find the rate of change of the $y$-coordinate as the point passes through the point $(2, 4)$.

4 A point is moving along the curve $y = 3\sqrt{x} - \dfrac{5}{\sqrt{x}}$ in such a way that the $x$-coordinate is increasing at a constant rate of 0.02 units per second. Find the rate of change of the $y$-coordinate when $x = 1$.

5 A point, $P$, travels along the curve $y = 3x + \dfrac{1}{\sqrt{x}}$ in such a way that the $x$-coordinate of $P$ is increasing at a constant rate of 0.5 units per second. Find the rate at which the $y$-coordinate of $P$ is changing when $P$ is at the point $(1, 4)$.

6 A point is moving along the curve $y = \dfrac{2}{x} + 5x$ in such a way that the $x$-coordinate is increasing at a constant rate of 0.02 units per second. Find the rate at which the $y$-coordinate is changing when $x = 2$, stating whether the $y$-coordinate is increasing or decreasing.

7 A point moves along the curve $y = \dfrac{8}{7 - 2x}$. As it passes through the point $P$, the $x$-coordinate is increasing at a rate of 0.125 units per second and the $y$-coordinate is increasing at a rate of 0.08 units per second. Find the possible $x$-coordinates of $P$.

8 A point, $P$, travels along the curve $y = \sqrt[3]{2x^2 - 3}$ in such a way that at time $t$ minutes the $x$-coordinate of $P$ is increasing at a constant rate of 0.012 units per minute. Find the rate at which the $y$-coordinate of $P$ is changing when $P$ is at the point $(1, -1)$.

**PS** 9 A point, $P(x, y)$, travels along the curve $y = x^3 - 5x^2 + 5x$ in such a way that the rate of change of $x$ is constant. Find the values of $x$ at the points where the rate of change of $y$ is double the rate of change of $x$.

229

## 8.5 Practical applications of connected rates of change

**WORKED EXAMPLE 8.12**

Oil is leaking from a pipeline under the sea and a circular patch is formed on the surface of the sea.

The radius of the patch increases at a rate of 2 metres per hour.

Find the rate at which the area is increasing when the radius of the patch is 25 metres.

**Answer**

We need to find $\dfrac{\mathrm{d}A}{\mathrm{d}t}$ when $r = 25$.

Radius increasing at a rate of 2 metres per hour, so $\dfrac{\mathrm{d}r}{\mathrm{d}t} = 2$.

Let $A$ = area of circular oil patch, in $\mathrm{m}^2$.

$A = \pi r^2$             **Differentiate with respect to $r$.**

$\dfrac{\mathrm{d}A}{\mathrm{d}r} = 2\pi r$

When $r = 25$, $\dfrac{\mathrm{d}A}{\mathrm{d}r} = 50\pi$

Using the chain rule, $\dfrac{\mathrm{d}A}{\mathrm{d}t} = \dfrac{\mathrm{d}A}{\mathrm{d}r} \times \dfrac{\mathrm{d}r}{\mathrm{d}t}$

$\qquad\qquad\qquad = 50\pi \times 2$

$\qquad\qquad\qquad = 100\pi$

The area is increasing at a rate of $100\pi\,\mathrm{m}^2$ per hour.

**WORKED EXAMPLE 8.13**

A solid sphere has radius $r$ cm, surface area $A\,\mathrm{cm}^2$ and volume $V\,\mathrm{cm}^3$.

The radius is increasing at a rate of $\dfrac{1}{5\pi}\,\mathrm{cm\,s^{-1}}$.

  **a**  Find the rate of increase of the surface area when $r = 3$.

  **b**  Find the rate of increase of the volume when $r = 5$.

**Answer**

  **a**  We need to find $\dfrac{\mathrm{d}A}{\mathrm{d}t}$ when $r = 3$.

      Radius increasing at a rate of $\dfrac{1}{5\pi}\,\mathrm{cm\,s^{-1}}$, so $\dfrac{\mathrm{d}r}{\mathrm{d}t} = \dfrac{1}{5\pi}$.

         $A = 4\pi r^2$           **Differentiate with respect to $r$.**

         $\dfrac{\mathrm{d}A}{\mathrm{d}r} = 8\pi r$

      When $r = 3$, $\dfrac{\mathrm{d}A}{\mathrm{d}r} = 24\pi$

      Using the chain rule, $\qquad \dfrac{\mathrm{d}A}{\mathrm{d}t} = \dfrac{\mathrm{d}A}{\mathrm{d}r} \times \dfrac{\mathrm{d}r}{\mathrm{d}t}$

                             $= 24\pi \times \dfrac{1}{5\pi}$

                       $= 4.8$

      The surface area is increasing at a rate of $4.8\,\mathrm{cm}^2\,\mathrm{s}^{-1}$.

**b**  We need to find $\dfrac{dV}{dt}$ when $r = 5$.

$$V = \frac{4}{3}\pi r^3$$ ............................................ Differentiate with respect to $r$.

$$\frac{dV}{dr} = 4\pi r^2$$

When $r = 5$, $\dfrac{dV}{dr} = 100\pi$

Using the chain rule, $\dfrac{dV}{dt} = \dfrac{dV}{dr} \times \dfrac{dr}{dt}$

$$= 100\pi \times \frac{1}{5\pi}$$

$$= 20$$

The volume is increasing at a rate of $20\,\text{cm}^3\,\text{s}^{-2}$.

---

**WORKED EXAMPLE 8.14**

Water is poured into the conical container shown, at a rate of $2\pi\,\text{cm}^3\text{s}^{-1}$.

After $t$ seconds, the volume of water in the container, $V\,\text{cm}^3$, is given by $V = \dfrac{1}{12}\pi h^3$, where $h$ cm is the height of the water in the container.

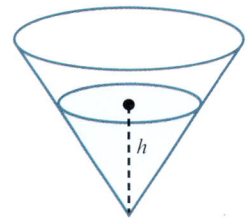

**a**  Find the rate of change of $h$ when $h = 5$.

**b**  Given that the container has radius $10\,$cm and height $20\,$cm, find the rate of change of $h$ when the container is half full. Give your answer correct to 3 significant figures.

**Answer**

**a**  We need to find $\dfrac{dh}{dt}$ when $h = 5$.

Volume increasing at a rate of $2\pi\,\text{cm}^3\text{s}^{-1}$, so $\dfrac{dV}{dt} = 2\pi$.

$$V = \frac{1}{12}\pi h^3$$ ............................................ Differentiate with respect to $h$.

$$\frac{dV}{dh} = \frac{1}{4}\pi h^2$$

When $h = 5$, $\dfrac{dV}{dh} = \dfrac{25\pi}{4}$

Using the chain rule, $\dfrac{dh}{dt} = \dfrac{dh}{dV} \times \dfrac{dV}{dt}$

$$= \frac{4}{25\pi} \times 2\pi$$

$$= 0.32$$

The height is increasing at a rate of $0.32\,\text{cm}\,\text{s}^{-1}$.

231

**b**  Volume when half full $= \dfrac{1}{2}\left[\dfrac{1}{12}\pi h^3\right] = \dfrac{1}{2}\left[\dfrac{1}{12}\pi(20)^3\right] = \dfrac{1000}{3}\pi$

Using $V = \dfrac{1}{12}\pi h^3$, $\dfrac{1}{12}\pi h^3 = \dfrac{1000}{3}\pi$

$$h^3 = 4000$$
$$h = 15.874$$

When $h = 15.874$, $\dfrac{\mathrm{d}V}{\mathrm{d}h} = \dfrac{1}{4}\pi(15.874)^2 = 197.9$

Using the chain rule, $\dfrac{\mathrm{d}h}{\mathrm{d}t} = \dfrac{\mathrm{d}h}{\mathrm{d}V} \times \dfrac{\mathrm{d}V}{\mathrm{d}t}$

$$= \dfrac{1}{197.9} \times 2\pi$$
$$= 0.0317\,\text{cm}\,\text{s}^{-1}$$

## EXERCISE 8E

**1**  A circle has radius $r$ cm and area $A$ cm$^2$.

The radius is increasing at a rate of $0.1\,\text{cm}\,\text{s}^{-1}$.

Find the rate of increase of $A$ when $r = 4$.

**2**  A sphere has radius $r$ cm and volume $V$ cm$^3$.

The radius is increasing at a rate of $\dfrac{1}{2\pi}\,\text{cm}\,\text{s}^{-1}$.

Find the rate of increase of the volume when $V = 36\pi$.

**3**  A cone has base radius $r$ cm and a fixed height of $30$ cm.

The radius of the base is increasing at a rate of $0.01\,\text{cm}\,\text{s}^{-1}$.

Find the rate of change of the volume when $r = 5$.

**4**  A square has side length $x$ cm and area $A$ cm$^2$.

The area is increasing at a constant rate of $0.03\,\text{cm}^2\,\text{s}^{-1}$.

Find the rate of increase of $x$ when $A = 25$.

**5**  A cube has sides of length $x$ cm and volume $V$ cm$^3$.

The volume is increasing at a rate of $1.5\,\text{cm}^3\,\text{s}^{-1}$.

Find the rate of increase of $x$ when $V = 8$.

**6**  A solid metal cuboid has dimensions $x$ cm by $x$ cm by $4x$ cm.

The cuboid is heated and the volume increases at a rate of $0.15\,\text{cm}^3\,\text{s}^{-1}$.

Find the rate of increase of $x$ when $x = 2$.

**7**  A closed circular cylinder has radius $r$ cm and surface area $A$ cm$^2$, where $A = 2\pi r^2 + \dfrac{400\pi}{r}$.

Given that the radius of the cylinder is increasing at a rate of $0.25\,\text{cm}\,\text{s}^{-1}$, find the rate of change of $A$ when $r = 10$.

8  The diagram shows a water container in the shape of a triangular prism of length 120 cm.

The vertical cross-section is an equilateral triangle.

Water is poured into the container at a rate of $24\,\text{cm}^3\text{s}^{-1}$.

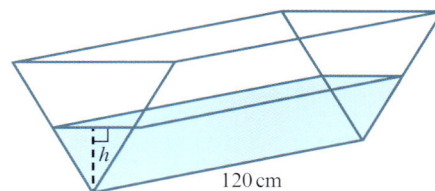

a  Show that the volume of water in the container, $V\,\text{cm}^3$, is given by $V = 40\sqrt{3}\,h^2$, where $h$ cm is the height of the water in the container.

b  Find the rate of change of $h$ when $h = 12$.

9  Water is poured into the hemispherical bowl of radius 5 cm at a rate of $3\pi\,\text{cm}^3\text{s}^{-1}$.

After $t$ seconds, the volume of water in the bowl, $V\,\text{cm}^3$, is given by

$V = 5\pi h^2 - \dfrac{1}{3}\pi h^3$, where $h$ cm is the height of the water in the bowl.

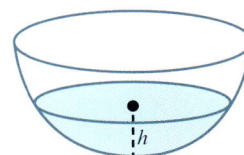

a  Find the rate of change of $h$ when $h = 1$.

b  Find the rate of change of $h$ when $h = 3$.

10  The diagram shows a right circular cone with radius 10 cm and height 30 cm.

The cone is initially completely filled with water.

Water leaks out of the cone through a small hole at the vertex at a rate of $4\,\text{cm}^3\text{s}^{-1}$.

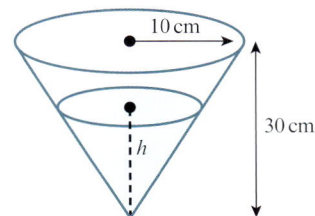

a  Show that the volume of water in the cone, $V\,\text{cm}^3$, when the height of the water is $h$ cm is given by the formula $V = \dfrac{\pi h^3}{27}$.

b  Find the rate of change of $h$ when $h = 20$.

11  Oil is poured onto a flat surface and a circular patch is formed.

The radius of the patch increases at a rate of $2\sqrt{r}\,\text{cm s}^{-1}$.

Find the rate at which the area is increasing when the circumference is $8\pi$ cm.

12  Paint is poured onto a flat surface and a circular patch is formed.

The area of the patch increases at a rate of $5\,\text{cm}^2\text{s}^{-1}$.

a  Find, in terms of $\pi$, the radius of the patch after 8 seconds.

b  Find, in terms of $\pi$, the rate of increase of the radius after 8 seconds.

**PS** 13  A cylindrical container of radius 8 cm and height 25 cm is completely filled with water.

The water is then poured at a constant rate from the cylinder into an empty inverted cone.

The cone has radius 15 cm and height 24 cm and its axis is vertical.

It takes 40 seconds for all of the water to be transferred.

a  If $V$ represents the volume of water, in $\text{cm}^3$, in the cone at time $t$ seconds, find $\dfrac{dV}{dt}$ in terms of $\pi$.

b  When the depth of the water in the cone is 10 cm, find:

i  the rate of change of the height of the water in the cone

ii  the rate of change of the horizontal surface area of the water in the cone.

233

# Checklist of learning and understanding

**Increasing and decreasing functions**

- $y = f(x)$ is increasing for a given interval of $x$ if $\dfrac{dy}{dx} > 0$ throughout the interval.

- $y = f(x)$ is decreasing for a given interval of $x$ if $\dfrac{dy}{dx} < 0$ throughout the interval.

**Stationary points**

- Stationary points (turning points) of a function $y = f(x)$ occur when $\dfrac{dy}{dx} = 0$.

**First derivative test for maximum and minimum points**

At a maximum point:

- $\dfrac{dy}{dx} = 0$

- the gradient is positive to the left of the maximum and negative to the right.

At a minimum point:

- $\dfrac{dy}{dx} = 0$

- the gradient is negative to the left of the minimum and positive to the right.

**Second derivative test for maximum and minimum points**

- If $\dfrac{dy}{dx} = 0$ and $\dfrac{d^2 y}{dx^2} < 0$, then the point is a maximum point.

- If $\dfrac{dy}{dx} = 0$ and $\dfrac{d^2 y}{dx^2} > 0$, then the point is a minimum point.

- If $\dfrac{dy}{dx} = 0$ and $\dfrac{d^2 y}{dx^2} = 0$, then the nature of the stationary point can be found using the first derivative test.

**Connected rates of change**

- When two variables, $x$ and $y$, both vary with a third variable, $t$, the three variables can be connected using the chain rule: $\dfrac{dy}{dt} = \dfrac{dy}{dx} \times \dfrac{dx}{dt}$.

- You may also need to use the rule: $\dfrac{dx}{dy} = \dfrac{1}{\dfrac{dy}{dx}}$.

**END-OF-CHAPTER REVIEW EXERCISE 8**

1   The volume of a spherical balloon is increasing at a constant rate of $40\,\text{cm}^3$ per second. Find the rate of increase of the radius of the balloon when the radius is $15\,\text{cm}$.

[The volume, $V$, of a sphere with radius $r$ is $V = \dfrac{4}{3}\pi r^3$.]   [4]

2   An oil pipeline under the sea is leaking oil and a circular patch of oil has formed on the surface of the sea. At midday the radius of the patch of oil is $50\,\text{m}$ and is increasing at a rate of 3 metres per hour. Find the rate at which the area of the oil is increasing at midday.   [4]

*Cambridge International AS & A Level Mathematics 9709 Paper 11 Q3 November 2012*

3   A curve has equation $y = 27x - \dfrac{4}{(x+2)^2}$. Show that the curve has a stationary point at $x = -\dfrac{8}{3}$ and determine its nature.   [5]

4   A watermelon is assumed to be spherical in shape while it is growing. Its mass, $M$ kg, and radius, $r$ cm, are related by the formula $M = kr^3$, where $k$ is a constant. It is also assumed that the radius is increasing at a constant rate of 0.1 centimetres per day. On a particular day the radius is $10\,\text{cm}$ and the mass is $3.2\,\text{kg}$. Find the value of $k$ and the rate at which the mass is increasing on this day.   [5]

*Cambridge International AS & A Level Mathematics 9709 Paper 11 Q4 June 2012*

5

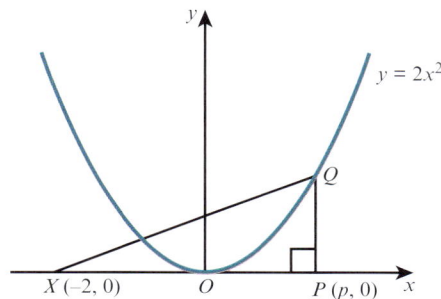

The diagram shows the curve $y = 2x^2$ and the points $X(-2, 0)$ and $P(p, 0)$. The point $Q$ lies on the curve and $PQ$ is parallel to the $y$-axis.

i   Express the area, $A$, of triangle $XPQ$ in terms of $p$.   [2]

The point $P$ moves along the $x$-axis at a constant rate of 0.02 units per second and $Q$ moves along the curve so that $PQ$ remains parallel to the $y$-axis.

ii   Find the rate at which $A$ is increasing when $p = 2$.   [3]

*Cambridge International AS & A Level Mathematics 9709 Paper 11 Q2 June 2015*

6

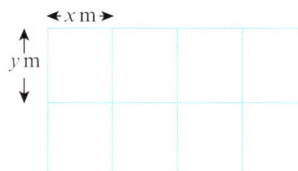

A farmer divides a rectangular piece of land into 8 equal-sized rectangular sheep pens as shown in the diagram. Each sheep pen measures $x$ m by $y$ m and is fully enclosed by metal fencing. The farmer uses 480 m of fencing.

i   Show that the total area of land used for the sheep pens, $A\,\text{m}^2$, is given by $A = 384x - 9.6x^2$.   [3]

ii   Given that $x$ and $y$ can vary, find the dimensions of each sheep pen for which the value of $A$ is a maximum. (There is no need to verify that the value of $A$ is a maximum.)   [3]

*Cambridge International AS & A Level Mathematics 9709 Paper 11 Q5 June 2016*

235

7  The variables $x$, $y$ and $z$ can take only positive values and are such that $z = 3x + 2y$ and $xy = 600$.

   i   Show that $z = 3x + \dfrac{1200}{x}$.      **[1]**

   ii  Find the stationary value of $z$ and determine its nature.      **[6]**

                                  *Cambridge International AS & A Level Mathematics 9709 Paper 11 Q6 June 2011*

8

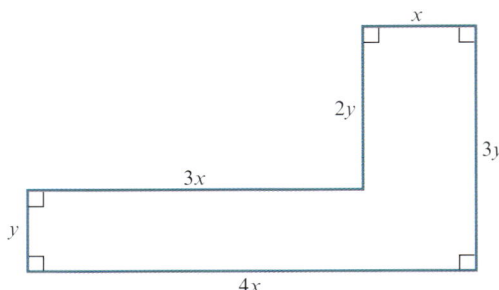

The diagram shows the dimensions in metres of an L-shaped garden. The perimeter of the garden is 48 m.

   i   Find an expression for $y$ in terms of $x$.      **[1]**

   ii  Given that the area of the garden is $A\,\text{m}^2$, show that $A = 48x - 8x^2$.      **[2]**

   iii Given that $x$ can vary, find the maximum area of the garden, showing that this is a maximum value rather than a minimum value.      **[4]**

                                 *Cambridge International AS & A Level Mathematics 9709 Paper 11 Q7 November 2011*

9  A curve has equation $y = \dfrac{8}{x} + 2x$.

   i   Find $\dfrac{dy}{dx}$ and $\dfrac{d^2y}{dx^2}$.      **[3]**

   ii  Find the coordinates of the stationary points and state, with a reason, the nature of each stationary point.      **[5]**

                               *Cambridge International AS & A Level Mathematics 9709 Paper 11 Q5 November 2015*

10

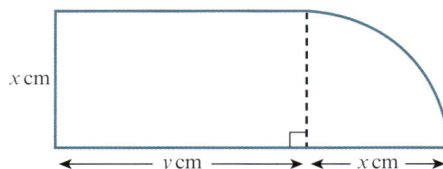

The diagram shows a metal plate consisting of a rectangle with sides $x$ cm and $y$ cm and a quarter-circle of radius $x$ cm. The perimeter of the plate is 60 cm.

   i   Express $y$ in terms of $x$.      **[2]**

   ii  Show that the area of the plate, $A\,\text{cm}^2$, is given by $A = 30x - x^2$.      **[2]**

Given that $x$ can vary,

   iii find the value of $x$ at which $A$ is stationary,      **[2]**

   iv find this stationary value of $A$, and determine whether it is a maximum or a minimum value.      **[2]**

                               *Cambridge International AS & A Level Mathematics 9709 Paper 11 Q8 November 2010*

**11** A curve has equation $y = x^3 + x^2 - 5x + 7$.

    **a** Find the set of values of $x$ for which the gradient of the curve is less than 3.   **[4]**

    **b** Find the coordinates of the two stationary points on the curve and determine the nature of each stationary point.   **[5]**

**12**

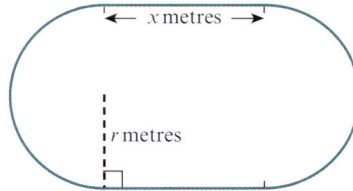

The inside lane of a school running track consists of two straight sections each of length $x$ metres, and two semicircular sections each of radius $r$ metres, as shown in the diagram. The straight sections are perpendicular to the diameters of the semicircular sections. The perimeter of the inside lane is 400 metres.

    **i** Show that the area, $A\,\text{m}^2$, of the region enclosed by the inside lane is given by $A = 400r - \pi r^2$.   **[4]**

    **ii** Given that $x$ and $r$ can vary, show that, when $A$ has a stationary value, there are no straight sections in the track. Determine whether the stationary value is a maximum or a minimum.   **[5]**

*Cambridge International AS & A Level Mathematics 9709 Paper 11 Q8 November 2013*

**13** The equation of a curve is $y = x^3 + px^2$, where $p$ is a positive constant.

    **i** Show that the origin is a stationary point on the curve and find the coordinates of the other stationary point in terms of $p$.   **[4]**

    **ii** Find the nature of each of the stationary points.   **[3]**

Another curve has equation $y = x^3 + px^2 + px$.

    **iii** Find the set of values of $p$ for which this curve has no stationary points.   **[3]**

*Cambridge International AS & A Level Mathematics 9709 Paper 11 Q9 June 2015*

# Chapter 9
# Integration

**In this chapter you will learn how to:**

- understand integration as the reverse process of differentiation, and integrate $(ax+b)^n$ (for any rational $n$ except $-1$), together with constant multiples, sums and differences
- solve problems involving the evaluation of a constant of integration
- evaluate definite integrals
- use definite integration to find the:
  - area of a region bounded by a curve and lines parallel to the axes, or between a curve and a line, or between two curves
  - volume of revolution about one of the axes.

| Where it comes from | What you should be able to do | Check your skills |
|---|---|---|
| IGCSE / O Level Mathematics | Substitute values for $x$ and $y$ into equations of the form $y = f(x) + c$ and solve to find $c$. | 1 a Given that the line $y = 5x + c$ passes through the point $(3, -4)$, find the value of $c$.<br><br>b Given that the curve $y = x^2 - 2x + c$ passes through the point $(-1, 2)$, find the value of $c$. |
| IGCSE / O Level Mathematics | Find the $x$-coordinates of the points where a curve crosses the $x$-axis. | 2 Find the $x$-coordinates of the points where the curve crosses the $x$-axis.<br><br>a $y = 3x^2 - 13x - 10$<br><br>b $y = 3\sqrt{x} - x$ |
| Chapter 7 | Differentiate constant multiples, sums and differences of expressions containing terms of the form $ax^n$. | 3 Find $\dfrac{dy}{dx}$.<br><br>a $y = 3x^8 - 13x - 10$<br><br>b $y = 5x^2 - 4x + 10\sqrt{x}$ |

## Why do we study integration?

In Chapters 7 and 8 you studied differentiation, which is the first basic tool of calculus. In this chapter you will learn about integration, which is the second basic tool of calculus. We often refer to integration as the reverse process of differentiation. It has many applications; for example, planning spacecraft flight paths, or modelling real-world behaviour for computer games.

Isaac Newton and Gottfried Wilhelm Leibniz formulated the principles of integration independently, in the 17th century, by thinking of an integral as an infinite sum of rectangles of infinitesimal width.

239

**WEB LINK**

Explore the *Calculus meets functions* station on the Underground Mathematics website.

## 9.1 Integration as the reverse of differentiation

In Chapter 7, you learnt about the process of obtaining $\dfrac{dy}{dx}$ when $y$ is known. We call this process differentiation.

You learnt the rule for differentiating power functions:

**KEY POINT 9.1**

If $y = x^n$, then $\dfrac{dy}{dx} = nx^{n-1}$.

Applying this rule to functions of the form $y = x^3 + c$, we obtain:

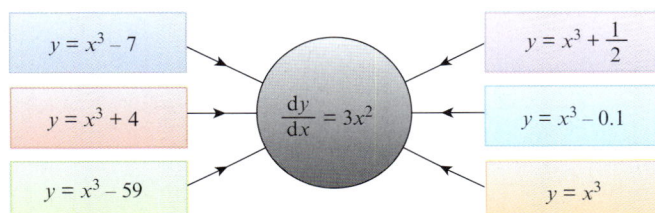

$y = x^3 - 7$    $y = x^3 + \dfrac{1}{2}$

$y = x^3 + 4$    $\dfrac{dy}{dx} = 3x^2$    $y = x^3 - 0.1$

$y = x^3 - 59$    $y = x^3$

This shows that there are an infinite number of functions that when differentiated give the answer $3x^2$. They are all of the form $y = x^3 + c$, where $c$ is some constant.

In this chapter you will learn about the reverse process of obtaining $y$ when $\dfrac{dy}{dx}$ is known.

We can call this reverse process **antidifferentiation**.

There is a seemingly unrelated problem that you will study in Section 9.6: what is the area under the graph of $y = 3x^2$? The process used to answer that question is known as **integration**. There is a remarkable theorem due to both Newton and Leibniz that says that *integration is essentially the same as antidifferentiation*. This is now known as the Fundamental theorem of Calculus.

Because of this theorem, we do not need to use the term antidifferentiation. So from now on, we will only talk about **integration**, whether we are reversing the process of differentiation or finding the area under a graph.

### EXPLORE 9.1

1   Find $\dfrac{dy}{dx}$ for each of the following functions.

   **a**   $y = \dfrac{1}{3}x^3 - 2$           **b**   $y = \dfrac{1}{6}x^6 + 8$           **c**   $y = \dfrac{1}{15}x^{15} + 1$

   **d**   $y = -\dfrac{1}{2}x^{-2} + 3$        **e**   $y = -\dfrac{1}{7}x^{-7} + 0.2$      **f**   $y = \dfrac{2}{3}x^{\frac{3}{2}} - \dfrac{5}{8}$

2   Discuss your results with those of your classmates and try to find a rule for obtaining $y$ if $\dfrac{dy}{dx} = x^n$.

3   Describe your rule, in words.

4   Discuss with your classmates whether your rule works for finding $y$ when $\dfrac{dy}{dx} = \dfrac{1}{x}$.

From the class discussion we can conclude that:

### KEY POINT 9.2

If $\dfrac{dy}{dx} = x^n$, then $y = \dfrac{1}{n+1}x^{n+1} + c$ (where $c$ is an arbitrary constant and $n \neq -1$).

You may find it easier to remember this in words:

'Increase the power $n$ by 1 to obtain the new power, then divide by the new power. Remember to add a constant $c$ at the end.'

Using function notation we write this rule as:

### KEY POINT 9.3

If $f'(x) = x^n$, then $f(x) = \dfrac{1}{n+1}x^{n+1} + c$ (where $c$ is an arbitrary constant and $n \neq -1$).

The special symbol $\displaystyle\int$ is used to denote integration.

240

When we need to integrate $x^3$, for example, we write:

$$\int x^3 \, dx = \frac{1}{4} x^4 + c$$

$\int x^3 \, dx$ is called the **indefinite integral** of $x^3$ with respect to $x$.

We call it 'indefinite' because it has infinitely many solutions.

Using this notation, we can write the rule for integrating powers as:

> 🔍 **KEY POINT 9.4**
>
> $\int x^n \, dx = \frac{1}{n+1} x^{n+1} + c$ (where $c$ is a constant and $n \neq -1$).

We write the rule for integrating constant multiples of a function as:

> 🔍 **KEY POINT 9.5**
>
> $\int k f(x) \, dx = k \int f(x) \, dx$, where $k$ is a constant.

We write the rule for integrating sums and differences of two functions as:

> 🔍 **KEY POINT 9.6**
>
> $\int \left[ f(x) \pm g(x) \right] dx = \int f(x) \, dx \pm \int g(x) \, dx$

**WORKED EXAMPLE 9.1**

Find $y$ in terms of $x$ for each of the following.

   **a** $\dfrac{dy}{dx} = x^3$          **b** $\dfrac{dy}{dx} = \dfrac{1}{x^2}$          **c** $\dfrac{dy}{dx} = x\sqrt{x}$

**Answer**

**a** $\dfrac{dy}{dx} = x^3$

$y = \dfrac{1}{3+1} x^{3+1} + c$

$\quad = \dfrac{1}{4} x^4 + c$

**b** $\dfrac{dy}{dx} = x^{-2}$

$y = \dfrac{1}{-2+1} x^{-2+1} + c$

$\quad = -x^{-1} + c$

$\quad = -\dfrac{1}{x} + c$

**c** $\dfrac{dy}{dx} = x^{\frac{3}{2}}$

$y = \dfrac{1}{\frac{3}{2}+1} x^{\frac{3}{2}+1} + c$

$\quad = \dfrac{1}{\left(\frac{5}{2}\right)} x^{\frac{5}{2}} + c$

$\quad = \dfrac{2}{5} x^{\frac{5}{2}} + c$

**WORKED EXAMPLE 9.2**

Find f($x$) in terms of $x$ for each of the following.

**a** $f'(x) = 4x^3 - \dfrac{2}{x^2} + 4$

**b** $f'(x) = 8x^2 - \dfrac{1}{2x^4} + 2x$

**c** $f'(x) = \dfrac{(x+3)(x-1)}{\sqrt{x}}$

**Answer**

**a** $f'(x) = 4x^3 - 2x^{-2} + 4x^0$ → Write in index form ready for integration.

$f(x) = \dfrac{4}{4}x^4 - \dfrac{2}{(-1)}x^{-1} + \dfrac{4}{1}x^1 + c$

$= x^4 + 2x^{-1} + 4x + c$

$= x^4 + \dfrac{2}{x} + 4x + c$

**b** $f'(x) = 8x^2 - \dfrac{1}{2}x^{-4} + 2x^1$ → Write in index form ready for integration.

$f(x) = \dfrac{8}{3}x^3 - \dfrac{1}{2(-3)}x^{-3} + \dfrac{2}{2}x^2 + c$

$= \dfrac{8}{3}x^3 + \dfrac{1}{6}x^{-3} + x^2 + c$

$= \dfrac{8}{3}x^3 + \dfrac{1}{6x^3} + x^2 + c$

**c** $f'(x) = \dfrac{x^2 + 2x - 3}{\sqrt{x}}$ → Write in index form ready for integration.

$= x^{\frac{3}{2}} + 2x^{\frac{1}{2}} - 3x^{-\frac{1}{2}}$

$f(x) = \dfrac{1}{\left(\frac{5}{2}\right)}x^{\frac{5}{2}} + \dfrac{2}{\left(\frac{3}{2}\right)}x^{\frac{3}{2}} - \dfrac{3}{\left(\frac{1}{2}\right)}x^{\frac{1}{2}} + c$

$= \dfrac{2}{5}x^{\frac{5}{2}} + \dfrac{4}{3}x^{\frac{3}{2}} - 6\sqrt{x} + c$

**WORKED EXAMPLE 9.3**

Find:

**a** $\displaystyle\int x(2x-1)(2x+3)\,dx$

**b** $\displaystyle\int \dfrac{4x^2 - 3\sqrt{x}}{x}\,dx$

**Answer**

**a** $\displaystyle\int x(2x-1)(2x+3)\,dx = \int (4x^3 + 4x^2 - 3x)\,dx$

$= \dfrac{4x^4}{4} + \dfrac{4x^3}{3} - \dfrac{3x^2}{2} + c$

$= x^4 + \dfrac{4x^3}{3} - \dfrac{3x^2}{2} + c$

$$\textbf{b} \quad \int \frac{4x^2 - 3\sqrt{x}}{x}\,dx = \int \left(4x - 3x^{-\frac{1}{2}}\right)dx$$

$$= \frac{4}{2}x^2 - \frac{3}{\left(\frac{1}{2}\right)}x^{\frac{1}{2}} + c$$

$$= 2x^2 - 6x^{\frac{1}{2}} + c$$

$$= 2x^2 - 6\sqrt{x} + c$$

## EXERCISE 9A

**1** Find $y$ in terms of $x$ for each of the following.

**a** $\dfrac{dy}{dx} = 15x^2$      **b** $\dfrac{dy}{dx} = 14x^6$      **c** $\dfrac{dy}{dx} = 12x^3$

**d** $\dfrac{dy}{dx} = \dfrac{3}{x^2}$      **e** $\dfrac{dy}{dx} = \dfrac{1}{2x^3}$      **f** $\dfrac{dy}{dx} = \dfrac{4}{\sqrt{x}}$

**2** Find $f(x)$ in terms of $x$ for each of the following.

**a** $f'(x) = 5x^4 - 2x^3 + 2$      **b** $f'(x) = 3x^5 + x^2 - 2x$

**c** $f'(x) = \dfrac{2}{x^3} + \dfrac{8}{x^2} + 6x$      **d** $f'(x) = \dfrac{9}{x^7} - \dfrac{3}{x^2} - 4$

**3** Find $y$ in terms of $x$ for each of the following.

**a** $\dfrac{dy}{dx} = x(x + 5)$      **b** $\dfrac{dy}{dx} = 2x^2(3x + 1)$      **c** $\dfrac{dy}{dx} = x(x + 2)(x - 8)$

**d** $\dfrac{dy}{dx} = \dfrac{x^4 - 2x + 5}{2x^3}$      **e** $\dfrac{dy}{dx} = \sqrt{x}(x - 3)^2$      **f** $\dfrac{dy}{dx} = \dfrac{5x^2 + 3x + 1}{\sqrt{x}}$

**4** Find each of the following.

**a** $\displaystyle\int 12x^5\,dx$      **b** $\displaystyle\int 20x^3\,dx$      **c** $\displaystyle\int 3x^{-2}\,dx$

**d** $\displaystyle\int \dfrac{4}{x^3}\,dx$      **e** $\displaystyle\int \dfrac{2}{3\sqrt{x}}\,dx$      **f** $\displaystyle\int \dfrac{5}{x\sqrt{x}}\,dx$

**5** Find each of the following.

**a** $\displaystyle\int (x + 1)(x + 4)\,dx$      **b** $\displaystyle\int (x - 3)^2\,dx$      **c** $\displaystyle\int (2\sqrt{x} - 1)^2\,dx$

**d** $\displaystyle\int \sqrt[3]{x}(x^2 + 1)\,dx$      **e** $\displaystyle\int \dfrac{x^2 - 1}{2x^2}\,dx$      **f** $\displaystyle\int \dfrac{x^3 + 6}{2x^3}\,dx$

**g** $\displaystyle\int \dfrac{x^2 + 2\sqrt{x}}{3x}\,dx$      **h** $\displaystyle\int \dfrac{x^4 - 10}{x\sqrt{x}}\,dx$      **i** $\displaystyle\int \left(2\sqrt{x} - \dfrac{3}{x^2\sqrt{x}}\right)^2\,dx$

243

## 9.2 Finding the constant of integration

The next two examples show how we can find the equation of a curve if we know the gradient function and the coordinates of a point on the curve.

**WORKED EXAMPLE 9.4**

A curve is such that $\dfrac{dy}{dx} = \dfrac{6x^5 - 18}{x^3}$ , and $(1, 6)$ is a point on the curve.

Find the equation of the curve.

**Answer**

$\dfrac{dy}{dx} = \dfrac{6x^5 - 18}{x^3}$        Write in index form ready for integration.

$\qquad = 6x^2 - 18x^{-3}$

$y = 2x^3 + 9x^{-2} + c$

$\qquad = 2x^3 + \dfrac{9}{x^2} + c$

When $x = 1$, $y = 6$.

$6 = 2(1)^3 + \dfrac{9}{(1)^2} + c$

$6 = 2 + 9 + c$

$c = -5$

The equation of the curve is $y = 2x^3 + \dfrac{9}{x^2} - 5$.

**WORKED EXAMPLE 9.5**

The function f is such that $f'(x) = 15x^4 - 6x$ and $f(-1) = 1$. Find $f(x)$.

**Answer**

$f'(x) = 15x^4 - 6x$

$f(x) = 3x^5 - 3x^2 + c$

Using $f(-1) = 1$ gives:

$\qquad\qquad 1 = 3(-1)^5 - 3(-1)^2 + c$

$\qquad\qquad 1 = -3 - 3 + c$

$\qquad\qquad c = 7$

$\therefore f(x) = 3x^5 - 3x^2 + 7$

**WORKED EXAMPLE 9.6**

A curve is such that $\dfrac{dy}{dx} = 6x + k$, where $k$ is a constant. The gradient of the normal to the curve at the point $(1, -3)$ is $\dfrac{1}{2}$. Find the equation of the curve.

**Answer**

$\dfrac{dy}{dx} = 6x + k$ $\qquad\qquad\qquad\qquad\qquad\qquad$ Integrate.

$y = 3x^2 + kx + c$

When $x = 1,\quad y = -3.$

$-3 = 3(1)^2 + k(1) + c$

$c + k = -6$ $\text{-----------------------}$ (1)

When $x = 1, \dfrac{dy}{dx} = 6(1) + k = 6 + k$

Gradient of normal $= \dfrac{1}{2}$ so gradient of tangent $= -2$

$6 + k = -2$

$k = -8$

Substituting for $k$ into (1) gives $c = 2.$

The equation of the curve is $y = 3x^2 - 8x + 2.$

245

**EXERCISE 9B**

1 Find the equation of the curve, given $\dfrac{dy}{dx}$ and a point $P$ on the curve.

    a $\dfrac{dy}{dx} = 3x^2 + 1,\ \ P = (1, 4)$ $\qquad\qquad\qquad$ b $\dfrac{dy}{dx} = 2x(3x - 1),\ \ P = (-1, 2)$

    c $\dfrac{dy}{dx} = \dfrac{4}{x^2},\qquad P = (4, 9)$ $\qquad\qquad\qquad$ d $\dfrac{dy}{dx} = \dfrac{2x^3 - 6}{x^2},\qquad P = (3, 7)$

    e $\dfrac{dy}{dx} = \dfrac{2}{\sqrt{x}} - 1,\ \ P = (4, 6)$ $\qquad\qquad\qquad$ f $\dfrac{dy}{dx} = \dfrac{(1 - \sqrt{x}\ )^2}{\sqrt{x}},\ P = (9, 5)$

2 A curve is such that $\dfrac{dy}{dx} = -\dfrac{k}{x^2}$, where $k$ is a constant. Given that the curve passes through the points $(6, 2.5)$ and $(-3, 1)$, find the equation of the curve.

3 A curve is such that $\dfrac{dy}{dx} = kx^2 - 12x + 5$, where $k$ is a constant. Given that the curve passes through the points $(1, -3)$ and $(3, 11)$, find the equation of the curve.

4 A curve is such that $\dfrac{dy}{dx} = kx^2 - \dfrac{6}{x^3}$, where $k$ is a constant. Given that the curve passes through the point $P(1, 6)$ and that the gradient of the curve at $P$ is 9, find the equation of the curve.

5   A curve is such that $\dfrac{dy}{dx} = 5x\sqrt{x} + 2$. Given that the curve passes through the point $(1, 3)$, find:

    a   the equation of the curve

    b   the equation of the tangent to the curve when $x = 4$.

6   A curve is such that $\dfrac{dy}{dx} = kx + 3$, where $k$ is a constant. The gradient of the normal to the curve at the point

    $(1, -2)$ is $-\dfrac{1}{7}$. Find the equation of the curve.

7   A function $y = f(x)$ has gradient function $f'(x) = 8 - 2x$. The maximum value of the function is $20$. Find $f(x)$ and sketch the graph of $y = f(x)$.

8   A curve is such that $\dfrac{dy}{dx} = 3x^2 + x - 10$. Given that the curve passes through the point $(2, -7)$ find:

    a   the equation of the curve

    b   the set of values of $x$ for which the gradient of the curve is positive.

**PS**   9   A curve is such that $\dfrac{d^2y}{dx^2} = 12x + 12$. The gradient of the curve at the point $(0, 4)$ is $10$.

    a   Express $y$ in terms of $x$.

    b   Show that the gradient of the curve is never less than $4$.

10   A curve is such that $\dfrac{d^2y}{dx^2} = -6x - 4$. Given that the curve has a minimum point at $(-2, -6)$, find the equation of the curve.

11   A curve $y = f(x)$ has a stationary point at $P(2, -13)$ and is such that $f'(x) = 2x^2 + 3x - k$, where $k$ is a constant.

    a   Find the $x$-coordinate of the other stationary point, $Q$, on the curve $y = f(x)$.

    b   Determine the nature of each of the stationary points $P$ and $Q$.

**PS**   12   A curve is such that $\dfrac{dy}{dx} = k + x$, where $k$ is a constant.

    a   Given that the tangents to the curve at the points where $x = 5$ and $x = 7$ are perpendicular, find the value of $k$.

    b   Given also that the curve passes through the point $(10, -8)$, find the equation of the curve.

13   A curve $y = f(x)$ has a stationary point at $(1, -1)$ and is such that $f''(x) = 2 + \dfrac{4}{x^3}$. Find $f'(x)$ and $f(x)$.

14   A curve is such that $\dfrac{d^2y}{dx^2} = 2x + 8$. Given that the curve has a minimum point at $(3, -49)$, find the coordinates of the maximum point.

15   A curve is such that $\dfrac{dy}{dx} = 3 - 2x$ and $(1, 11)$ is a point on the curve.

    a   Find the equation of the curve.

    b   A line with gradient $\dfrac{1}{5}$ is a normal to the curve at the point $(4, 5)$. Find the equation of this normal.

16   A curve is such that $\dfrac{dy}{dx} = 3\sqrt{x} - 6$ and the point $P(1, 6)$ is a point on the curve.

    a   Find the equation of the curve.

    b   Find the coordinates of the stationary point on the curve and determine its nature.

**17** A curve is such that $\dfrac{d^2y}{dx^2} = 2 - \dfrac{12}{x^3}$. The curve has a stationary point at $P$ where $x = 1$. Given that the curve passes through the point $(2, 5)$, find the coordinates of the stationary point $P$ and determine its nature.

**18** A curve is such that $\dfrac{dy}{dx} = 2x - 5$ and the point $P(3, -4)$ is a point on the curve. The normal to the curve at $P$ meets the curve again at $Q$.

   **a**  Find the equation of the curve.

   **b**  Find the equation of the normal to the curve at $P$.

   **c**  Find the coordinates of $Q$.

## 9.3 Integration of expressions of the form $(ax + b)^n$

In Chapter 7 you learnt that:

$$\frac{d}{dx}\left[\frac{1}{3 \times 7}(3x - 1)^7\right] = (3x - 1)^6$$

Hence, $\displaystyle\int (3x - 1)^6 \, dx = \frac{1}{3 \times 7}(3x - 1)^7 + c$

This leads to the general rule:

> 🔍 **KEY POINT 9.7**
>
> If $n \neq -1$ and $a \neq 0$, then $\displaystyle\int (ax + b)^n \, dx = \frac{1}{a(n + 1)}(ax + b)^{n + 1} + c$

It is very important to note that this rule *only* works for powers of linear functions.

For example, $\displaystyle\int (ax^2 + b)^6 \, dx$ is not equal to $\dfrac{1}{3a}(ax^2 + b)^3 + c$. (Try differentiating the latter expression to see why.)

> **WORKED EXAMPLE 9.7**
>
> Find:
>
>   **a**  $\displaystyle\int (2x - 3)^4 \, dx$       **b**  $\displaystyle\int \frac{20}{(1 - 4x)^6} \, dx$       **c**  $\displaystyle\int \frac{5}{\sqrt{2x + 7}} \, dx$
>
> **Answer**
>
>   **a**  $\displaystyle\int (2x - 3)^4 \, dx = \frac{1}{2(4 + 1)}(2x - 3)^{4 + 1} + c$
>
>         $= \dfrac{1}{10}(2x - 3)^5 + c$
>
>   **b**  $\displaystyle\int \frac{20}{(1 - 4x)^6} \, dx = 20\int (1 - 4x)^{-6} \, dx$
>
>         $= \dfrac{20}{(-4)(-6 + 1)}(1 - 4x)^{-6 + 1} + c$
>
>         $= (1 - 4x)^{-5} + c$
>
>         $= \dfrac{1}{(1 - 4x)^5} + c$

c $\quad \int \dfrac{5}{\sqrt{2x+7}}\, dx = 5 \int (2x+7)^{-\frac{1}{2}}\, dx$

$\qquad\qquad = \dfrac{5}{2\left(-\frac{1}{2}+1\right)}(2x+7)^{-\frac{1}{2}+1} + c$

$\qquad\qquad = 5\sqrt{2x+7} + c$

## EXERCISE 9C

**1** Find:

**a** $\displaystyle\int (2x-7)^8\, dx$

**b** $\displaystyle\int (3x+1)^5\, dx$

**c** $\displaystyle\int 2(5x-2)^8\, dx$

**d** $\displaystyle\int 3(1-2x)^5\, dx$

**e** $\displaystyle\int \sqrt[3]{5-4x}\, dx$

**f** $\displaystyle\int \sqrt{(2x+1)^3}\, dx$

**g** $\displaystyle\int \dfrac{2}{\sqrt{3x-2}}\, dx$

**h** $\displaystyle\int \left(\dfrac{2}{2x+1}\right)^3\, dx$

**i** $\displaystyle\int \dfrac{5}{4(7-2x)^5}\, dx$

**2** Find the equation of the curve, given $\dfrac{dy}{dx}$ and a point $P$ on the curve.

**a** $\dfrac{dy}{dx} = (2x-1)^3,\ P = \left(\dfrac{3}{2},\, 4\right)$

**b** $\dfrac{dy}{dx} = \sqrt{2x+5},\quad P = (2,2)$

**c** $\dfrac{dy}{dx} = \dfrac{1}{\sqrt{x-2}},\quad P = (3,7)$

**d** $\dfrac{dy}{dx} = \dfrac{4}{(3-2x)^2},\ P = (2,4)$

**3** A curve is such that $\dfrac{dy}{dx} = k(x-5)^3$, where $k$ is a constant. The gradient of the normal to the curve at the point $(4, 2)$ is $\dfrac{1}{12}$. Find the equation of the curve.

**4** A curve is such that $\dfrac{dy}{dx} = \dfrac{5}{\sqrt{2x-3}}$.

Given that the curve passes through the point $P(2, 1)$, find:

**a** the equation of the normal to the curve at $P$

**b** the equation of the curve.

**5** A curve is such that $\dfrac{dy}{dx} = \dfrac{12}{\sqrt{3x+1}} - 4x - 2$.

**a** Show that the curve has a stationary point when $x = 1$ and determine its nature.

**b** Given that the curve passes through the point $(0, 13)$, find the equation of the curve.

**PS** **6** A curve is such that $\dfrac{dy}{dx} = \dfrac{4}{\sqrt{2x+k}}$, where $k$ is a constant. The point $P(3, 2)$ lies on the curve and the normal to the curve at $P$ is $x + 4y = 11$. Find the equation of the curve.

## 9.4 Further indefinite integration

In this section we use the concept that integration is the reverse process of differentiation to help us integrate some more complicated expressions.

> **KEY POINT 9.8**
>
> If $\dfrac{d}{dx}[F(x)] = f(x)$, then $\displaystyle\int f(x)\,dx = F(x) + c$

**WORKED EXAMPLE 9.8**

**a** Show that $\dfrac{d}{dx}\left[\left(3x^2 - 4\right)^8\right] = 48x\left(3x^2 - 4\right)^7$.    **b** Hence, find $\displaystyle\int 6x\left(3x^2 - 4\right)^7 dx$.

**Answer**

**a** Let $y = (3x^2 - 4)^8$  · · · · · · · · · · · · · · · · · · Use the chain rule.

$\dfrac{dy}{dx} = (6x)(8)(3x^2 - 4)^{8-1}$

$\qquad = 48x(3x^2 - 4)^7$

**b** $\displaystyle\int 6x(3x^2 - 4)^7\,dx = \dfrac{1}{8}\int 48x(3x^2 - 4)^7\,dx$

$\qquad\qquad\qquad\qquad = \dfrac{1}{8}(3x^2 - 4)^8 + c$

**EXERCISE 9D**

1   **a**   Differentiate $(x^2 + 2)^4$ with respect to $x$.

    **b**   Hence, find $\displaystyle\int x(x^2 + 2)^3\,dx$.

2   **a**   Differentiate $(2x^2 - 1)^5$ with respect to $x$.

    **b**   Hence, find $\displaystyle\int x(2x^2 - 1)^4\,dx$.

3   **a**   Given that $y = \dfrac{1}{x^2 - 5}$, show that $\dfrac{dy}{dx} = \dfrac{kx}{(x^2 - 5)^2}$, and state the value of $k$.

    **b**   Hence, find $\displaystyle\int \dfrac{4x}{(x^2 - 5)^2}\,dx$.

4   **a**   Differentiate $\dfrac{1}{4 - 3x^2}$ with respect to $x$.

    **b**   Hence, find $\displaystyle\int \dfrac{3x}{(4 - 3x^2)^2}\,dx$.

5   **a**   Differentiate $(x^2 - 3x + 5)^6$ with respect to $x$.

    **b**   Hence, find $\displaystyle\int 2(2x - 3)(x^2 - 3x + 5)^5\,dx$.

6   **a**   Differentiate $(\sqrt{x} + 3)^8$ with respect to $x$.

    **b**   Hence, find $\displaystyle\int \dfrac{(\sqrt{x} + 3)^7}{\sqrt{x}}\,dx$.

**7**  **a**  Differentiate $(2x\sqrt{x} - 1)^5$ with respect to $x$.

    **b**  Hence, find $\int 3\sqrt{x}(2x\sqrt{x} - 1)^4 \, \mathrm{d}x$.

## 9.5 Definite integration

In the remaining sections of this chapter, you will be learning how to find areas and volumes of various shapes. To do this, you will be using a technique known as definite integration, which is an extension of the indefinite integrals you have been using up to now. In this section, you will learn this technique, before going on to apply it in the next section.

Recall that

$$\int x^3 \, \mathrm{d}x = \frac{1}{4}x^4 + c,$$

where $c$ is an arbitrary constant, is called the indefinite integral of $x^3$ with respect to $x$.

We can integrate a function between two specified limits.

We write the integral of the function $x^3$ with respect to $x$ between the limits $x = 2$ and $x = 4$ as:

$$\int_2^4 x^3 \, \mathrm{d}x$$

The method for evaluating this integral is:

$$\int_2^4 x^3 \, \mathrm{d}x = \left[ \frac{1}{4}x^4 + c \right]_2^4$$

$$= \left( \frac{1}{4} \times 4^4 + c \right) - \left( \frac{1}{4} \times 2^4 + c \right)$$

$$= 60$$

> **TIP**
>
> The limits of integration are always written either to the right of the integral sign, as printed, or directly below and above it. They should never be written to the left of the integral sign.

Note that the '$c$'s cancel out, so the process can be simplified to:

$$\int_2^4 x^3 \, \mathrm{d}x = \left[ \frac{1}{4}x^4 \right]_2^4$$

$$= \left( \frac{1}{4} \times 4^4 \right) - \left( \frac{1}{4} \times 2^4 \right)$$

$$= 60$$

$\int_2^4 x^3 \, \mathrm{d}x$ is called the **definite integral** of $x^3$ with respect to $x$ between the limits 2 and 4.

Hence, we can write the evaluation of a definite integral as:

> **KEY POINT 9.9**
>
> $$\int_a^b f(x) \, \mathrm{d}x = \left[ F(x) \right]_a^b = F(b) - F(a)$$

The following rules for definite integrals may also be used.

### KEY POINT 9.10

$$\int_a^b k\text{f}(x)\,\mathrm{d}x = k\int_a^b \text{f}(x)\,\mathrm{d}x, \text{ where } k \text{ is a constant}$$

$$\int_a^b \left[\text{f}(x) \pm \text{g}(x)\right]\mathrm{d}x = \int_a^b \text{f}(x)\,\mathrm{d}x \pm \int_a^b \text{g}(x)\,\mathrm{d}x$$

$$\int_a^b \text{f}(x)\,\mathrm{d}x = -\int_b^a \text{f}(x)\,\mathrm{d}x$$

### KEY POINT 9.11

$$\int_a^b \text{f}(x)\,\mathrm{d}x + \int_b^c \text{f}(x)\,\mathrm{d}x = \int_a^c \text{f}(x)\,\mathrm{d}x$$

### WORKED EXAMPLE 9.9

Evaluate:

a $\displaystyle\int_1^2 \frac{6x^4 - 1}{x^2}\,\mathrm{d}x$
b $\displaystyle\int_0^3 \sqrt{5x + 1}\,\mathrm{d}x$
c $\displaystyle\int_{-2}^1 \frac{8}{\left(5 - 2x\right)^2}\,\mathrm{d}x$

**Answer**

a $\displaystyle\int_1^2 \frac{6x^4 - 1}{x^2}\,\mathrm{d}x = \int_1^2 \left(6x^2 - x^{-2}\right)\mathrm{d}x$

$$= \left[\frac{6}{3}x^3 + x^{-1}\right]_1^2$$

$$= \left(2(2)^3 + (2)^{-1}\right) - \left(2(1)^3 + (1)^{-1}\right)$$

$$= \left(16 + \frac{1}{2}\right) - (2 + 1)$$

$$= 13\frac{1}{2}$$

b $\displaystyle\int_0^3 \sqrt{5x + 1}\,\mathrm{d}x = \int_0^3 (5x + 1)^{\frac{1}{2}}\,\mathrm{d}x$

$$= \left[\frac{1}{(5)\left(\frac{3}{2}\right)}(5x + 1)^{\frac{3}{2}}\right]_0^3$$

$$= \left[\frac{2}{15}(5x + 1)^{\frac{3}{2}}\right]_0^3$$

$$= \left(\frac{2}{15} \times 16^{\frac{3}{2}}\right) - \left(\frac{2}{15} \times 1^{\frac{3}{2}}\right)$$

$$= \left(\frac{128}{15}\right) - \left(\frac{2}{15}\right)$$

$$= 8\frac{2}{5}$$

c $\displaystyle\int_{-2}^{1} \frac{8}{(5-2x)^2}\,dx = \int_{-2}^{1} 8(5-2x)^{-2}\,dx$

$\displaystyle = \left[ \frac{8}{(-2)(-1)}(5-2x)^{-1} \right]_{-2}^{1}$

$\displaystyle = \left[ \frac{4}{5-2x} \right]_{-2}^{1}$

$\displaystyle = \left( \frac{4}{3} \right) - \left( \frac{4}{9} \right)$

$\displaystyle = \frac{8}{9}$

## EXERCISE 9E

1  Evaluate:

a $\displaystyle\int_{1}^{2} 3x^2\,dx$

b $\displaystyle\int_{1}^{3} \frac{4}{x^3}\,dx$

c $\displaystyle\int_{-1}^{1} (2x-3)\,dx$

d $\displaystyle\int_{0}^{3} \left(10-x^2\right)dx$

e $\displaystyle\int_{-1}^{2} \left(4x^2-2x\right)dx$

f $\displaystyle\int_{2}^{4} \left(2-\frac{6}{x^2}\right)dx$

2  Evaluate:

a $\displaystyle\int_{1}^{2} \left(3x^2-2+\frac{1}{x^2}\right)dx$

b $\displaystyle\int_{-2}^{-1} \left(\frac{8-x^2}{x^2}\right)dx$

c $\displaystyle\int_{1}^{2} (x+3)(7-2x)\,dx$

d $\displaystyle\int_{0}^{1} \sqrt{x}(1-x)\,dx$

e $\displaystyle\int_{1}^{2} \frac{(3-x)(8+x)}{x^4}\,dx$

f $\displaystyle\int_{1}^{4} \left(3\sqrt{x}+\frac{2}{\sqrt{x}}\right)dx$

3  Evaluate:

a $\displaystyle\int_{-1}^{0} (2x+3)^3\,dx$

b $\displaystyle\int_{0}^{4} \sqrt{2x+1}\,dx$

c $\displaystyle\int_{1}^{2} \sqrt{(x-1)^3}\,dx$

d $\displaystyle\int_{-1}^{1} \frac{6}{(x-2)^2}\,dx$

e $\displaystyle\int_{2}^{3} \frac{9}{(2x-3)^3}\,dx$

f $\displaystyle\int_{-2}^{2} \frac{4}{\sqrt{5-2x}}\,dx$

4  a  Given that $y = \dfrac{2}{x^2+5}$, find $\dfrac{dy}{dx}$.

   b  Hence, evaluate $\displaystyle\int_{0}^{2} \frac{2x}{\left(x^2+5\right)^2}\,dx$.

5  a  Given that $y = \left(x^3-2\right)^5$, find $\dfrac{dy}{dx}$.

   b  Hence, evaluate $\displaystyle\int_{0}^{1} x^2\left(x^3-2\right)^4\,dx$.

6  a  Given that $y = \dfrac{\left(\sqrt{x}+1\right)^5}{10}$, find $\dfrac{dy}{dx}$.

   b  Hence, evaluate $\displaystyle\int_{1}^{4} \frac{\left(\sqrt{x}+1\right)^4}{\sqrt{x}}\,dx$.

## 9.6 Area under a curve

Consider the area bounded by the curve $y = x^2$, the $x$-axis and the lines $x = 2$ and $x = 5$.

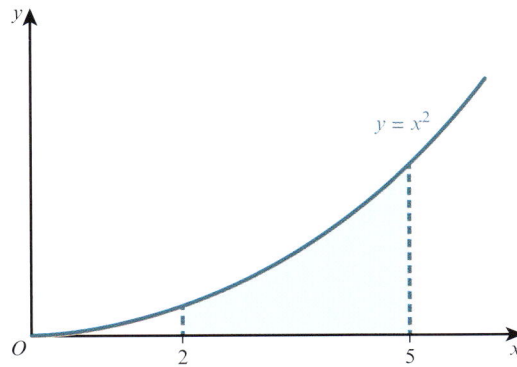

The area, $A$, of the region can be approximated by a series of rectangular strips of thickness $\delta x$ (corresponding to a small increase in $x$) and height $y$ (corresponding to the height of the function).

The approximation for $A$ is then $\sum y \delta x$.

As $\delta x \to 0$, then $A \to \displaystyle\int_2^5 y \, dx$.

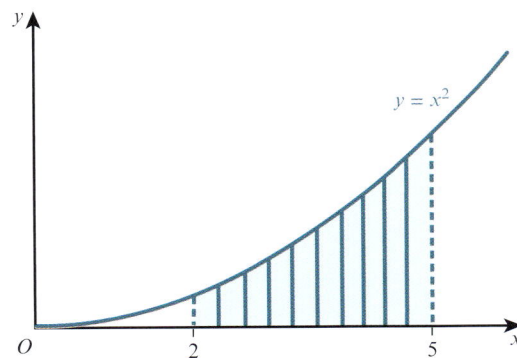

This leads to the general rule:

### KEY POINT 9.12

If $y = f(x)$ is a function with $y \geqslant 0$, then the area, $A$, bounded by the curve $y = f(x)$, the $x$-axis and the lines $x = a$ and $x = b$ is given by the formula $A = \displaystyle\int_a^b y \, dx$.

253

**WORKED EXAMPLE 9.10**

Find the area of the shaded region.

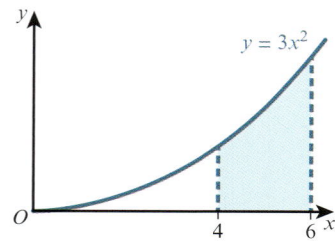

**Answer**

$$\text{Area} = \int_4^6 3x^2 \, dx$$

$$= \left[ \frac{3}{3} x^3 \right]_4^6$$

$$= (216) - (64)$$

$$= 152 \, \text{units}^2$$

In Worked example 9.10, the required area is above the $x$-axis.

If the required area lies below the $x$-axis, then $\int_a^b f(x) \, dx$ will have a negative value. This is because the integral is summing the $y$ values, and these are all negative.

**WORKED EXAMPLE 9.11**

Find the area of the shaded region.

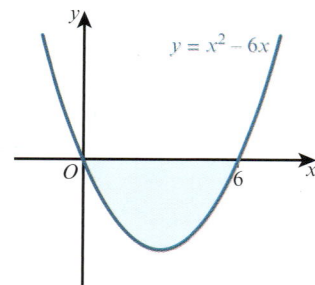

**Answer**

$$\int_0^6 \left( x^2 - 6x \right) dx = \left[ \frac{1}{3} x^3 - \frac{6}{2} x^2 \right]_0^6$$

$$= (72 - 108) - (0 - 0)$$

$$= -36$$

Area is $36 \, \text{units}^2$.

The required region could consist of a section above the $x$-axis and a section below the $x$-axis.

If this happens we must evaluate each area separately.

This is illustrated in Worked example 9.12.

**WORKED EXAMPLE 9.12**

Find the total area of the shaded regions.

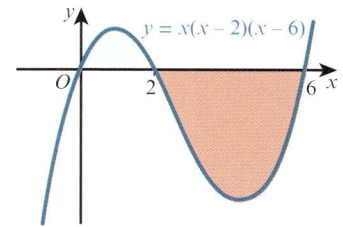

**Answer**

$$\int_0^2 x(x-2)(x-6)\,dx = \int_0^2 \left(x^3 - 8x^2 + 12x\right)dx$$

$$= \left[\frac{1}{4}x^4 - \frac{8}{3}x^3 + 6x^2\right]_0^2$$

$$= \left(\frac{1}{4}(2)^4 - \frac{8}{3}(2)^3 + 6(2)^2\right) - \left(\frac{1}{4}(0)^4 - \frac{8}{3}(0)^3 + 6(0)^2\right)$$

$$= \left(6\frac{2}{3}\right) - (0)$$

$$= 6\frac{2}{3}$$

$$\int_2^6 x(x-2)(x-6)\,dx = \int_2^6 \left(x^3 - 8x^2 + 12x\right)dx$$

$$= \left[\frac{1}{4}x^4 - \frac{8}{3}x^3 + 6x^2\right]_2^6$$

$$= \left(\frac{1}{4}(6)^4 - \frac{8}{3}(6)^3 + 6(6)^2\right) - \left(\frac{1}{4}(2)^4 - \frac{8}{3}(2)^3 + 6(2)^2\right)$$

$$= (-36) - \left(6\frac{2}{3}\right)$$

$$= -42\frac{2}{3}$$

Hence, the total area of the shaded regions $= 6\frac{2}{3} + 42\frac{2}{3} = 49\frac{1}{3}$ units$^2$.

## Area enclosed by a curve and the $y$-axis

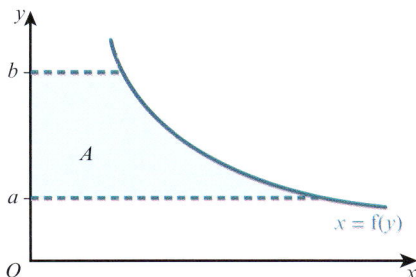

---

🔍 **KEY POINT 9.13**

If $x = f(y)$ is a function with $x \geq 0$, then the area, $A$, bounded by the curve $x = f(y)$, the $y$-axis and the lines $y = a$ and $y = b$ is given by the formula $A = \displaystyle\int_a^b x\,dy$ when $x \geq 0$.

---

**WORKED EXAMPLE 9.13**

Find the area of the shaded region.

**Answer**

$$\text{Area} = \int_0^4 x\,dy$$

$$= \int_0^4 \left(4y - y^2\right) dy$$

$$= \left[\frac{4}{2}y^2 - \frac{1}{3}y^3\right]_0^4$$

$$= \left(2(4)^2 - \frac{1}{3}(4)^3\right) - \left(2(0)^2 - \frac{1}{3}(0)^3\right)$$

$$= 10\tfrac{2}{3}$$

Area is $10\tfrac{2}{3}$ units$^2$.

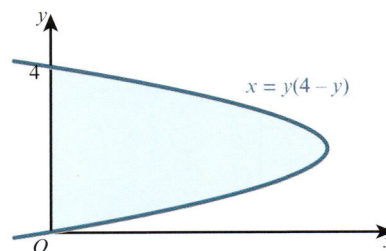

$x = y(4 - y)$

---

256

---

1   Find the area of each shaded region.

**a**

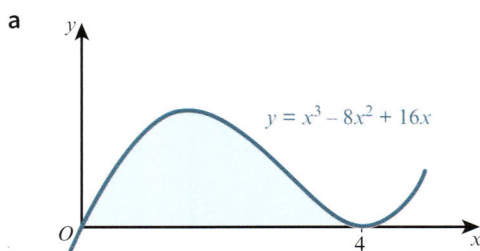

$y = x^3 - 8x^2 + 16x$

**b**

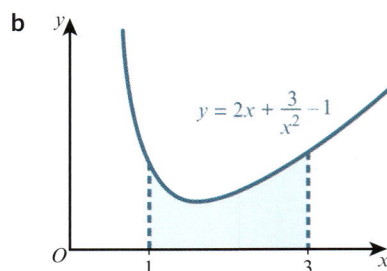

$y = 2x + \dfrac{3}{x^2} - 1$

**c**

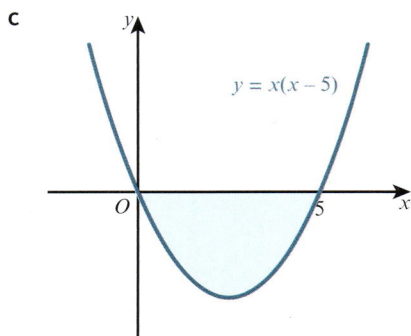

$y = x(x - 5)$

**d**

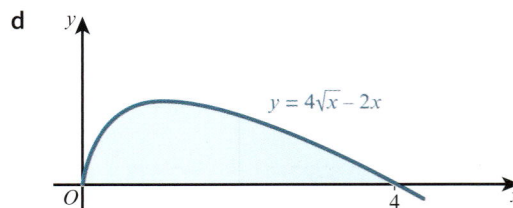

$y = 4\sqrt{x} - 2x$

**2**

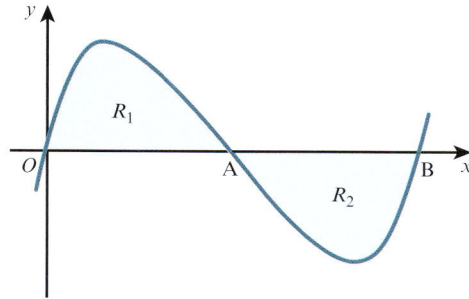

The diagram shows the curve $y = x(x - 2)(x - 4)$ that crosses the $x$-axis at the points $O(0, 0)$, $A(2, 0)$ and $B(4, 0)$.

Show by integration that the area of the shaded region $R_1$ is the same as the area of the shaded region $R_2$.

**3** Sketch the curve and find the total area bounded by the curve and the $x$-axis for each of these functions.

    **a**   $y = x(x - 3)(x + 1)$              **b**   $y = x\left(x^2 - 9\right)$

    **c**   $y = x(2x - 1)(x + 2)$         **d**   $y = (x - 1)(x + 1)(x - 4)$

**4** Sketch the curve and find the enclosed area for each of the following.

    **a**   $y = x^4 - 6x^2 + 9$, the $x$-axis and the lines $x = 0$ and $x = 1$

    **b**   $y = 2x + \dfrac{5}{x^2}$, the $x$-axis and the lines $x = 1$ and $x = 2$

    **c**   $y = 5 + \dfrac{8}{x^3}$, the $x$-axis and the lines $x = 2$ and $x = 5$

    **d**   $y = 3\sqrt{x}$, the $x$-axis and the lines $x = 1$ and $x = 4$

    **e**   $y = \dfrac{4}{\sqrt{x}}$, the $x$-axis and the lines $x = 1$ and $x = 9$

    **f**   $y = \sqrt{2x + 3}$, the $x$-axis and the line $x = 3$

**5** Sketch the curve and find the enclosed area for each of the following.

    **a**   $y = x^3$, the $y$-axis and the lines $y = 8$ and $y = 27$

    **b**   $x = y^2 + 1$, the $y$-axis and the lines $y = -1$ and $y = 2$

**6**

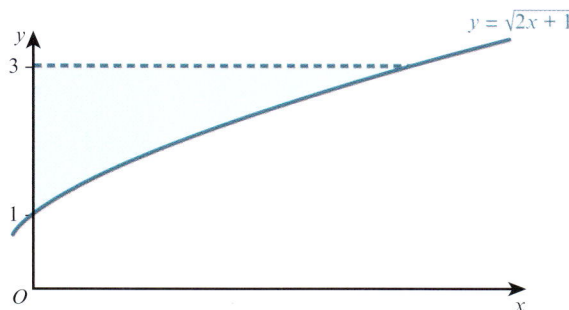

The diagram shows the curve $y = \sqrt{2x + 1}$. The shaded region is bounded by the curve, the $y$-axis and the line $y = 3$. Find the area of the shaded region.

7

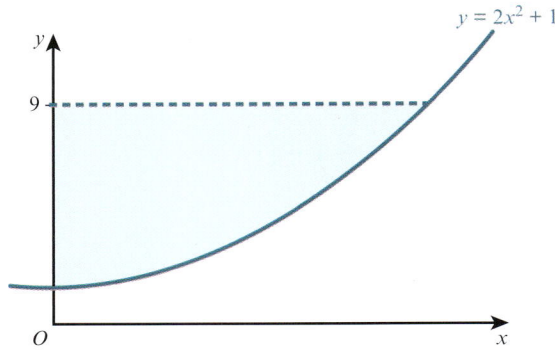

Find the area of the region bounded by the curve $y = 2x^2 + 1$, the line $y = 9$ and the $y$-axis.

8  **a**  Find the area of the region enclosed by the curve $y = \dfrac{12}{x^2}$, the $x$-axis and the lines $x = 1$ and $x = 4$.

   **b**  The line $x = p$ divides the region in part **a** into two parts of equal area. Find the value of $p$.

9

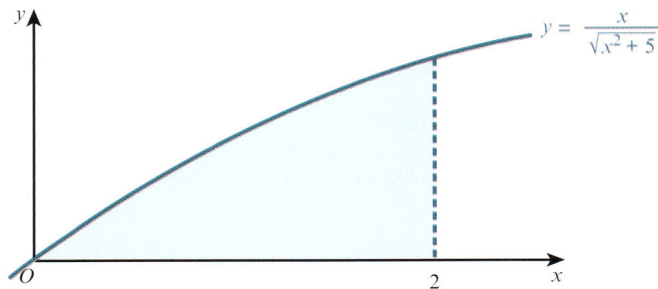

   **a**  Show that $\dfrac{\mathrm{d}}{\mathrm{d}x}\left(\sqrt{x^2 + 5}\right) = \dfrac{x}{\sqrt{x^2 + 5}}$.

   **b**  Use your result from part **a** to evaluate the area of the shaded region.

10

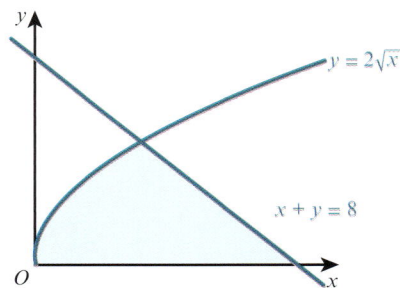

Find the shaded area enclosed by the curve $y = 2\sqrt{x}$, the line $x + y = 8$ and the $x$-axis.

11  The tangent to the curve $y = 8x - x^2$ at the point $(2, 12)$ cuts the $x$-axis at the point $P$.

   **a**  Find the coordinates of $P$.

   **b**  Find the area of the shaded region.

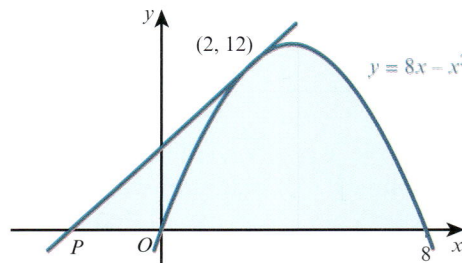

**12**

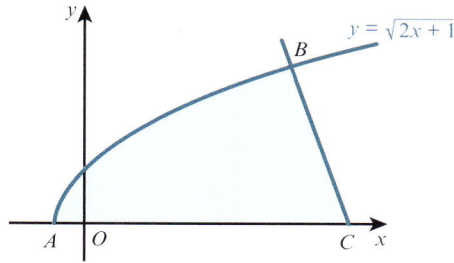

The diagram shows the curve $y = \sqrt{2x + 1}$ that intersects the $x$-axis at $A$. The normal to the curve at $B(4, 3)$ meets the $x$-axis at $C$. Find the area of the shaded region.

**PS** **13**

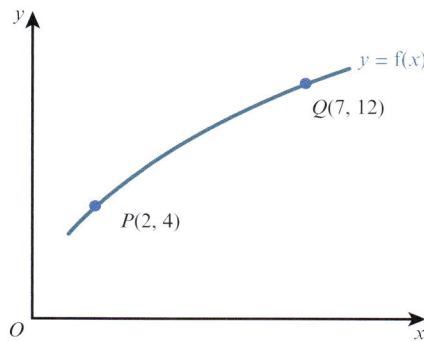

The figure shows part of the curve $y = \mathrm{f}(x)$. The points $P(2, 4)$ and $Q(7, 12)$ lie on the curve. Given that $\displaystyle\int_2^7 y \, \mathrm{d}x = 42$, find the value of $\displaystyle\int_4^{12} x \, \mathrm{d}y$.

**PS** **14**

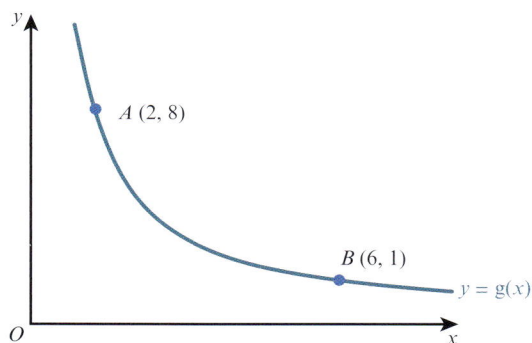

The figure shows part of the curve $y = \mathrm{g}(x)$. The points $A(2, 8)$ and $B(6, 1)$ lie on the curve. Given that $\displaystyle\int_2^6 y \, \mathrm{d}x = 16$, find the value of $\displaystyle\int_1^8 x \, \mathrm{d}y$.

259

**WEB**

Try the following resources on the Underground Mathematics website:

- *What else do you know?*
- *Slippery areas.*

## 9.7 Area bounded by a curve and a line or by two curves

The following example shows a possible method for finding the area enclosed by a curve and a straight line.

**WORKED EXAMPLE 9.14**

The diagram shows the curve $y = -x^2 + 8x - 5$ and the line $y = x + 1$ that intersect at the points $(1, 2)$ and $(6, 7)$.

Find the area of the shaded region.

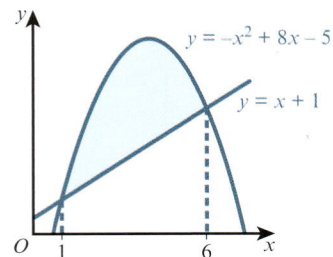

**Answer**

Area = area under curve − area of trapezium

$$= \int_1^6 \left(-x^2 + 8x - 5\right) dx - \frac{1}{2} \times (2 + 7) \times 5$$

$$= \left[-\frac{1}{3} x^3 + 4x^2 - 5x\right]_1^6 - 22\frac{1}{2}$$

$$= \left(-\frac{1}{3}(6)^3 + 4(6)^2 - 5(6)\right) - \left(-\frac{1}{3}(1)^3 + 4(1)^2 - 5(1)\right) - 22\frac{1}{2}$$

$$= 20\frac{5}{6} \text{ units}^2$$

There is an alternative method for finding the shaded area in Worked example 9.14.

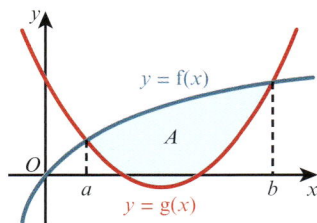

If two functions, $f(x)$ and $g(x)$, intersect at $x = a$ and $x = b$, then the area, $A$, enclosed between the two curves is given by:

**KEY POINT 9.14**

$$A = \int_a^b f(x)\, dx - \int_a^b g(x)\, dx \qquad \text{or} \qquad A = \int_a^b \left[f(x) - g(x)\right] dx$$

So for the area enclosed by $y = -x^2 + 8x - 5$ and $y = x + 1$:

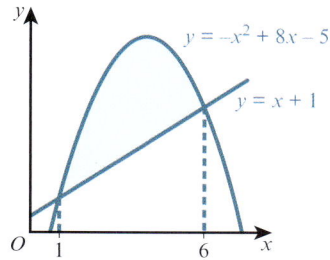

Using $f(x) = -x^2 + 8x - 5$ and $g(x) = x + 1$ gives:

$$\text{Area} = \int_1^6 f(x)\,dx - \int_1^6 g(x)\,dx$$

$$= \int_1^6 \left(-x^2 + 8x - 5\right)dx - \int_1^6 (x + 1)\,dx$$

$$= \int_1^6 \left(-x^2 + 7x - 6\right)dx$$

$$= \left[-\frac{1}{3}x^3 + \frac{7}{2}x^2 - 6x\right]_1^6$$

$$= \left(-\frac{1}{3}(6)^3 + \frac{7}{2}(6)^2 - 6(6)\right) - \left(-\frac{1}{3}(1)^3 + \frac{7}{2}(1)^2 - 6(1)\right)$$

$$= 20\tfrac{5}{6} \text{ units}^2$$

This alternative method is the easiest method to use in the next example.

**WORKED EXAMPLE 9.15**

The diagram shows the curve $y = x^2 - 6x - 2$ and the line $y = 2x - 9$, which intersect when $x = 1$ and $x = 7$.

Find the area of the shaded region.

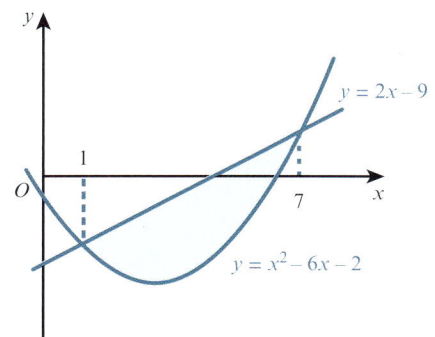

**Answer**

$$\text{Area} = \int_1^7 (2x - 9)\,dx - \int_1^7 \left(x^2 - 6x - 2\right)dx$$

$$= \int_1^7 \left(-x^2 + 8x - 7\right)dx$$

$$= \left[-\frac{1}{3}x^3 + 4x^2 - 7x\right]_1^7$$

$$= \left(-\frac{1}{3}(7)^3 + 4(7)^2 - 7(7)\right) - \left(-\frac{1}{3}(1)^3 + 4(1)^2 - 7(1)\right)$$

$$= 36 \text{ units}^2$$

**1**

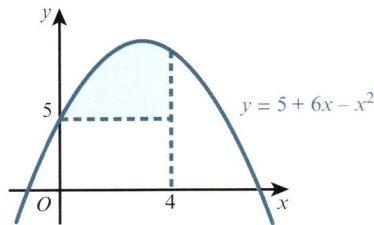

Find the area of the region bounded by the curve $y = 5 + 6x - x^2$, the line $x = 4$ and the line $y = 5$.

**2**

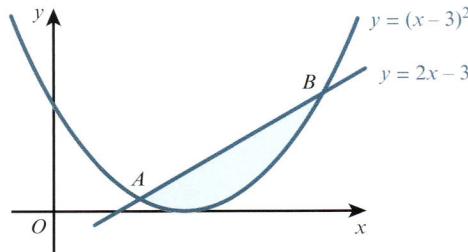

The diagram shows the curve $y = (x - 3)^2$ and the line $y = 2x - 3$ that intersect at points $A$ and $B$. Find the area of the shaded region.

**3**

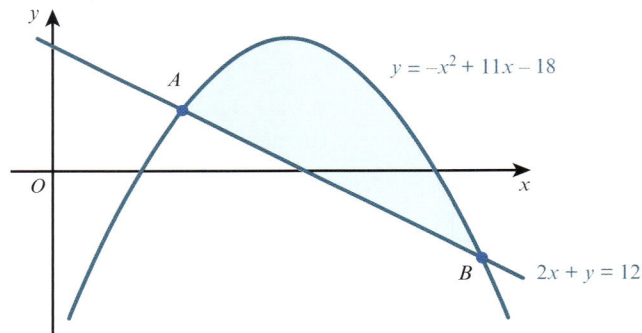

The diagram shows the curve $y = -x^2 + 11x - 18$ and the line $2x + y = 12$. Find the area of the shaded region.

**4** Sketch the following curves and lines and find the area enclosed between their graphs.

   **a** $y = x^2 - 3$ and $y = 6$

   **b** $y = -x^2 + 12x - 20$ and $y = 2x + 1$

   **c** $y = x^2 - 4x + 4$ and $2x + y = 12$

**5** Sketch the curves $y = x^2$ and $y = x(2 - x)$ and find the area enclosed between the two curves.

**6**

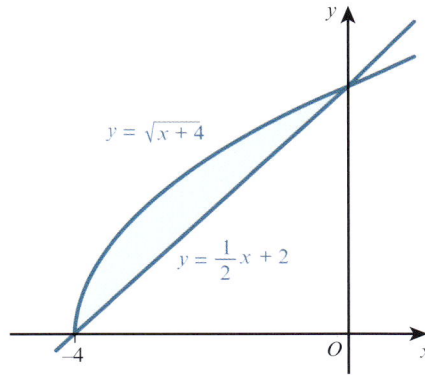

The diagram shows the curve $y = \sqrt{x+4}$ and the line $y = \dfrac{1}{2}x + 2$ meeting at the points $(-4, 0)$ and $(0, 2)$.

Find the area of the shaded region.

**7**

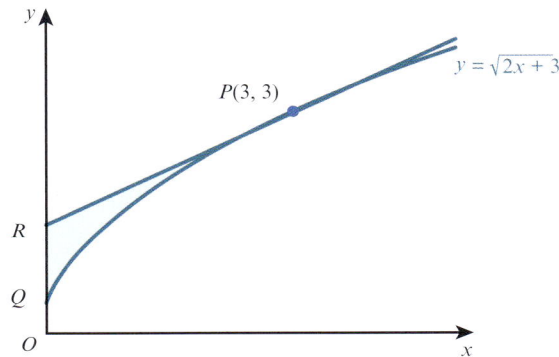

The curve $y = \sqrt{2x+3}$ meets the $y$-axis at the point $Q$.

The tangent at the point $P(3, 3)$ to this curve meets the $y$-axis at the point $R$.

**a**  Find the equation of the tangent to the curve at $P$.

**b**  Find the exact value of the area of the shaded region $PQR$.

**8**

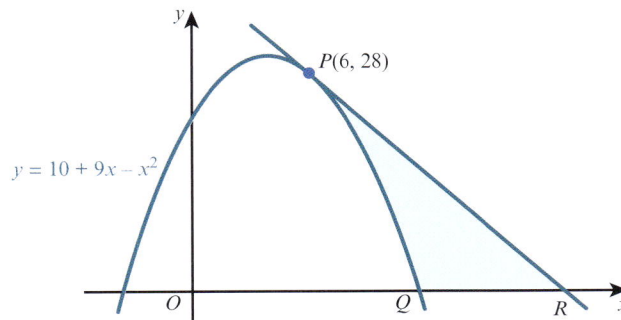

The diagram shows the curve $y = 10 + 9x - x^2$. Points $P(6, 28)$ and $Q(10, 0)$ lie on the curve. The tangent at $P$ intersects the $x$-axis at $R$.

**a**  Find the equation of the tangent to the curve at $P$.

**b**  Find the area of the shaded region.

**9**

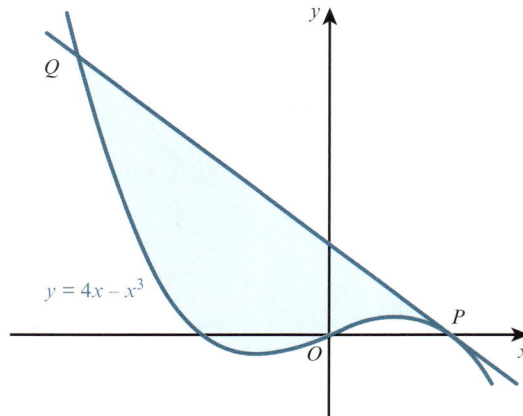

The diagram shows the curve $y = 4x - x^3$.

The point $P$ has coordinates $(2, 0)$ and the point $Q$ has coordinates $(-4, 48)$.

a   Find the equation of the tangent to the curve at $P$.

b   Find the area of the shaded region.

**10**

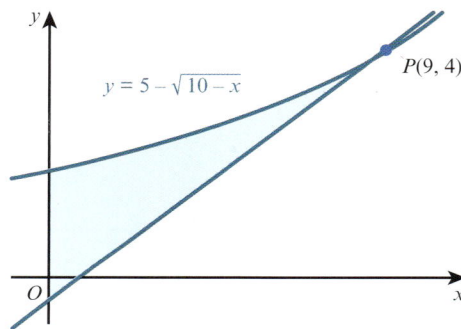

The diagram shows part of the curve $y = 5 - \sqrt{10 - x}$ and the tangent to the curve at $P(9, 4)$.

a   Find the equation of the tangent to the curve at $P$.

b   Find the area of the shaded region. Give your answer correct to 3 significant figures.

## 9.8 Improper integrals

In this section, we will consider what happens if some part of a definite integral becomes infinite. These are known as **improper integrals**, and we will look at two different types.

### Type 1

These are definite integrals that have either one limit infinite or both limits infinite.

Examples of these are $\displaystyle\int_1^\infty \frac{1}{x^2}\,\mathrm{d}x$ and $\displaystyle\int_{-\infty}^{-2} \frac{1}{x^3}\,\mathrm{d}x.$

> **KEY POINT 9.15**
>
> We can evaluate integrals of the form $\displaystyle\int_a^\infty \mathrm{f}(x)\,\mathrm{d}x$ by replacing the infinite limit with a finite
>
> value, $X$, and then taking the limit as $X \to \infty$, provided the limit exists.

We can evaluate integrals of the form $\displaystyle\int_{-\infty}^{b} f(x)\,dx$ by replacing the infinite limit with a finite value, $X$, and then taking the limit as $X \to -\infty$, provided the limit exists.

**WORKED EXAMPLE 9.16**

Show that the improper integral $\displaystyle\int_{1}^{\infty} \frac{1}{x^2}\,dx$ has a finite value and find this value.

**Answer**

$$\int_{1}^{X} \frac{1}{x^2}\,dx = \int_{1}^{X} x^{-2}\,dx \qquad\qquad\qquad \text{Write the integral with an upper limit } X.$$

$$= \left[ -x^{-1} \right]_{1}^{X}$$

$$= \left( -\frac{1}{X} \right) - \left( -\frac{1}{1} \right)$$

$$= 1 - \frac{1}{X}$$

As $X \to \infty$, $\dfrac{1}{X} \to 0$

$$\therefore \int_{1}^{\infty} \frac{1}{x^2}\,dx = 1 - 0 = 1$$

Hence, the improper integral $\displaystyle\int_{1}^{\infty} \frac{1}{x^2}\,dx$ has a finite value of 1.

265

**WORKED EXAMPLE 9.17**

The diagram shows part of the curve $y = \dfrac{1}{(1-x)^3}$.

Show that as $p \to -\infty$, the shaded area tends to a finite value and find this value.

**Answer**

$$\text{Area} = \int_{p}^{0} \frac{1}{(1-x)^3}\,dx$$

$$= \int_{p}^{0} (1-x)^{-3}\,dx$$

$$= \left[ \frac{1}{(-2)(-1)} (1-x)^{-2} \right]_{p}^{0}$$

$$= \left[ \frac{1}{2(1-x)^2} \right]_{p}^{0}$$

$$= \frac{1}{2} - \frac{1}{2(1-p)^2}$$

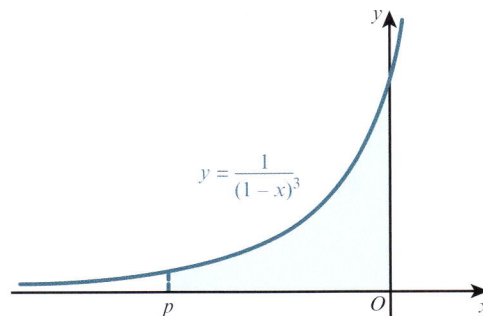

As $p \to -\infty$, $\dfrac{1}{2(1-p)^2} \to 0$.

Hence, as $p \to -\infty$, the shaded area tends to a finite value of $\dfrac{1}{2}$.

## Type 2

These are integrals where the function to be integrated approaches an infinite value (or approaches ± infinity) at either or both end points in the interval (of integration).

For example, $\displaystyle\int_{-1}^{1} \dfrac{1}{x^2}\, dx$ is an invalid integral because $\dfrac{1}{x^2}$ is not defined when $x = 0$.

However, $\displaystyle\int_{0}^{1} \dfrac{1}{x^2}\, dx$ is an improper integral because $\dfrac{1}{x^2}$ tends to infinity as $x \to 0$ and it is well-defined everywhere else in the interval of integration.

For this section we will consider only those improper integrals where the function is not defined at one end of the interval.

> ### KEY POINT 9.17
>
> We can evaluate integrals of the form $\displaystyle\int_{a}^{b} f(x)\, dx$ where $f(x)$ is not defined when $x = a$ can be evaluated by replacing the limit $a$ with an $X$ and then taking the limit as $X \to a$, provided the limit exists.

> ### KEY POINT 9.18
>
> We can evaluate integrals of the form $\displaystyle\int_{a}^{b} f(x)\, dx$ where $f(x)$ is not defined when $x = b$ by replacing the limit $b$ with an $X$ and then taking the limit as $X \to b$, provided the limit exists.

### WORKED EXAMPLE 9.18

Find the value, if it exists, of $\displaystyle\int_{0}^{2} \dfrac{5}{x^2}\, dx$.

**Answer**

The function $f(x) = \dfrac{5}{x^2}$ is not defined when $x = 0$.

$$\int_{X}^{2} \dfrac{5}{x^2}\, dx = \int_{X}^{2} 5x^{-2}\, dx \qquad\qquad \text{Write the integral with a lower limit } X.$$

$$= \left[ -5x^{-1} \right]_{X}^{2}$$

$$= \left( -\dfrac{5}{2} \right) - \left( -\dfrac{5}{X} \right)$$

$$= \dfrac{5}{X} - \dfrac{5}{2}$$

As $X \to 0$, $\dfrac{5}{X}$ tends to infinity.

Hence, $\displaystyle\int_0^2 \dfrac{5}{x^2}\,dx$ is undefined.

**WORKED EXAMPLE 9.19**

The diagram shows part of the curve $y = \dfrac{3}{\sqrt{2-x}}$.

Show that as $p \to 2$ the shaded area tends to a finite value and find this value.

$y = \dfrac{3}{\sqrt{2-x}}$

**Answer**

$$\text{Area} = \int_0^p \frac{3}{\sqrt{2-x}}\,dx$$

$$= \int_0^p 3(2-x)^{-\frac{1}{2}}\,dx$$

$$= \left[ \frac{3}{\left(\frac{1}{2}\right)(-1)}(2-x)^{\frac{1}{2}} \right]_0^p$$

$$= \left[ -6\sqrt{2-x} \right]_0^p$$

$$= \left(-6\sqrt{2-p}\right) - \left(-6\sqrt{2}\right)$$

$$= 6\sqrt{2} - 6\sqrt{2-p}$$

As $p \to 2$, $\displaystyle\int_0^p \dfrac{3}{\sqrt{2-x}}\,dx \to 6\sqrt{2}$.

Hence, as $p \to 2$ the shaded area tends to a finite value of $6\sqrt{2}$.

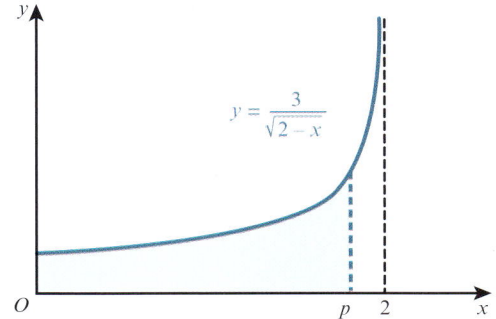

267

**EXERCISE 9H**

1  Show that each of the following improper integrals has a finite value and, in each case, find this value.

a  $\displaystyle\int_1^\infty \frac{2}{x^2}\,dx$

b  $\displaystyle\int_4^\infty \frac{4}{x^5}\,dx$

c  $\displaystyle\int_{-\infty}^{-2} \frac{10}{x^3}\,dx$

d  $\displaystyle\int_4^\infty \frac{4}{x\sqrt{x}}\,dx$

e  $\displaystyle\int_0^{25} \frac{5}{\sqrt{x}}\,dx$

f  $\displaystyle\int_4^8 \frac{4}{\sqrt{x-4}}\,dx$

g  $\displaystyle\int_0^3 \frac{3}{\sqrt{3-x}}\,dx$

h  $\displaystyle\int_2^\infty \frac{1}{(1-x)^2}\,dx$

i  $\displaystyle\int_1^\infty \left( \frac{2}{x^2} + \frac{4}{(x+2)^3} \right)dx$

2

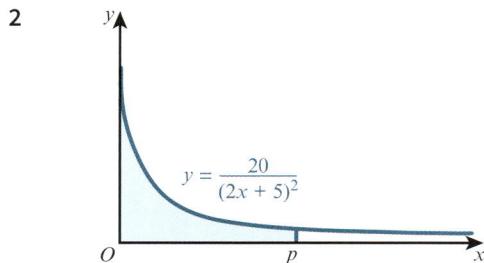

The diagram shows part of the curve $y = \dfrac{20}{(2x+5)^2}$.

Show that as $p \to \infty$, the shaded area tends to the value 2.

3 Show that none of the following improper integrals exists.

a $\displaystyle\int_4^\infty \frac{6}{\sqrt{x}}\,dx$

b $\displaystyle\int_0^\infty \frac{4}{x\sqrt{x}}\,dx$

c $\displaystyle\int_0^9 \frac{12}{x^2\sqrt{x}}\,dx$

d $\displaystyle\int_5^\infty \frac{2}{\sqrt{x+4}}\,dx$

e $\displaystyle\int_{\frac{1}{2}}^2 \frac{5}{(2x-1)^2}\,dx$

f $\displaystyle\int_0^{25} \left(\sqrt{x} + \frac{1}{x^2}\right)dx$

## 9.9 Volumes of revolution

Consider the area bounded by the curve $y = x^2$, the $x$-axis, and the lines $x = 2$ and $x = 5$.

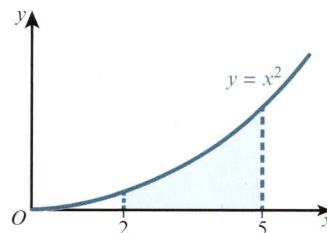

When this area is rotated about the $x$-axis through $360°$ a **solid of revolution** is formed.

The volume of this solid is called a **volume of revolution**.

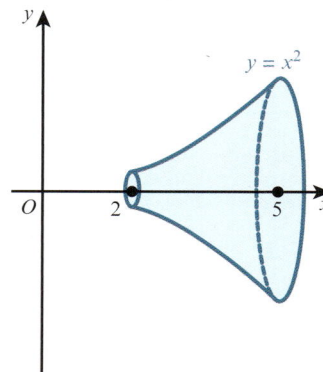

We can approximate the volume, $V$, of the solid by a series of cylindrical discs of thickness $\delta x$ (corresponding to a small increase in $x$) and radius $y$ (corresponding to the height of the function).

268

The volume of each cylindrical disc is $\pi y^2 \delta x$. An approximation for $V$ is then $\sum \pi y^2 \delta x$.

As $\delta x \to 0$, then $V \to \displaystyle\int_2^5 \pi y^2 \, dx$.

This leads to a general formula:

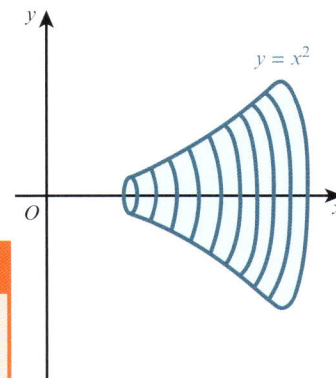

### KEY POINT 9.19

The volume, $V$, obtained when the function $y = f(x)$ is rotated through $360°$ about the $x$-axis between the boundary values $x = a$ and $x = b$ is given by the formula $V = \displaystyle\int_a^b \pi y^2 \, dx$.

### WORKED EXAMPLE 9.20

Find the volume obtained when the shaded region is rotated through $360°$ about the $x$-axis.

**Answer**

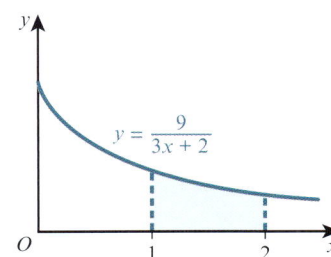

$$\text{Volume} = \pi \int_1^2 y^2 \, dx = \pi \int_1^2 \left( \frac{9}{3x + 2} \right)^2 dx$$

$$= \pi \int_1^2 81(3x + 2)^{-2} \, dx$$

$$= \pi \left[ \frac{81}{3(-1)} (3x + 2)^{-1} \right]_1^2$$

$$= \pi \left[ \frac{-27}{(3x + 2)} \right]_1^2$$

$$= \pi \left[ \left( \frac{-27}{8} \right) - \left( \frac{-27}{5} \right) \right]$$

$$= \frac{81\pi}{40} \ \text{units}^3$$

Sometimes a curve is rotated about the $y$-axis. In this case the general rule is:

### KEY POINT 9.20

The volume, $V$, obtained when the function $x = f(y)$ is rotated through $360°$ about the $y$-axis between the boundary values $y = a$ and $y = b$ is given by the formula $V = \displaystyle\int_a^b \pi x^2 \, dy$.

269

**WORKED EXAMPLE 9.21**

Find the volume obtained when the shaded region is rotated through 360° about the $y$-axis.

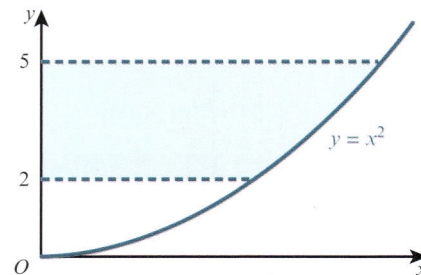

**Answer**

$$\text{Volume} = \pi \int_{2}^{5} x^2 \, dy = \pi \int_{2}^{5} y \, dy \qquad \cdots\cdots\cdots \text{Using the given } y = x^2.$$

$$= \pi \left[ \frac{y^2}{2} \right]_{2}^{5}$$

$$= \pi \left[ \left( \frac{25}{2} \right) - \left( \frac{4}{2} \right) \right]$$

$$= \frac{21\pi}{2} \text{ units}^3$$

**WORKED EXAMPLE 9.22**

Find the volume of the solid obtained when the shaded region is rotated through 360° about the $x$-axis.

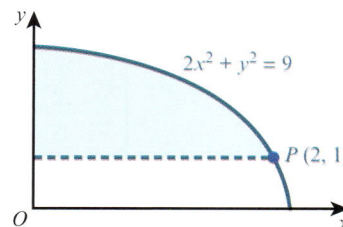

**Answer**

When the shaded region is rotated about the $x$-axis, a solid with a cylindrical hole is formed.

The radius of the cylindrical hole is 1 unit and the length of the hole is 2 units.

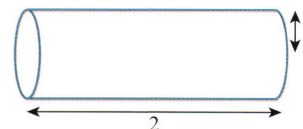

$$\text{Volume of solid} = \pi \int_{0}^{2} y^2 \, dx - \text{volume of cylinder}$$

$$= \pi \int_{0}^{2} \left( 9 - 2x^2 \right) dx - \pi \times r^2 \times h$$

$$= \pi \left[ 9x - \frac{2}{3} x^3 \right]_{0}^{2} - \pi \times 1^2 \times 2$$

$$= \pi \left[ \left( 18 - \frac{16}{3} \right) - (0 - 0) \right] - 2\pi$$

$$= \frac{32\pi}{3} \text{ units}^3$$

1   Find the volume obtained when the shaded region is rotated through 360° about the $x$-axis.

**a**

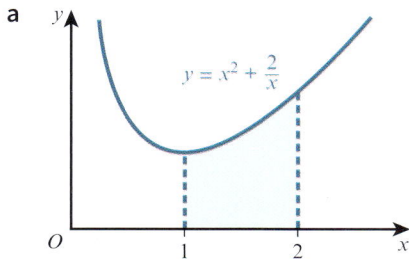

$y = x^2 + \dfrac{2}{x}$

**b**

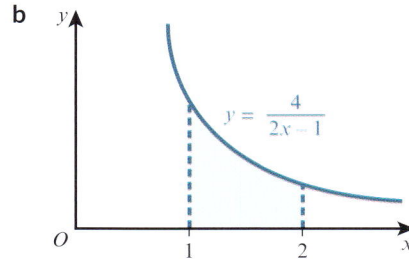

$y = \dfrac{4}{2x - 1}$

**c**

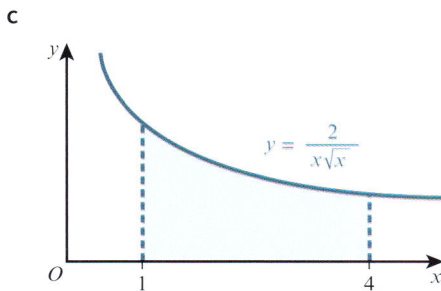

$y = \dfrac{2}{x\sqrt{x}}$

**d**

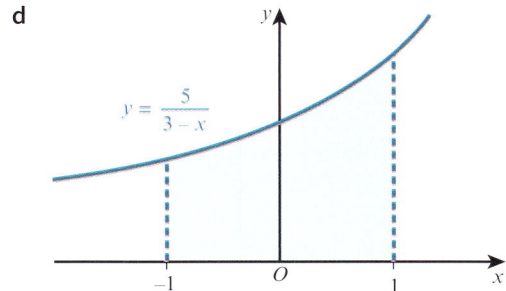

$y = \dfrac{5}{3 - x}$

2   Find the volume obtained when the shaded region is rotated through 360° about the $y$-axis.

**a**

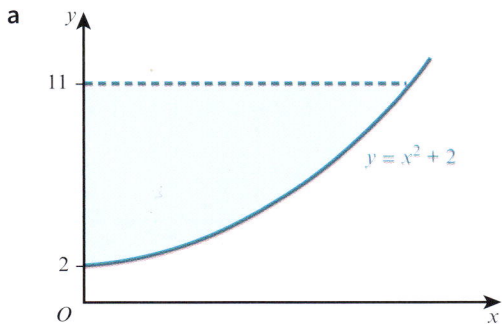

$y = x^2 + 2$

**b**

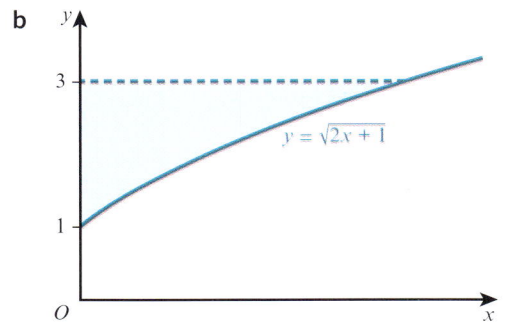

$y = \sqrt{2x + 1}$

3   The diagram shows part of the curve $y = \dfrac{a}{x}$, where $a > 0$. The volume obtained when the shaded region is rotated through 360° about the $x$-axis is $18\pi$. Find the value of $a$.

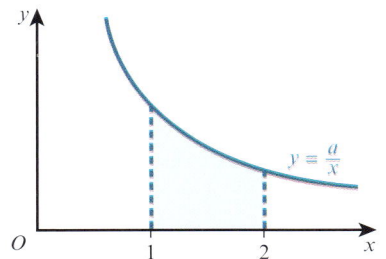

$y = \dfrac{a}{x}$

4   The diagram shows part of the curve $y = \sqrt{x^3 + 4x^2 + 3x + 2}$. Find the volume obtained when the shaded region is rotated through 360° about the $x$-axis.

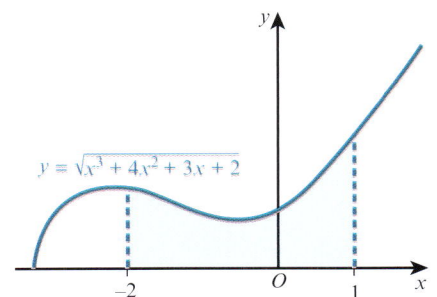

$y = \sqrt{x^3 + 4x^2 + 3x + 2}$

5   The diagram shows part of the line $3x + 8y = 24$. Rotating the shaded region through $360°$ about the $x$-axis would give a cone of base radius 3 and perpendicular height 8.

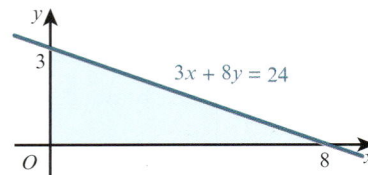

Find the volume of the cone using:

a   integration

b   the formula for the volume of a cone.

6   a   Sketch the graph of $y = (x - 2)^2$.

b   Find the volume of the solid formed when the enclosed region bounded by the curve, the $x$-axis and the $y$-axis is rotated through $360°$ about the $x$-axis.

7   The diagram shows part of the curve $y = 5\sqrt{x} - x$.

The curve meets the $x$-axis at $O$ and $P$.

a   Find the coordinates of $P$.

b   Find the volume obtained when the shaded region is rotated through $360°$ about the $x$-axis.

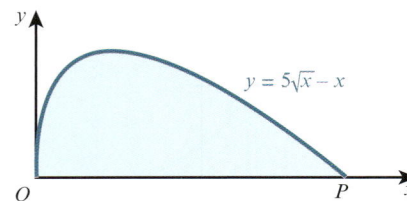

8   The diagram shows part of the curve $x = \dfrac{9}{y^2} - 1$ that intercepts the $y$-axis at the point $P$. The shaded region is bounded by the curve, the $y$-axis and the line $y = 1$.

a   Find the coordinates of $P$.

b   Find the volume obtained when the shaded region is rotated through $360°$ about the $y$-axis.

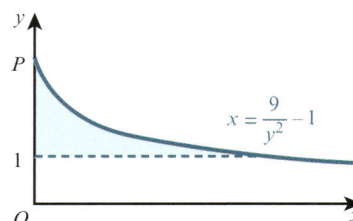

9   The diagram shows part of the curve $y = 3x + \dfrac{2}{x}$.

The line $y = 7$ intersects the curve at the points $P$ and $Q$.

a   Find the coordinates of $P$ and $Q$.

b   Find the volume obtained when the shaded region is rotated through $360°$ about the $x$-axis.

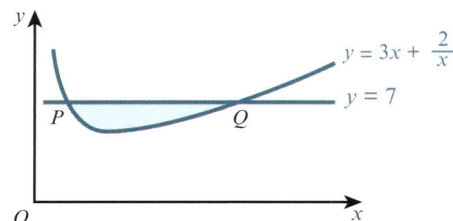

10  The diagram shows part of the curve $y = \dfrac{2}{2x + 1}$. The shaded area is rotated through $360°$ about the $x$-axis between $x = 0$ and $x = p$.

Show that as $p \to \infty$, the volume approaches the value $2\pi$.

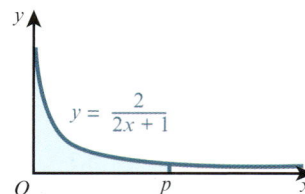

11  The diagram shows part of the curve $y = \sqrt{25 - x^2}$. The point $P(4, 3)$ lies on the curve.

a   Find the volume obtained when the shaded region is rotated through $360°$ about the $y$-axis.

b   Find the volume obtained when the shaded region is rotated through $360°$ about the $x$-axis.

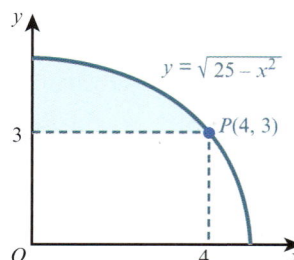

272

**12** The diagram shows the curve $y = \sqrt{4-x}$ and the line $x + 2y = 4$ that intersect at the points $(4, 0)$ and $(0, 2)$.

    **a** Find the volume obtained when the shaded region is rotated through $360°$ about the $x$-axis.

    **b** Find the volume obtained when the shaded region is rotated through $360°$ about the $y$-axis.

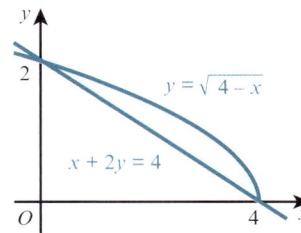

**13** A mathematical model for the inside of a bowl is obtained by rotating the curve $x^2 + y^2 = 100$ through $360°$ about the $y$-axis between $y = -8$ and $y = 0$. Each unit of $x$ and $y$ represents $1\,\text{cm}$.

    **a** Find the volume of the bowl.

    The bowl is filled with water to a depth of $3\,\text{cm}$.

    **b** Find the volume of water in the bowl.

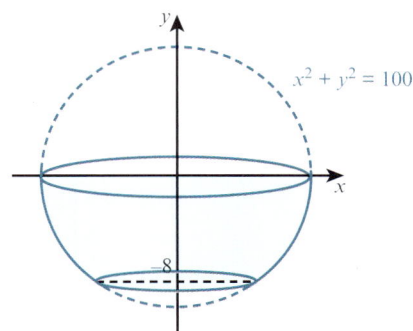

**(P) 14** Use integration to prove that the volume, $V\,\text{cm}^3$, of a sphere with radius $r\,\text{cm}$ is given by the formula $V = \dfrac{4}{3}\pi r^3$.

# Checklist of learning and understanding

**Integration as the reverse of differentiation**

- If $\dfrac{d}{dx}[F(x)] = f(x)$, then $\displaystyle\int f(x)\,dx = F(x) + c$.

**Integration formulae**

- $\displaystyle\int x^n\,dx = \dfrac{1}{n+1}x^{n+1} + c$ (where $c$ is a constant and $n \neq -1$)

- $\displaystyle\int (ax+b)^n\,dx = \dfrac{1}{a(n+1)}(ax+b)^{n+1} + c$ $(n \neq -1$ and $a \neq 0)$

**Rules for indefinite integration**

- $\displaystyle\int k\,f(x)\,dx = k\int f(x)\,dx$, where $k$ is a constant

- $\displaystyle\int [f(x) \pm g(x)]\,dx = \int f(x)\,dx \pm \int g(x)\,dx$

**Rules for definite integration**

- If $\displaystyle\int f(x)\,dx = F(x) + c$, then $\displaystyle\int_a^b f(x)\,dx = \Big[F(x)\Big]_a^b = F(b) - F(a)$.

- $\displaystyle\int_a^b k\,f(x)\,dx = k\int_a^b f(x)\,dx$, where $k$ is a constant.

- $\displaystyle\int_a^b \left[ \mathrm{f}(x) \pm \mathrm{g}(x) \right] \mathrm{d}x = \int_a^b \mathrm{f}(x)\,\mathrm{d}x \ \pm \int_a^b \mathrm{g}(x)\,\mathrm{d}x$

- $\displaystyle\int_a^b \mathrm{f}(x)\,\mathrm{d}x = -\int_b^a \mathrm{f}(x)\,\mathrm{d}x$

**Area under a curve**

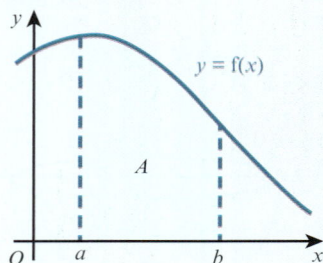

- The area, $A$, bounded by the curve $y = \mathrm{f}(x)$, the $x$-axis and the lines $x = a$ and $x = b$ is given by the formula:

$$A = \int_a^b y\,\mathrm{d}x \ \text{when} \ y \geqslant 0 \qquad\qquad (\text{or} \ A = \int_a^b \mathrm{f}(x)\,\mathrm{d}x \ \text{when} \ \mathrm{f}(x) \geqslant 0 ).$$

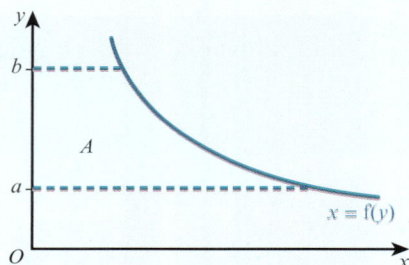

- The area, $A$, bounded by the curve $x = \mathrm{f}(y)$, the $y$-axis and the lines $y = a$ and $y = b$ is given by the formula:

$$A = \int_a^b x\,\mathrm{d}y \ \text{when} \ x \geqslant 0 \qquad\qquad (\text{or} \ A = \int_a^b \mathrm{f}(y)\,\mathrm{d}y \ \text{when} \ \mathrm{f}(y) \geqslant 0 ).$$

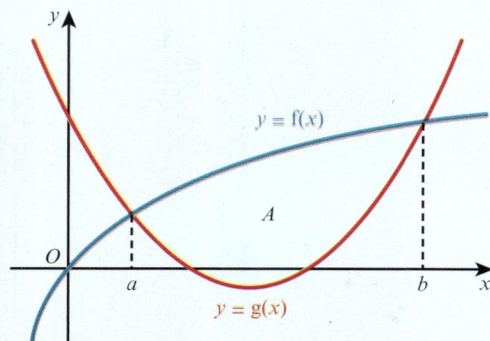

- The area, $A$, enclosed between $y = \mathrm{f}(x)$ and $y = \mathrm{g}(x)$ is given by the formula:

$$A = \int_a^b \left[\, \mathrm{f}(x) - \mathrm{g}(x) \,\right] \mathrm{d}x$$

where $a$ and $b$ are the $x$-coordinates of the points of intersection of the functions f and g.

## Improper integrals

- Integrals of the form $\displaystyle\int_a^\infty \mathrm{f}(x)\,\mathrm{d}x$ can be evaluated by replacing the infinite limit with a finite value, $X$, and then taking the limit as $X \to \infty$, provided the limit exists.

- Integrals of the form $\displaystyle\int_{-\infty}^b \mathrm{f}(x)\,\mathrm{d}x$ can be evaluated by replacing the infinite limit with a finite value, $X$, and then taking the limit as $X \to -\infty$, provided the limit exists.

- Integrals of the form $\displaystyle\int_a^b \mathrm{f}(x)\,\mathrm{d}x$ where $\mathrm{f}(x)$ is not defined when $x = a$ can be evaluated by replacing the limit $a$ with an $X$ and then taking the limit as $X \to a$, provided the limit exists.

- Integrals of the form $\displaystyle\int_a^b \mathrm{f}(x)\,\mathrm{d}x$ where $\mathrm{f}(x)$ is not defined when $x = b$ can be evaluated by replacing the limit $b$ with an $X$ and then taking the limit as $X \to b$, provided the limit exists.

## Volume of revolution

- The volume, $V$, obtained when the function $y = \mathrm{f}(x)$ is rotated through $360°$ about the $x$-axis between the boundary values $x = a$ and $x = b$ is given by the formula $V = \displaystyle\int_a^b \pi y^2\,\mathrm{d}x$.

- The volume, $V$, obtained when the function $x = \mathrm{f}(y)$ is rotated through $360°$ about the $y$-axis between the boundary values $y = a$ and $y = b$ is given by the formula $V = \displaystyle\int_a^b \pi x^2\,\mathrm{d}y$.

1   The function f is such that $f'(x) = 12x^3 + 10x$ and $f(-1) = 1$.

   Find $f(x)$.   [3]

2   Find $\displaystyle \int \left(5x - \frac{2}{x}\right)^2 \, dx$.   [3]

3   A curve is such that $\dfrac{dy}{dx} = \dfrac{6}{x^2} - 5x$ and the point $(3, 5.5)$ lies on the curve. Find the equation of the curve.   [4]

4   A curve has equation $y = f(x)$. It is given that $f'(x) = \dfrac{3}{\sqrt{x+2}} - \dfrac{8}{x^3}$ and that $f(2) = 3$. Find $f(x)$.   [5]

5

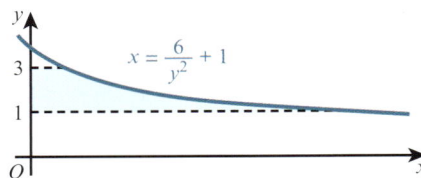

   The diagram shows part of the curve $x = \dfrac{6}{y^2} + 1$. The shaded region is bounded by the curve, the $y$-axis, and

   the lines $y = 1$ and $y = 3$. Find the volume, in terms of $\pi$, when this shaded region is rotated through $360°$
   about the $y$-axis.   [5]

6   A function is defined for $x \in \mathbb{R}$ and is such that $f'(x) = 6x - 6$. The range of the function is given
   by $f(x) \geqslant 5$.

   a   State the value of $x$ for which $f(x)$ has a stationary value.   [1]

   b   Find an expression for $f(x)$ in terms of $x$.   [4]

7

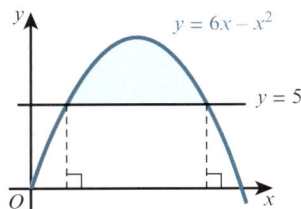

   The diagram shows the curve $y = 6x - x^2$ and the line $y = 5$. Find the area of the shaded region.   [6]

   *Cambridge International AS & A Level Mathematics 9709 Paper 11 Q4 June 2010*

8   a   Sketch the curve $y = (x - 3)^2 + 2$.   [1]

   b   The region enclosed by the curve, the $x$-axis, the $y$-axis, the line $x = 3$ is rotated through $360°$ about the
       $x$-axis. Find the volume obtained, giving your answer in terms of $\pi$.   [6]

**9**

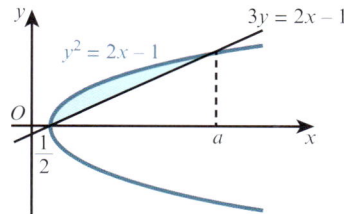

The diagram shows the curve $y^2 = 2x - 1$ and the straight line $3y = 2x - 1$.

The curve and straight line intersect at $x = \dfrac{1}{2}$ and $x = a$, where $a$ is a constant.

i   Show that $a = 5$. [2]

ii  Find, showing all necessary working, the area of the shaded region. [6]

*Cambridge International AS & A Level Mathematics 9709 Paper 11 Q8 November 2012*

**10**

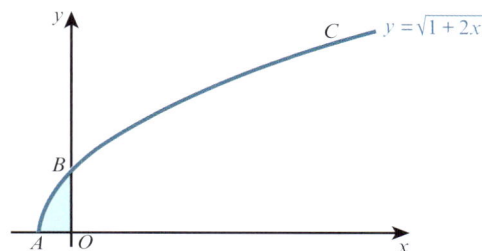

The diagram shows the curve $y = \sqrt{1 + 2x}$ meeting the $x$-axis at $A$ and the $y$-axis at $B$.
The $y$-coordinate of the point $C$ on the curve is 3.

i   Find the coordinates of $B$ and $C$. [2]

ii  Find the equation of the normal to the curve at $C$. [4]

iii Find the volume obtained when the shaded region is rotated through $360°$ about the $y$-axis. [5]

*Cambridge International AS & A Level Mathematics 9709 Paper 11 Q10 November 2011*

**11**

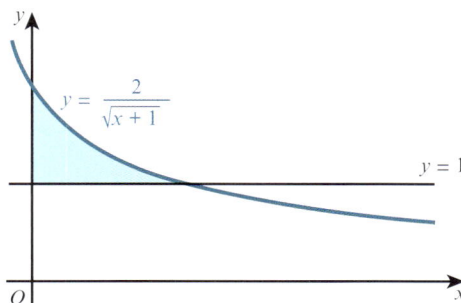

The diagram shows the line $y = 1$ and part of the curve $y = \dfrac{2}{\sqrt{x + 1}}$.

i   Show that the equation $y = \dfrac{2}{\sqrt{x + 1}}$ can be written in the form $x = \dfrac{4}{y^2} - 1$. [1]

ii  Find $\displaystyle\int_1^2 \left( \dfrac{4}{y^2} - 1 \right) dy$. Hence find the area of the shaded region. [5]

iii The shaded region is rotated through $360°$ about the $y$-axis. Find the exact value of the volume of revolution obtained. [5]

*Cambridge International AS & A Level Mathematics 9709 Paper 11 Q11 June 2012*

277

12  A curve has equation $y = f(x)$ and is such that $f'(x) = 3x^{\frac{1}{2}} + 3x^{-\frac{1}{2}} - 10$.

   i  By using the substitution $u = x^{\frac{1}{2}}$, or otherwise, find the values of $x$ for which the curve $y = f(x)$ has stationary points. **[4]**

   ii  Find $f''(x)$ and hence, or otherwise, determine the nature of each stationary point. **[3]**

   iii  It is given that the curve $y = f(x)$ passes through the point $(4, -7)$. Find $f(x)$. **[4]**

*Cambridge International AS & A Level Mathematics 9709 Paper 11 Q9 June 2013*

13

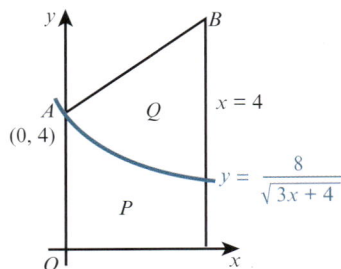

The diagram shows part of the curve $y = \dfrac{8}{\sqrt{3x+4}}$. The curve intersects the $y$-axis at $A(0, 4)$. The normal to the curve at $A$ intersects the line $x = 4$ at the point $B$.

   i  Find the coordinates of $B$. **[5]**

   ii  Show, with all necessary working, that the areas of the regions $P$ and $Q$ are equal. **[6]**

*Cambridge International AS & A Level Mathematics 9709 Paper 11 Q10 June 2015*

14

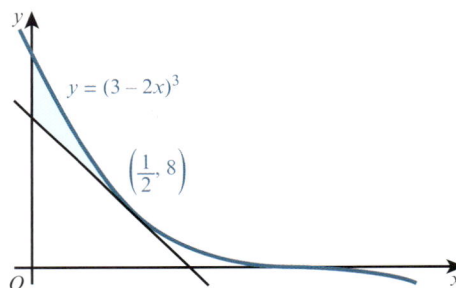

The diagram shows the curve $y = (3 - 2x)^3$ and the tangent to the curve at the point $\left(\dfrac{1}{2}, 8\right)$.

   i  Find the equation of this tangent, giving your answer in the form $y = mx + c$. **[5]**

   ii  Find the area of the shaded region. **[6]**

*Cambridge International AS & A Level Mathematics 9709 Paper 11 Q10 November 2013*

15

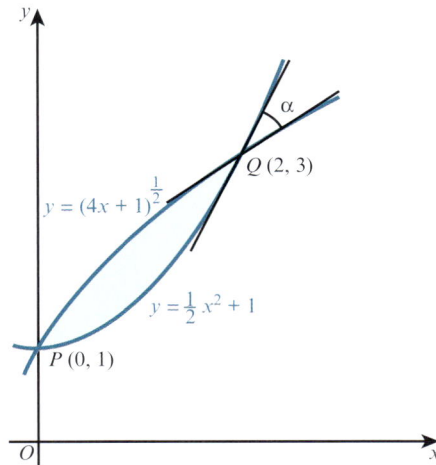

The diagram shows parts of the curves $y = (4x + 1)^{\frac{1}{2}}$ and $y = \frac{1}{2}x^2 + 1$ intersecting at points $P(0, 1)$ and

$Q(2, 3)$. The angle between the tangents to the curves at $Q$ is $\alpha$.

i   Find $\alpha$, giving your answer in degrees correct to 3 significant figures.                    [6]

ii  Find by integration the area of the shaded region.                                                [6]

*Cambridge International AS & A Level Mathematics 9709 Paper 11 Q11 November 2014*

1   A curve is such that $\dfrac{dy}{dx} = 2x^2 - 3$. Given that the curve passes through the point $(-3, -2)$, find the equation

of the curve. [4]

2   A curve is such that $\dfrac{dy}{dx} = 2 - 8(3x + 4)^{-\frac{1}{2}}$.

i   A point $P$ moves along the curve in such a way that the $x$-coordinate is increasing at a constant rate of
0.3 units per second. Find the rate of change of the $y$-coordinate as $P$ crosses the $y$-axis. [2]

The curve intersects the $y$-axis where $y = \dfrac{4}{3}$.

ii  Find the equation of the curve. [4]

*Cambridge International AS & A Level Mathematics 9709 Paper 11 Q4 June 2016*

3   A curve is such that $\dfrac{dy}{dx} = 3x^{\frac{1}{2}} - 6$ and the point $(9, 2)$ lies on the curve.

i   Find the equation of the curve. [4]

ii  Find the $x$-coordinate of the stationary point on the curve and determine the nature of the stationary
point. [3]

*Cambridge International AS & A Level Mathematics 9709 Paper 11 Q6 June 2010*

4   A curve is such that $\dfrac{dy}{dx} = \dfrac{3}{(1 + 2x)^2}$ and the point $\left(1, \dfrac{1}{2}\right)$ lies on the curve.

i   Find the equation of the curve. [4]

ii  Find the set of values of $x$ for which the gradient of the curve is less than $\dfrac{1}{3}$. [3]

*Cambridge International AS & A Level Mathematics 9709 Paper 11 Q7 June 2011*

280

5

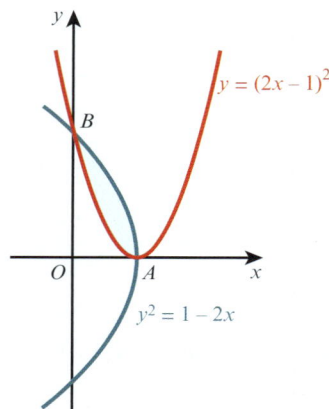

The diagram shows parts of the curves $y = (2x - 1)^2$ and $y^2 = 1 - 2x$, intersecting at points $A$ and $B$.

i   State the coordinates of $A$. [1]

ii  Find, showing all necessary working, the area of the shaded region. [6]

*Cambridge International AS & A Level Mathematics 9709 Paper 11 Q7 November 2016*

6   A curve has equation $y = f(x)$ and it is given that $f'(x) = 3x^{\frac{1}{2}} - 2x^{-\frac{1}{2}}$.
    The point $A$ is the only point on the curve at which the gradient is $-1$.

    i   Find the $x$-coordinate of $A$.                                                      [3]

    ii  Given that the curve also passes through the point $(4, 10)$, find the $y$-coordinate of $A$, giving your answer
        as a fraction.                                                                       [6]

        *Cambridge International AS & A Level Mathematics 9709 Paper 11 Q10 November 2016*

7

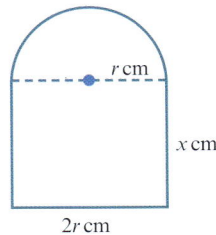

The diagram shows a metal plate. The plate has a perimeter of $50\,\text{cm}$ and consists of a rectangle of width
$2r\,\text{cm}$ and height $x\,\text{cm}$, and a semicircle of radius $r\,\text{cm}$.

    a   Show that the area, $A\,\text{cm}^2$, of the plate is given by $A = 50r - 2r^2 - \frac{1}{2}\pi r^2$.   [4]

    Given that $x$ and $r$ can vary:

    b   show that $A$ has a stationary value when $r = \dfrac{50}{4 + \pi}$                    [4]

    c   find this stationary value of $A$ and determine the nature of this stationary value.  [2]

8   A line has equation $y = 2x + c$ and a curve has equation $y = 8 - 2x - x^2$.
    i   For the case where the line is a tangent to the curve, find the value of the constant $c$.   [3]

    ii  For the case where $c = 11$, find the $x$-coordinates of the points of intersection of the line and the curve.
        Find also, by integration, the area of the region between the line and the curve.     [7]

        *Cambridge International AS & A Level Mathematics 9709 Paper 11 Q11 June 2014*

9   The equation of a curve is $y = \dfrac{9}{2 - x}$.

    i   Find an expression for $\dfrac{dy}{dx}$ and determine, with a reason, whether the curve has any stationary points.   [3]

    ii  Find the volume obtained when the region bounded by the curve, the coordinate axes and the line $x = 1$ is
        rotated through $360°$ about the $x$-axis.                                            [4]

    iii Find the set of values of $k$ for which the line $y = x + k$ intersects the curve at two distinct points.   [4]

        *Cambridge International AS & A Level Mathematics 9709 Paper 11 Q11 November 2010*

281

10  A function f is defined as $f(x) = \dfrac{4}{2x+1}$ for $x \geqslant 0$.

   a  Find an expression, in terms of $x$, for $f'(x)$ and explain how your answer shows that f is a decreasing function.    **[3]**

   b  Find an expression, in terms of $x$, for $f^{-1}(x)$ and find the domain of $f^{-1}$.    **[4]**

   c  On a diagram, sketch the graph of $y = f(x)$ and the graph of $y = f^{-1}(x)$, making clear the relationship between the two graphs.    **[4]**

11  A curve is such that $\dfrac{dy}{dx} = x^{\frac{1}{2}} - x^{-\frac{1}{2}}$. The curve passes through the point $\left(4, \dfrac{2}{3}\right)$.

   i  Find the equation of the curve.    **[4]**

   ii  Find $\dfrac{d^2y}{dx^2}$.    **[2]**

   iii  Find the coordinates of the stationary point and determine its nature.    **[5]**

         *Cambridge International AS & A Level Mathematics 9709 Paper 11 Q12 June 2014*

12  The function f is defined for $x > 0$ and is such that $f'(x) = 2x - \dfrac{2}{x^2}$. The curve $y = f(x)$ passes through the point $P(2, 6)$.

   i  Find the equation of the normal to the curve at $P$.    **[3]**

   ii  Find the equation of the curve.    **[4]**

   iii  Find the $x$-coordinate of the stationary point and state with a reason whether this point is a maximum or a minimum.    **[4]**

         *Cambridge International AS & A Level Mathematics 9709 Paper 11 Q9 November 2014*

13  The point $P(3, 5)$ lies on the curve $y = \dfrac{1}{x-1} - \dfrac{9}{x-5}$.

   i  Find the $x$-coordinate of the point where the normal to the curve at $P$ intersects the $x$-axis.    **[5]**

   ii  Find the $x$-coordinate of each of the stationary points on the curve and determine the nature of each stationary point, justifying your answers.    **[6]**

         *Cambridge International AS & A Level Mathematics 9709 Paper 11 Q11 November 2016*

14

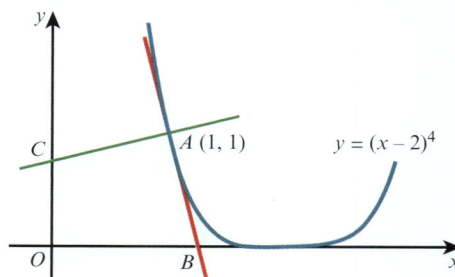

The diagram shows part of the curve $y = (x-2)^4$ and the point $A(1, 1)$ on the curve. The tangent at $A$ cuts the $x$-axis at $B$ and the normal at $A$ cuts the $y$-axis at $C$.

   i  Find the coordinates of $B$ and $C$.    **[6]**

   ii  Find the distance $AC$, giving your answer in the form $\dfrac{\sqrt{a}}{b}$, where $a$ and $b$ are integers.    **[2]**

   iii  Find the area of the shaded region.    **[4]**

         *Cambridge International AS & A Level Mathematics 9709 Paper 11 Q10 June 2013*

**15**

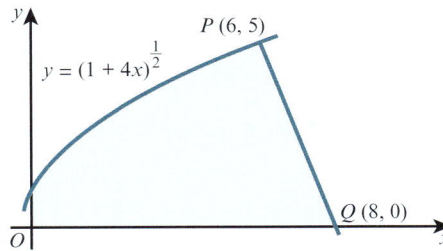

The diagram shows part of the curve $y = (1 + 4x)^{\frac{1}{2}}$ and a point $P(6, 5)$ lying on the curve. The line $PQ$ intersects the $x$-axis at $Q(8, 0)$.

**i** Show that $PQ$ is a normal to the curve. [5]

**ii** Find, showing all necessary working, the exact volume of revolution obtained when the shaded region is rotated through $360°$ about the $x$-axis. [7]

[In part **ii** you may find it useful to apply the fact that the volume, $V$, of a cone of base radius $r$ and vertical height $h$, is given by $V = \dfrac{1}{3} \pi r^2 h$.]

*Cambridge International AS & A Level Mathematics 9709 Paper 11 Q11 November 2015*

**Time allowed is 1 hour 50 minutes (75 marks).**

1   It is given that $f(x) = 2x - \dfrac{5}{x^3}$, for $x > 0$. Show that f is an increasing function. [2]

2   The graph of $y = x^3 - 3$ is transformed by applying a translation of $\begin{pmatrix} 2 \\ 0 \end{pmatrix}$ followed by a reflection in the $x$-axis.

   Find the equation of the resulting graph in the form $y = ax^3 + bx^2 + cx + d$. [3]

3   Prove the identity $\dfrac{1 - \tan^2 x}{1 + \tan^2 x} \equiv 2\cos^2 x - 1$. [4]

4   a   Find the first three terms in the expansion of $(3 - 2x)^7$, in ascending powers of $x$. [3]

   b   Find the coefficient of $x^2$ in the expansion of $(1 + 5x)(3 - 2x)^7$. [3]

5

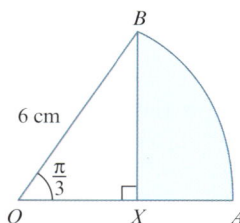

The diagram shows sector $OAB$ of a circle with centre $O$, radius $6\,\text{cm}$ and sector angle $\dfrac{\pi}{3}$ radians. The point $X$ lies on the line $OA$ and $BX$ is perpendicular to $OA$.

   a   Find the exact area of the shaded region. [4]

   b   Find the exact perimeter of the shaded region. [3]

6   A circle has centre $(3, -2)$ and passes through the point $P(5, -6)$.

   a   Find the equation of the circle. [3]

   b   Find the equation of the tangent to the circle at the point $P$, giving your answer in the form $ax + by = c$. [4]

7   a   The sum, $S_n$, of the first $n$ terms of an arithmetic progression is given by $S_n = 11n - 4n^2$. Find the first term and the common difference. [3]

   b   The first term of a geometric progression is $2\frac{1}{4}$ and the fourth term is $\dfrac{1}{12}$. Find:

      i   the common ratio [3]

      ii   the sum to infinity. [2]

8   The equation of a curve is $y = 3 + 12x - 2x^2$.

   a   Express $3 + 12x - 2x^2$ in the form $a - 2(x + b)^2$, where $a$ and $b$ are constants to be found. [3]

   b   Find the coordinates of the stationary point on the curve. [2]

   c   Find the set of values of $x$ for which $y \leqslant -5$. [3]

9  The function $f : x \mapsto 6 - 5\cos x$ is defined for the domain $0 \leqslant x \leqslant 2\pi$.

   a  Find the range of f.                                                              [1]

   b  Sketch the graph of $y = f(x)$.                                                    [2]

   c  Solve the equation $f(x) = 3$.                                                     [3]

   The function $g : x \mapsto 6 - 5\cos x$ is defined for the domain $0 \leqslant x \leqslant \pi$.

   d  Find $g^{-1}(x)$.                                                                  [2]

10 A curve has equation $y = \dfrac{6}{9 - 2x}$ and $A(3, 2)$ is a point on the curve.

   a  Find the equation of the normal to the curve at the point $A$.                     [5]

   b  A point $P(x, y)$ moves along the curve in such a way that the $y$-coordinate is increasing at a
      constant rate of 0.05 units per second. Find the rate of increase of the $x$-coordinate when $x = 4$.  [5]

11 A curve has equation $y = \dfrac{16}{x} - x^2$.

   a  Find $\dfrac{dy}{dx}$ and $\dfrac{d^2y}{dx^2}$ in terms of $x$.                     [3]

   b  Find the coordinates of the stationary point on the curve and determine its nature.  [4]

   c  Find the volume of the solid formed when the region enclosed by the curve, the $x$-axis, and the
      lines $x = 1$ and $x = 2$ is rotated about the $x$-axis.                           [5]

# Answers

## 1 Quadratics

### Prerequisite knowledge

1  a  $-4, 3$    b  $3$    c  $-\frac{1}{3}, 6$
2  a  $x > 2$    b  $x \geqslant -2$
3  a  $x = 2, y = 3$    b  $x = -2, y = -5$
4  a  $2\sqrt{5}$    b  $5$    c  $4\sqrt{2}$

### Exercise 1A

1  a  $-5, 2$    b  $3, 4$
   c  $-2, 8$    d  $-3, -\frac{4}{5}$
   e  $-\frac{5}{2}, 1\frac{1}{3}$    f  $-\frac{1}{5}, \frac{3}{2}$

2  a  $-1, 6$    b  $-1, 4$
   c  $-\frac{3}{4}, 1$    d  $\pm 2$
   e  $-\frac{1}{2}, 1$    f  $-\frac{1}{2}, 0$

3  a  $-2, \frac{5}{3}$    b  $-3, 2$
   c  $\pm 3$    d  $4$
   e  $-\frac{2}{3}, \frac{1}{2}$    f  $-5, \frac{1}{2}$

4  a  $-5, 3$    b  $\frac{5}{2}, 3$
   c  $1, 3$    d  $-4, -\frac{1}{2}$
   e  $-5, 3$    f  $2, 3, 4, 5$

5  a  Proof    b  $20\,\text{cm}, 21\,\text{cm}, 29\,\text{cm}$

6  $5\frac{1}{2}$

7  $-\frac{2}{3}, \frac{1}{2}, 4, 6, 7$

### Exercise 1B

1  a  $(x-3)^2 - 9$    b  $(x+4)^2 - 16$
   c  $\left(x - \frac{3}{2}\right)^2 - \frac{9}{4}$    d  $\left(x + \frac{15}{2}\right)^2 - \frac{225}{4}$
   e  $(x+2)^2 + 4$    f  $(x-2)^2 - 12$
   g  $\left(x + \frac{7}{2}\right)^2 - \frac{45}{4}$    h  $\left(x - \frac{3}{2}\right)^2 + \frac{7}{4}$

2  a  $2(x-3)^2 + 1$    b  $3(x-2)^2 - 13$
   c  $2\left(x + \frac{5}{4}\right)^2 - \frac{33}{8}$    d  $2\left(x + \frac{7}{4}\right)^2 - \frac{9}{8}$

3  a  $4 - (x-2)^2$    b  $16 - (x-4)^2$
   c  $\frac{25}{4} - \left(x + \frac{3}{2}\right)^2$    d  $\frac{61}{4} - \left(x - \frac{5}{2}\right)^2$

4  a  $15 - 2(x+2)^2$    b  $21 - 2(x+3)^2$
   c  $15 - 2(x-1)^2$    d  $\frac{49}{12} - 3\left(x - \frac{5}{6}\right)^2$

5  a  $(3x-1)^2 - 4$    b  $(2x+5)^2 + 5$
   c  $(5x+4)^2 - 20$    d  $(3x-7)^2 + 12$

6  a  $-9, 1$    b  $-6, 2$    c  $-5, 7$
   d  $2, 7$    e  $-6, 3$    f  $-10, 1$

7  a  $-2 \pm \sqrt{11}$    b  $5 \pm \sqrt{23}$    c  $-4 \pm \sqrt{17}$
   d  $1 \pm \sqrt{\frac{7}{2}}$    e  $\frac{-3 \pm \sqrt{3}}{2}$    f  $2 \pm \sqrt{\frac{11}{2}}$

8  $3 \pm \sqrt{10}$

9  $\sqrt{19} - 2$

10  $-\frac{8}{3}, 1, \frac{1}{6}(-5 - \sqrt{97}), \frac{1}{6}(\sqrt{97} - 5)$

11  a  $\frac{9000\sqrt{3}}{49} \approx 318\,\text{m}$    b  $\frac{9000\sqrt{3}}{98} \approx 159\,\text{m}$

### Exercise 1C

1  a  $-0.29, 10.29$    b  $-5.24, -0.76$
   c  $-4.19, 1.19$    d  $-3.39, 0.89$
   e  $-1.39, -0.36$    f  $-1.64, 0.24$

2  $4.93$

3  $3.19$

4  $-0.217, 9.22$

5  $x = \dfrac{b \pm \sqrt{b^2 - 4ac}}{2a}$; the solutions each increase by $\dfrac{b}{a}$.

### Exercise 1D

1  a  $(-3, 9), (2, 4)$    b  $\left(-8, \frac{7}{2}\right), (2, 1)$
   c  $(-10, 0), (8, 6)$    d  $(-2, -7), (1, 2)$
   e  $(2, -2), (10, 2)$    f  $(-1, -3), (2, 1)$
   g  $(2, 4)$    h  $(-3, 1), (9, 7)$
   i  $(2, 2), (10, -2)$    j  $(-5, -24), (5, 1)$
   k  $(-6, -2), \left(8, 1\frac{1}{2}\right)$    l  $(4, -6), (12, 10)$
   m  $\left(-3, -4\frac{1}{3}\right), (4, -9)$    n  $(-1, 3), (3, 1)$
   o  $(6, -2), (18, -1)$

2  a  $9$ and $17$    b  $13 - \sqrt{19}$ and $13 + \sqrt{19}$

3  $2\frac{1}{2}\,\text{cm}$ and $5\frac{2}{5}\,\text{cm}$

4  $3\frac{1}{2}\,\text{cm}$ and $9\,\text{cm}$

286

**5**   7 cm and 11 cm

**6**   $x = 4\frac{1}{2}$ and $y = 16$ or $x = 16$ and $y = 4\frac{1}{2}$

**7**   $r = 5, h = 13$

**8**   **a**  $(-3, 5)$ and $(2, 0)$   **b**  $5\sqrt{2}$

**9**   **a**  $(-2, 1)$ and $(3, -1)$   **b**  $\left(\frac{1}{2}, 0\right)$

**10**  $2\sqrt{53}$

**11**  $7x + y = 0$

**12**  $(2, 3)$

**13**  $y = -2x - 3$

**14**  **a**  $2, 8$   **b**  $\dfrac{N}{2} + \dfrac{D}{2N}, \dfrac{N}{2} - \dfrac{D}{2N}$

## Exercise 1E

**1**   **a**  $\pm 2, \pm 3$   **b**  $-1, 2$

    **c**  $\pm\sqrt{5}, \pm 1$   **d**  $\pm\dfrac{\sqrt{2}}{2}, \pm\sqrt{5}$

    **e**  $\pm 1$   **f**  $1, \dfrac{1}{2}$

    **g**  $\pm\sqrt{3}$   **h**  No solutions

    **i**  $\pm 2$   **j**  $-\dfrac{1}{2}, 1$

    **k**  $\pm\dfrac{3}{2}$   **l**  $-1, 2$

**2**   **a**  $4, 6\frac{1}{4}$   **b**  $4$

    **c**  $\dfrac{1}{9}, 6\frac{1}{4}$   **d**  $\dfrac{4}{25}$

    **e**  $\dfrac{1}{4}, 1\frac{9}{16}$   **f**  $\dfrac{1}{9}, 25$

**3**   **a**  $x - 6\sqrt{x} + 8 = 0$   **b**  $(4, 4), (16, 8)$

    **c**  $4\sqrt{10}$

**4**   $a = 2, b = -9, c = 7$

**5**   $a = 2, b = -40, c = 128$

## Exercise 1F

**1**   **a**  ∪-shaped curve, minimum point: $(3, -1)$, axes crossing points: $(2, 0), (4, 0), (0, 8)$

    **b**  ∪-shaped curve, minimum point: $\left(-2\frac{1}{2}, -20\frac{1}{4}\right)$, axes crossing points: $(-7, 0), (2, 0), (0, -14)$

    **c**  ∪-shaped curve, minimum point: $\left(-1\frac{3}{4}, -21\frac{1}{8}\right)$, axes crossing points: $(-5, 0), \left(1\frac{1}{2}, 0\right), (0, -15)$

    **d**  ∩-shaped curve, maximum point: $\left(\frac{1}{2}, 12\frac{1}{4}\right)$, axes crossing points: $(-3, 0), (4, 0), (0, 12)$

**2**   **a**  $2(x - 2)^2 - 3$

    **b**  $x = 2$

**3**   **a**  $\dfrac{53}{4} - \left(x - \dfrac{5}{2}\right)^2$

    **b**  $\left(2\frac{1}{2}, 13\frac{1}{4}\right)$, maximum

**4**   **a**  $2\left(x + \dfrac{9}{4}\right)^2 - \dfrac{49}{8}$

    **b**  $\left(-2\frac{1}{4}, -6\frac{1}{8}\right)$, minimum

**5**   $-4\frac{1}{4}$ when $x = 3\frac{1}{2}$

**6**   **a**  $\dfrac{9}{8} - 2\left(x - \dfrac{1}{4}\right)^2$

    **b**  ∩-shaped curve, maximum point: $\left(\frac{1}{4}, 1\frac{1}{8}\right)$, axes crossing points: $\left(-\frac{1}{2}, 0\right), (1, 0), (0, 1)$

**7**   Proof

**8**   A: $y = (x - 4)^2 + 2$ or $x^2 - 8x + 18$

    B: $y = 4(x + 2)^2 - 6$ or $4x^2 + 16x + 10$

    C: $y = 8 - \dfrac{1}{2}(x - 2)^2$ or $6 + 2x - \dfrac{1}{2}x^2$

**9**   **a**

| | |
|---|---|
| A | $y = x^2 - 6x + 13$ |
| B | $y = x^2 - 6x + 5$ |
| C | $y = -x^2 + 6x - 5$ |
| D | $y = -x^2 + 6x - 13$ |
| E | $y = x^2 + 6x + 13$ |
| F | $y = x^2 + 6x + 5$ |
| G | $y = -x^2 - 6x - 5$ |
| H | $y = -x^2 - 6x - 13$ |

    **b**  Student's own answers

**10**  $y = 3x^2 - 6x - 24$

**11**  $y = 5 + 3x - \dfrac{1}{2}x^2$

**12**  Proof

## Exercise 1G

1  a  $0 \leqslant x \leqslant 3$  b  $x < -2$ or $x > 3$

   c  $4 \leqslant x \leqslant 6$  d  $-\dfrac{3}{2} < x < 2$

   e  $-6 \leqslant x \leqslant 5$  f  $x < -\dfrac{1}{2}$ or $x > \dfrac{1}{3}$

2  a  $x \leqslant -5$ or $x \geqslant 5$  b  $-5 \leqslant x \leqslant -2$

   c  $x < -7$ or $x > 1$  d  $-\dfrac{3}{2} \leqslant x \leqslant \dfrac{2}{7}$

   e  $\dfrac{4}{3} < x < \dfrac{5}{2}$  f  $x < -4$ or $x > \dfrac{1}{2}$

3  a  $-9 < x < 4$  b  $x < 7$ or $x > 8$

   c  $-12 \leqslant x \leqslant 1$  d  $-3 < x < 2$

   e  $x < -4$ or $x > 1$  f  $-\dfrac{1}{2} < x < \dfrac{3}{5}$

   g  $x \leqslant -9$ or $x \geqslant 1$  h  $x < -2$ or $x > 5$

   i  $-\dfrac{7}{2} < x < \dfrac{5}{3}$

4  $-3 < x < \dfrac{5}{2}$

5  a  $5 \leqslant x < 7$  b  $-7 \leqslant x < 1$

   c  $x < -2$ or $x \geqslant 3$

6  $x < -5$ or $x > 8$

7  a  $1 < x \leqslant \dfrac{3}{2}$  b  $-1 < x < 0$

   c  $-1 \leqslant x < 1$ or $x \geqslant 5$

   d  $-3 \leqslant x < 2$ or $x \geqslant 5$

   e  $-5 \leqslant x < -2$ or $1 \leqslant x < 2$

   f  $x < -4$ or $\dfrac{1}{2} \leqslant x < 5$

## Exercise 1H

1  a  Two equal roots  b  Two distinct roots
   c  Two distinct roots  d  Two equal roots
   e  No real roots  f  Two distinct roots

2  No real roots

3  $b = -2, c = -35$

4  a  $k = \pm4$  b  $k = 4$ or $k = 1$

   c  $k = \dfrac{1}{4}$  d  $k = 0$ or $k = 2$

   e  $k = 0$ or $k = -\dfrac{8}{9}$  f  $k = -10$ or $k = 14$

5  a  $k > -13$  b  $k < \dfrac{57}{8}$

   c  $k < 2$  d  $k < \dfrac{1}{2}$

   e  $k > \dfrac{3}{2}$  f  $k < \dfrac{25}{16}$

6  a  $k > \dfrac{1}{2}$  b  $k > \dfrac{13}{12}$

   c  $k > \dfrac{26}{5}$  d  $k > -\dfrac{39}{8}$

   e  $5 - \sqrt{21} < k < 5 + \sqrt{21}$
   f  $7 - 2\sqrt{10} < k < 7 + 2\sqrt{10}$

7  $k = \dfrac{p^2}{20}$

8  $k \leqslant \dfrac{25}{8}$

9  Proof

10  Proof

11  $k \leqslant -2\sqrt{2}$

## Exercise 1I

1  $-5, -9$

2  $-1, 7$

3  $5$

4  a  $\pm10$  b  $(2, 4), (-2, -4)$

5  $-6, -2, (-1, 12), (1, 4)$

6  $k < -2$ or $k > 6$

7  $k < -4\sqrt{3}$ or $k > 4\sqrt{3}$

8  $k < 6$

9  $-3 < m < 1$

10  $k > 6$

11  $\dfrac{1}{2}$

12  Proof

13  Proof

## End-of-chapter review exercise 1

1  $\left(\dfrac{1}{2}, 0\right)$

2  a  $\left(3x - \dfrac{5}{2}\right)^2 - \dfrac{25}{4}$  b  $-\dfrac{1}{3} < x < 2$

3  $x = \pm2, x = \pm\dfrac{3}{2}$

4  $x < -9 - 2\sqrt{3}$ or $x > -9 + 2\sqrt{3}$

5  $k < 1$ or $k > 2$

**6**  **a**  $\left(1\frac{1}{2}, -2\right)$  **b**  $k = -4$ or $k = -20$

**7**  **a**  Proof  **b**  (6, 29)
    **c**  $k = 1, C = (2, 5)$

**8**  **a**  Proof  **b**  (2, 1), (5, 7),
    **c**  $2 < x < 5$

**9**  **a**  $25 - (x - 5)^2$  **b**  (5, 25)
    **c**  $x \leqslant 1$ or $x \geqslant 9$

**10**  **i**  $3\sqrt{5}, \left(-\frac{1}{2}, 5\right)$  **ii**  $k = 3$ or $11$

**11**  **i**  $\left(2\frac{1}{2}, 2\frac{1}{2}\right)$  **ii**  $m = -8, (-2, 16)$

**12**  **i**  $2(x - 1)^2 - 1, (1, -1)$  **ii**  $\left(-\frac{1}{2}, 3\frac{1}{2}\right)$

    **iii**  $y - 3 = -\frac{1}{5}(x - 2)$

# 2 Functions

## Prerequisite knowledge

**1**  10

**2**  $3 - 2x$

**3**  $f^{-1}(x) = \dfrac{x - 4}{5}$

**4**  $2(x - 3)^2 - 13$

## Exercise 2A

**1**  **a**  function, one-one  **b**  function, many-one
    **c**  function, one-one  **d**  function, one-one
    **e**  function, one-one  **f**  function, one-one
    **g**  function, one-one  **h**  not a function

**2**  **a**
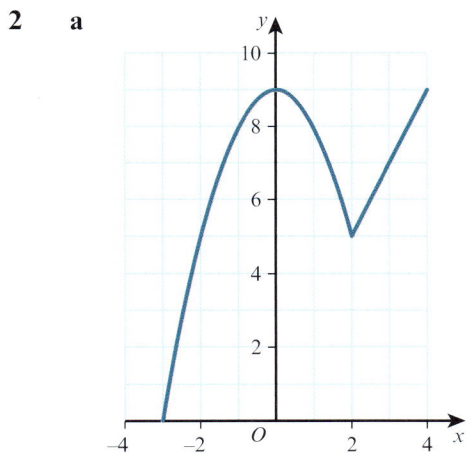
    **b**  Many-one

**3**  **a**
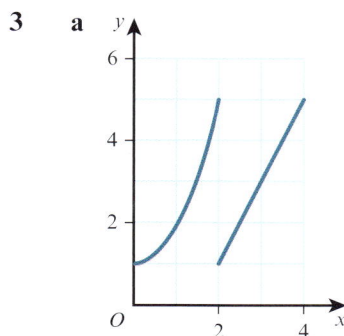
    **b**  each input does not have a unique output

**4**  **a**  domain: $x \in \mathbb{R}, -1 \leqslant x \leqslant 5$
       range: $f(x) \in \mathbb{R}, -8 \leqslant f(x) \leqslant 8$
    **b**  domain: $x \in \mathbb{R}, -3 \leqslant x \leqslant 2$
       range: $f(x) \in \mathbb{R}, -7 \leqslant f(x) \leqslant 20$

**5**  **a**  $f(x) > 12$  **b**  $-13 \leqslant f(x) \leqslant -3$
    **c**  $-1 \leqslant f(x) \leqslant 9$  **d**  $2 \leqslant f(x) \leqslant 32$
    **e**  $\dfrac{1}{32} \leqslant f(x) \leqslant 16$  **f**  $\dfrac{3}{2} \leqslant f(x) \leqslant 12$

**6**  **a**  $f(x) \geqslant -2$  **b**  $3 \leqslant f(x) \leqslant 28$
    **c**  $f(x) \leqslant 3$  **d**  $-5 \leqslant f(x) \leqslant 7$

**7**  **a**  $f(x) \geqslant 5$  **b**  $f(x) \geqslant -7$
    **c**  $-17 \leqslant f(x) \leqslant 8$  **d**  $f(x) \geqslant 1$

**8**  **a**  $f(x) \geqslant -20$  **b**  $f(x) \geqslant -6\frac{1}{3}$

**9**  **a**  $f(x) \leqslant 23$  **b**  $f(x) \leqslant 5$

**10**  **a**
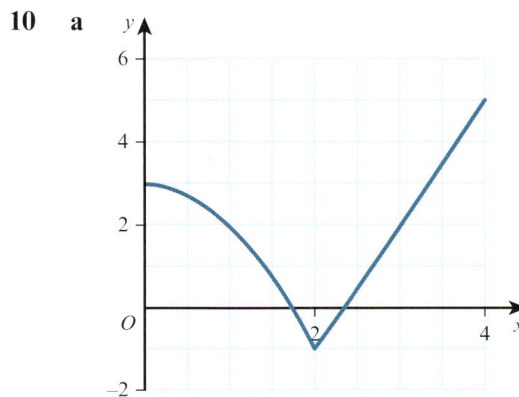
    **b**  $-1 \leqslant f(x) \leqslant 5$

**11**  $f(x) \geqslant k - 9$

**12**  $g(x) \leqslant \dfrac{a^2}{8} + 5$

**13**  $a = 2$

**14**  $a = 1$ or $a = -5$

**15**  **a**  $2(x - 2)^2 - 3$  **b**  $k = 4$
    **c**  $x \in \mathbb{R}, -3 \leqslant x \leqslant 5$

289

**16** **a** domain: $x \in \mathbb{R}$
range: $\mathrm{f}(x) \in \mathbb{R}$

**b** domain: $x \in \mathbb{R}$
range: $\mathrm{f}(x) \in \mathbb{R}, \mathrm{f}(x) \geqslant 2$

**c** domain: $x \in \mathbb{R}$
range: $\mathrm{f}(x) \in \mathbb{R}, \mathrm{f}(x) > 0$

**d** domain: $x \in \mathbb{R}, x \neq 0,$
range: $\mathrm{f}(x) \in \mathbb{R}, \mathrm{f}(x) \neq 0$

**e** domain: $x \in \mathbb{R}, x \neq 2,$
range: $\mathrm{f}(x) \in \mathbb{R}, \mathrm{f}(x) \neq 0$

**f** domain: $x \in \mathbb{R}, x \geqslant 3,$
range: $\mathrm{f}(x) \in \mathbb{R}, \mathrm{f}(x) \geqslant -2$

## Exercise 2B

**1** **a** 7 **b** 3 **c** 231

**2** **a** hk **b** kh **c** hh

**3** **a** $a = 3, b = -12$ **b** $5\frac{7}{9}$

**4** **a** $-\dfrac{6}{x + 1}$ **b** $-4$

**5** **a** $(2x + 5)^2 - 2$ **b** $-4\frac{1}{2}$ or $-\frac{1}{2}$

**6** $\dfrac{1}{2}$ or $3\frac{1}{2}$

**7** $-\dfrac{4}{3}$ or $0$

**8** $-9$

**9** $\dfrac{x + 2}{4x + 9}$

**10** **a** fg **b** gf **c** gg
**d** ff **e** gfg **f** fgf

**11** Proof

**12** $\pm 4$

**13** $k \geqslant -\dfrac{19}{2}$

**14** Proof

**15** **a** $2(x + 1)^2 - 10$ **b** $-1$

**16** **a** $x \leqslant -1$ or $x \geqslant 3$ **b** $(x - 1)^2 + 3$
**c** $\mathrm{f}(x) \geqslant 3$

**17** **a** $4x^2 + 2x - 6$ **b** $\mathrm{fg}(x) \geqslant -6\frac{1}{4}$

**18** **a** $\mathrm{ff}(x) = \dfrac{2(x + 1)}{x + 3}$ for $x \in \mathbb{R}, x \neq -3$
**b** Proof **c** $-2$ or $1$

**19** **a** PQ($x$), domain is $x \in \mathbb{R}$,
range is $\mathrm{f}(x) \in \mathbb{R}, \mathrm{f}(x) \geqslant -1$
**b** QP($x$), domain is $x \in \mathbb{R}$,
range is $\mathrm{f}(x) \in \mathbb{R}, \mathrm{f}(x) \geqslant 1$

**c** RR($x$), domain is $x \in \mathbb{R}, x \neq 0,$
range is $\mathrm{f}(x) \in \mathbb{R}, \mathrm{f}(x) \neq 0$

**d** QPR($x$), domain is $x \in \mathbb{R}, x \neq 0,$
range is $\mathrm{f}(x) \in \mathbb{R}, \mathrm{f}(x) > 1$

**e** RQQ($x$), domain is $x \in \mathbb{R}, x \neq -4,$
range is $\mathrm{f}(x) \in \mathbb{R}, \mathrm{f}(x) \neq 0$

**f** PS($x$), domain is $x \in \mathbb{R}, x \geqslant -1,$
range is $\mathrm{f}(x) \in \mathbb{R}, \mathrm{f}(x) \geqslant -1$

**g** SP($x$), domain is $x \in \mathbb{R}, x \geqslant -1,$
range is $\mathrm{f}(x) \in \mathbb{R}, \mathrm{f}(x) \geqslant -1$

## Exercise 2C

**1** **a** $\mathrm{f}^{-1}(x) = \dfrac{x + 8}{5}$ **b** $\mathrm{f}^{-1}(x) = \sqrt{x - 3}$

**c** $\mathrm{f}^{-1}(x) = 5 + \sqrt{x - 3}$ **d** $\mathrm{f}^{-1}(x) = \dfrac{3x + 8}{x}$

**e** $\mathrm{f}^{-1}(x) = \dfrac{7 - 2x}{x - 1}$ **f** $\mathrm{f}^{-1}(x) = 2 + \sqrt[3]{x + 1}$

**2** **a** Domain is $x \geqslant -4$, range is $\mathrm{f}^{-1}(x) \geqslant -2$
**b** $\mathrm{f}^{-1}(x) = -2 + \sqrt{x + 4}$

**3** **a** $\mathrm{f}^{-1}(x) = \dfrac{5 - x}{2x}$ **b** $x \leqslant 1$

**4** **a** $\mathrm{f}^{-1}(x) = -1 + \sqrt[3]{x + 4}$ **b** $x \geqslant -3$

**5** **a** g is one-one for $x \geqslant 3$, since vertex $= (2, 2)$
**b** $\mathrm{g}^{-1}(x) = 2 + \sqrt{\dfrac{x - 2}{2}}$

**6** **a** $-3$ **b** $\mathrm{f}^{-1}(x) = -3 + \sqrt{\dfrac{x + 32}{2}}$

**7** **a** $\mathrm{f}(x) \geqslant -9$
**b** No inverse since it is not one-one

**8** **a** $k = 3$
**b** **i** $\mathrm{f}^{-1}(x) = 3 + \sqrt{9 - x}$
**ii** Domain is $x \leqslant 9$, range is $3 \leqslant \mathrm{f}^{-1}(x) \leqslant 7$

**9** **a** $\mathrm{f}(x) = \dfrac{1}{5 - x}$ **b** Domain is $x \leqslant 4\frac{2}{3}$

**10** $a = 5, b = 12$

**11** **a** $\mathrm{f}^{-1}(x) = \dfrac{x + 1}{3}, \mathrm{g}^{-1}(x) = \dfrac{4x + 3}{2x}$
**b** Proof

**12** **a** $\mathrm{f}^{-1}(x) = \dfrac{1}{2}\left(1 + \sqrt[3]{x + 3}\right)$
**b** Domain is $-2 \leqslant x \leqslant 122$

**13** **a** $\mathrm{f}(x) = (x - 5)^2 - 25$
**b** $\mathrm{f}^{-1}(x) = 5 + \sqrt{x + 25}$, domain is $x \geqslant -25$

**14** **a** $f^{-1}(x) = \dfrac{x+1}{x}$ **b** Proof

**c** $\dfrac{1 \pm \sqrt{5}}{2}$

**15** **b** and **c**

**16** **a** $\dfrac{7-x}{6}$

**b** **i** $\dfrac{14-x}{6}$ **ii** $\dfrac{7-x}{6}$

**c** $(fg)^{-1}(x) = g^{-1} f^{-1}(x)$

## Exercise 2D

**1** **a**

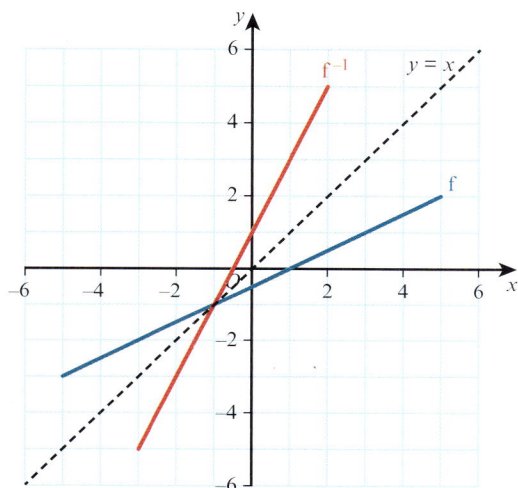

**b**

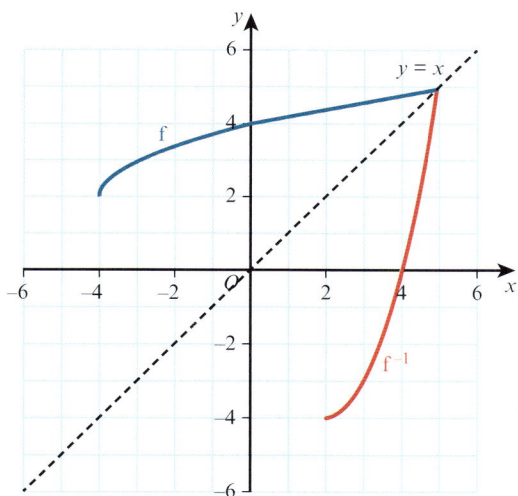

**c**

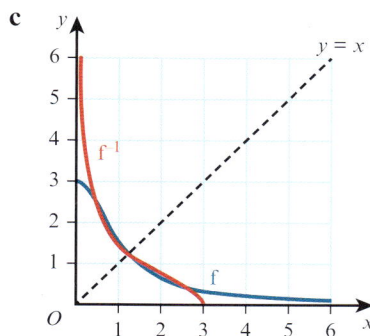

**d** $f^{-1}$ does not exist since f is not one-one

**2** **a** $f^{-1}(x) = \dfrac{x+1}{2}$

**b** Domain is $-3 \leqslant x \leqslant 5$, range is $-1 \leqslant x \leqslant 3$

**c**

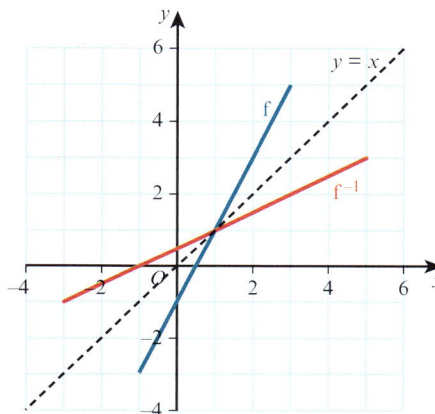

**3** **a** $0 < f(x) \leqslant 2$ **b** $f^{-1}(x) = \dfrac{4-2x}{x}$

**c** Domain is $0 < x \leqslant 2$, range is $f^{-1}(x) \geqslant 0$

**d**

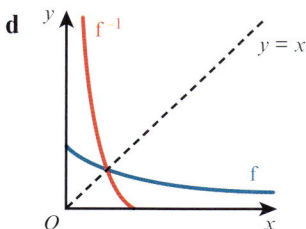

**4** **a** Symmetrical about $y = x$

**b** Not symmetrical about $y = x$

**c** Symmetrical about $y = x$

**d** Symmetrical about $y = x$

**5** **a** Proof **b** $d = -a$

## Exercise 2E

1   **a** $y = 2x^2 + 4$        **b** $y = 5\sqrt{x} - 2$

   **c** $y = 7x^2 - 2x + 1$    **d** $y = x^2 + 1$

   **e** $y = \dfrac{2}{x + 5}$        **f** $y = \dfrac{x - 3}{x - 2}$

   **g** $y = (x + 1)^2 + x + 1$   **h** $y = 3(x - 2)^2 + 1$

2   **a** Translation $\begin{pmatrix} 0 \\ 4 \end{pmatrix}$     **b** Translation $\begin{pmatrix} 0 \\ -5 \end{pmatrix}$

   **c** Translation $\begin{pmatrix} -1 \\ 0 \end{pmatrix}$     **d** Translation $\begin{pmatrix} 2 \\ 0 \end{pmatrix}$

   **e** Translation $\begin{pmatrix} 1 \\ 0 \end{pmatrix}$     **f** Translation $\begin{pmatrix} 2 \\ 4 \end{pmatrix}$

3   **a**

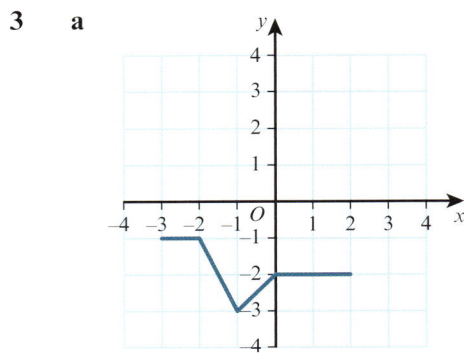

   **b**

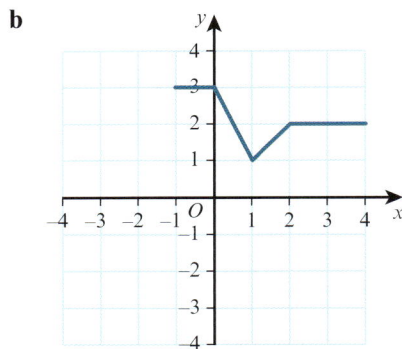

   **c**

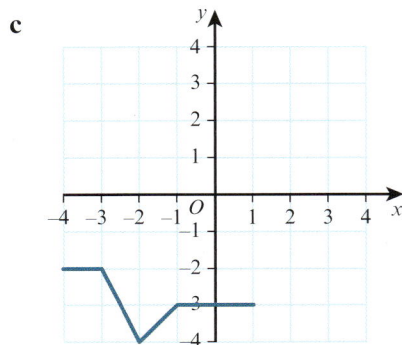

4   **a**

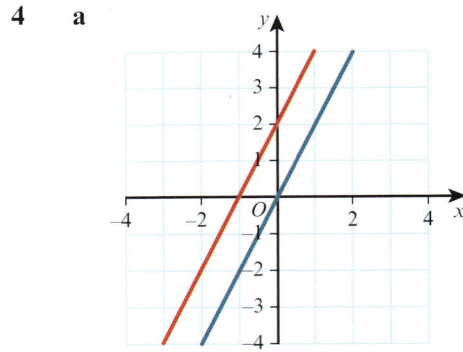

   **b** $a = 2$        **c** $b = -1$

5   $y = (x + 1)(x - 4)(x - 7)$

6   $y = x^2 - 6x + 8$

7   $a = 2, \ b = -3, \ c = 1$

## Exercise 2F

1   **a**

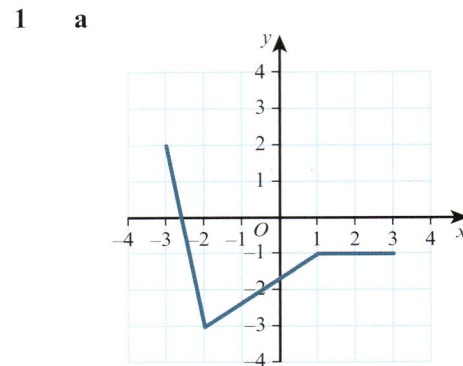

   **b**

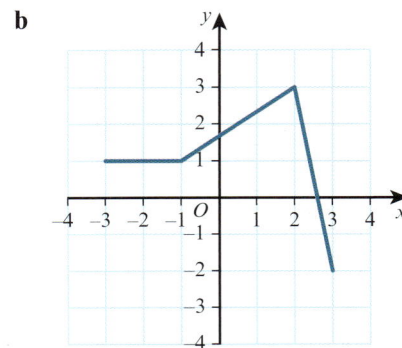

2   **a** $y = -5x^2$        **b** $y = 2x^4$

   **c** $y = 2x^2 + 3x + 1$    **d** $y = 3x^2 - 2x - 5$

3   **a** Reflection in the $x$-axis

   **b** Reflection in the $y$-axis

   **c** Reflection in the $x$-axis

   **d** Reflection in the $x$-axis

## Exercise 2G

**1 a**

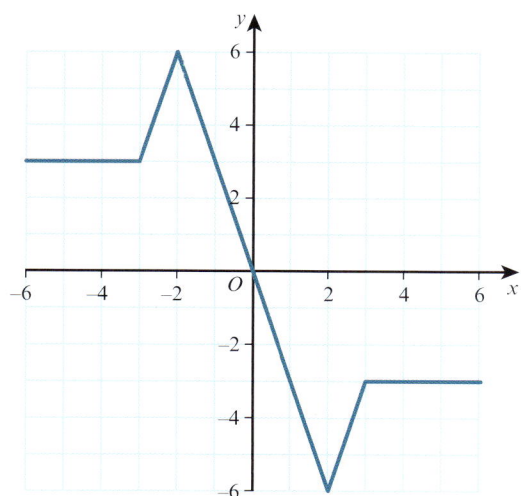

**b**

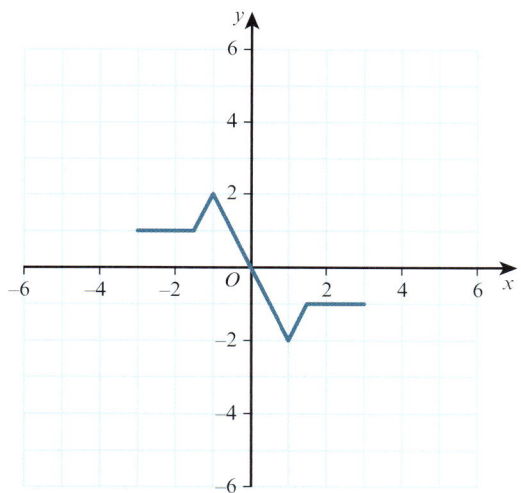

**2 a** $y = 6x^2$      **b** $y = 3x^3 - 3$

  **c** $y = 2^{x-1} + 2$      **d** $y = \dfrac{1}{2}x^2 - 4x + 10$

  **e** $y = 162x^3 - 108x$

**3 a** Stretch parallel to the $x$-axis with stretch factor $\dfrac{1}{2}$

  **b** Stretch parallel to the $y$-axis with stretch factor 3

  **c** Stretch parallel to the $y$-axis with stretch factor 2

  **d** Stretch parallel to the $x$-axis with stretch factor $\dfrac{1}{3}$

## Exercise 2H

**1 a**

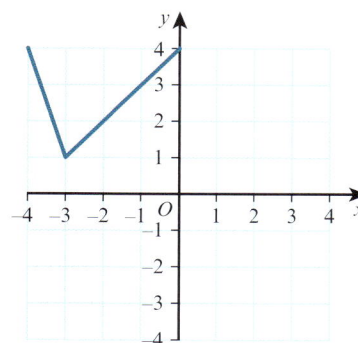

**b**

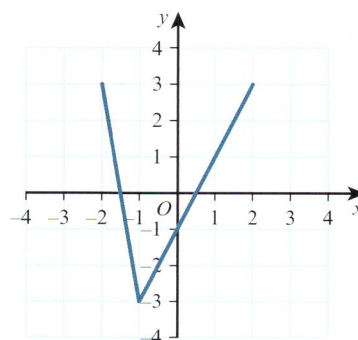

**c**

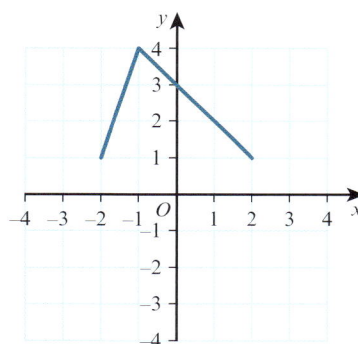

**d**

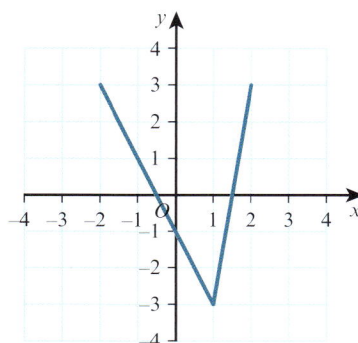

e

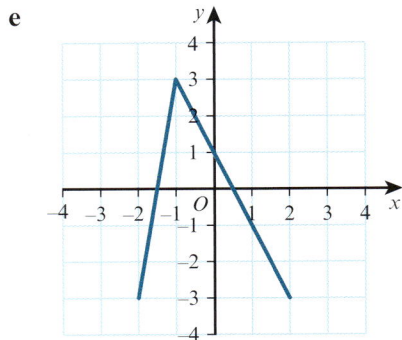

f

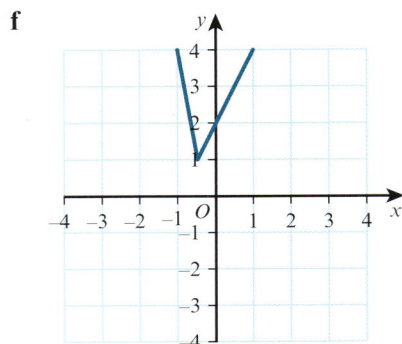

294

g

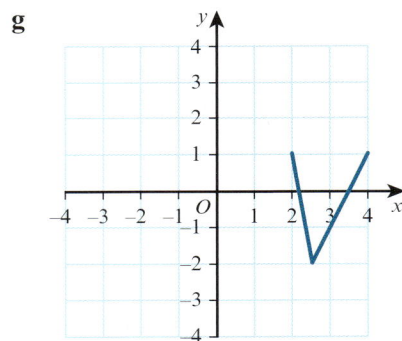

h

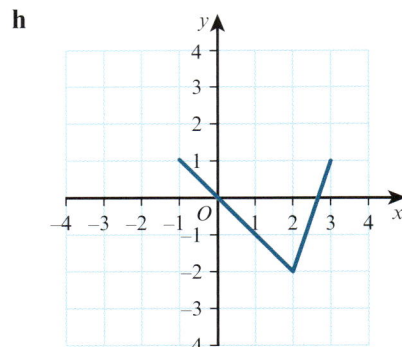

**2**  **a**  $y = 2f(-x)$          **b**  $y = 2 - f(x)$
    **c**  $y = 2f(x-1) + 1$
**3**  **a**  $y = 3(x-1)^2$          **b**  $y = 3(x-1)^2$

**4**  **a**  $y = \dfrac{1}{4}(x-5)^2$          **b**  $y = \left(\dfrac{1}{2}x - 5\right)^2$

  **c**

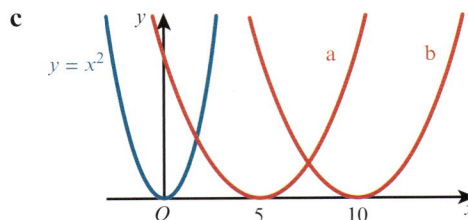

**5**  **a**  $y = 2x^2 - 8$          **b**  $y = -x^2 + 4x - 5$
**6**  **a**  $y = 2g(-x)$          **b**  $y = 3 - f(x-2)$
**7**  **a**  Stretch parallel to the $y$-axis with stretch

    factor $\dfrac{1}{2}$ followed by a translation $\begin{pmatrix} 0 \\ 3 \end{pmatrix}$

  **b**  Reflection in the $x$-axis followed by a

    translation $\begin{pmatrix} 0 \\ 2 \end{pmatrix}$

  **c**  Translation $\begin{pmatrix} 6 \\ 0 \end{pmatrix}$ followed by a stretch parallel

    to the $x$-axis with stretch factor $\dfrac{1}{2}$
  **d**  Stretch parallel to the $y$-axis with stretch

    factor 2 followed by a translation $\begin{pmatrix} 0 \\ -8 \end{pmatrix}$

**8**  **a**  Translation $\begin{pmatrix} -5 \\ 0 \end{pmatrix}$ followed by a stretch

    parallel to the $y$-axis with stretch factor $\dfrac{1}{2}$

  **b**  Translation $\begin{pmatrix} -1 \\ 0 \end{pmatrix}$, stretch parallel to the

    $y$-axis with stretch factor $\dfrac{1}{2}$, reflection in the

    $x$-axis, translation $\begin{pmatrix} 0 \\ -2 \end{pmatrix}$

  **c**  Translation $\begin{pmatrix} 3 \\ 0 \end{pmatrix}$, stretch parallel to the

    $y$-axis with stretch factor 2, reflection in the

    $x$-axis, translation $\begin{pmatrix} 0 \\ 4 \end{pmatrix}$

**9**  **a**  $y = -\sqrt{\dfrac{1}{2}x - 1} + 3$   **b**  $y = -\sqrt{\dfrac{1}{2}(x-1)} - 3$

**10**  **a**  $y = 3[(-x+4)^2 + 2] = 3(4-x)^2 + 6$
   **b**  $y = 3[(-(x+4))^2] + 2 = 3(x+4)^2 + 2$

**11** Translation $\begin{pmatrix} 2 \\ 0 \end{pmatrix}$ followed by reflection in the $y$-axis or reflection in the $y$-axis followed by translation $\begin{pmatrix} -2 \\ 0 \end{pmatrix}$

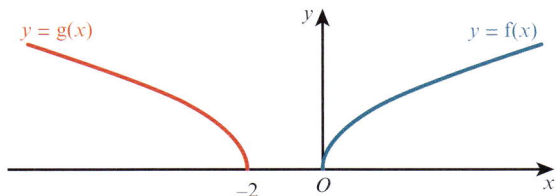

$y = g(x)$     $y = f(x)$

**12** Translation $\begin{pmatrix} -10 \\ 0 \end{pmatrix}$ followed by a stretch parallel to the $x$-axis with stretch factor $\dfrac{1}{2}$ or stretch parallel to the $x$-axis with stretch factor $\dfrac{1}{2}$ followed by translation $\begin{pmatrix} -5 \\ 0 \end{pmatrix}$

## End-of-chapter review exercise 2

**1** $\dfrac{25}{4} - 9\left(x - \dfrac{7}{6}\right)^2$

**2** **a**

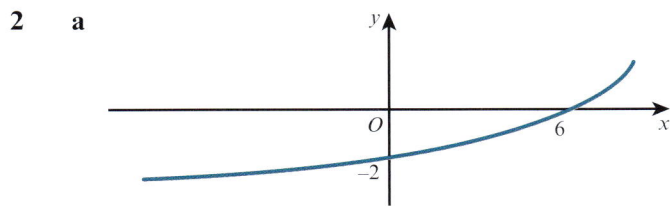

**b** Translation $\begin{pmatrix} -3 \\ 0 \end{pmatrix}$ followed by a reflection in the $y$-axis or reflection in the $y$-axis followed by translation $\begin{pmatrix} 3 \\ 0 \end{pmatrix}$

**3** **a**

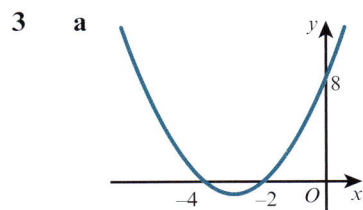

**b** $y = 3x^2 + 6x$

**4** **a** $f^{-1}: x \mapsto \sqrt{x+2}$ for $x \geqslant -2$

**b**

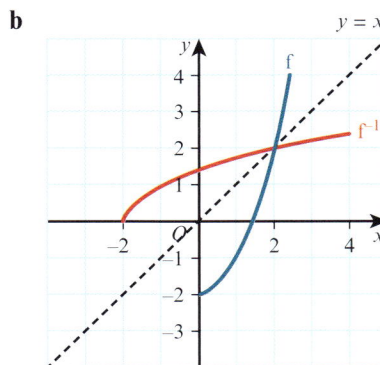

**5** **i** $-(x-3)^2 + 4$     **ii** 3

**iii** $f^{-1}(x) = 3 + \sqrt{4-x}$, domain is $x \leqslant 0$

**6** **i** $(x-2)^2 - 4 + k$

**ii** $f(x) \geqslant k - 4$

**iii** $p = 2$

**iv** $f^{-1}(x) = 2 + \sqrt{x + 4 - k}$, domain is $x \geqslant k - 4$

**7** **i** $-5 \leqslant f(x) \leqslant 4$

**ii**

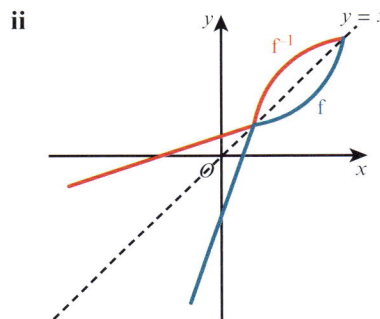

**iii** $f^{-1}(x) = \begin{cases} \dfrac{1}{3}(x+2) & \text{for } -5 \leqslant x \leqslant 1 \\[2mm] 5 - \dfrac{4}{x} & \text{for } 1 < x \leqslant 4 \end{cases}$

**8** **i** $4(x-3)^2 - 25$, vertex is $(3, -25)$

**ii** $g(x) \geqslant -9$

**iii** $g^{-1}(x) = 3 - \dfrac{1}{2}\sqrt{x+25}$, domain is $x \geqslant -9$

**9** **i** $2(x-3)^2 - 5$     **ii** 3

**iii** $f(x) \geqslant 27$

**iv** $f^{-1}(x) = 3 + \sqrt{\dfrac{x+5}{2}}$, domain is $x \geqslant 27$

**10** **i** $(x-1)^2 - 16$     **ii** $-16$

**iii** $p = 6$, $q = 10$     **iv** $f^{-1}(x) = 1 + \sqrt{x+16}$

**11 i** $2(x-3)^2 - 11$ **ii** $f \geqslant -11$
**iii** $-1 < x < 7$ **iv** $k = 22$

**12 i** $fg(x) = 2x^2 - 3$, $gf(x) = 4x^2 + 4x - 1$
**ii** $a = -1$ **iii** $b = 2$
**iv** $\frac{1}{2}(x^2 - 3)$ **v** $h^{-1}(x) = -\sqrt{x+2}$

**13 i** $2(x-2)^2 + 2$ **ii** $2 \leqslant f(x) \leqslant 10$
**iii** $2 \leqslant x \leqslant 10$
**iv** $f(x)$: half parabola from $(0, 10)$ to $(2, 2)$;
$g(x)$: line through $O$ at $45°$;
$f^{-1}(x)$: reflection of $f(x)$ in $g(x)$
**v** $f^{-1}(x) = 2 - \sqrt{\frac{1}{2}(x-2)}$

# 3 Coordinate geometry

## Prerequisite knowledge

**1** $\left(-4\frac{1}{2}, -2\right)$, $13$

**2 a** $-\frac{1}{6}$ **b** $6$

**3 a** $\frac{2}{3}$ **b** $-5$
**c** $7\frac{1}{2}$

**4 a** $(x-4)^2 - 21$ **b** $4 - \sqrt{21}$, $4 + \sqrt{21}$

## Exercise 3A

**1 a** $PQ = 5\sqrt{5}$, $QR = 4\sqrt{5}$, $PR = 3\sqrt{5}$,
right-angled triangle
**b** $PQ = \sqrt{197}$, $QR = \sqrt{146}$, $PR = 3\sqrt{5}$,
not right angled

**2** $17$ units$^2$

**3** $a = 3$ or $a = -9$

**4** $b = 3$ or $b = -5\frac{4}{5}$

**5** $a = 2$, $b = -1$

**6 a** $(-2, -1)$ **b** $(-1, 9)$
**c** $2\sqrt{41}$, $2\sqrt{101}$

**7** $k = 4$

**8** $38\frac{1}{2}$ units$^2$

**9** $k = 2$

**10** $(-2, 6)$

**11 a** $(5, 2)$ **b** $8\sqrt{2}$

**12** $A(-5, 5)$, $B(7, 3)$, $C(-3, -3)$

## Exercise 3B

**1 a** $\frac{1}{5}, \frac{1}{6}$ **b** Not collinear

**2** Proof

**3** $-\frac{2}{5}, \frac{5}{2}$

**4** $(7, -1)$

**5** $k = \frac{5}{7}$

**6** $k = 2$ or $k = 3$

**7** $(0, -26)$

**8 a** $1$ **b** $5$

**9** $a = 10$, $b = 4$

**10 a** $\frac{1}{2}$ **b** $-2$
**c** $a = 6$ or $a = -4$

**11 a** $(6, 6)$ **b** $a = -4$, $b = 16$, $c = 11$
**c** $4\sqrt{145}$ **d** $100$

## Exercise 3C

**1 a** $y = 2x + 1$ **b** $y = -3x - 1$
**c** $2x + 3y = 1$

**2 a** $2y = 3x - 3$ **b** $9x + 5y = 2$
**c** $2x - 3y = 9$

**3 a** $y = 3x + 4$ **b** $x + 2y = -8$
**c** $x + 2y = 8$ **d** $3x + 2y = 18$

**4 a** $y = 2x + 2$ **b** $5x + 3y = 9$
**c** $7x + 3y = -6$

**5** $(8, 2)$

**6 a** $y = \frac{3}{2}x + 8$ **b** $(0, 8)$
**c** $39$

**7 a** $(6, 3)$ **b** $y = -\frac{2}{3}x + 7$

**8 a** $y = \frac{4}{3}x + 10$ **b** $\left(-7\frac{1}{2}, 0\right)$, $(0, 10)$
**c** $12\frac{1}{2}$

**9 a** $2y = 5x + 33$ **b** $33$

**10** $E(4, 6)$, $F(10, 3)$

**11** 10

**12** $(14, -2)$

**13** **a** $y = -3x + 2$      **b** $(-1, 5)$
     **c** $5\sqrt{10}$, $4\sqrt{10}$      **d** 100

**14** **a** **i** $y = 4\frac{1}{2}$      **ii** $x + y = 7$
     **b** $\left(2\frac{1}{2}, 4\frac{1}{2}\right)$

**15** **a** $y = 2x - 7$      **b** $\left(4\frac{2}{5}, 1\frac{4}{5}\right)$

**16** $x + y = 8$, $3x + y = 3$. Other solutions possible.

## Exercise 3D

**1** **a** $(0, 0)$, 4      **b** $(0, 0)$, $\dfrac{3\sqrt{2}}{2}$
     **c** $(0, 2)$, 5      **d** $(5, -3)$, 2
     **e** $(-7, 0)$, $3\sqrt{2}$      **f** $(3, -4)$, $\dfrac{3\sqrt{10}}{2}$
     **g** $(4, -10)$, $\sqrt{6}$      **h** $\left(3\frac{1}{2}, 2\frac{1}{2}\right)$, 10

**2** **a** $x^2 + y^2 = 64$      **b** $(x - 5)^2 + (y + 2)^2 = 16$
     **c** $(x + 1)^2 + (y - 3)^2 = 7$
     **d** $\left(x - \dfrac{1}{2}\right)^2 + \left(y + \dfrac{3}{2}\right)^2 = \dfrac{25}{4}$

**3** $(x - 2)^2 + (y - 5)^2 = 25$

**4** $(x + 2)^2 + (y - 2)^2 = 52$

**5**
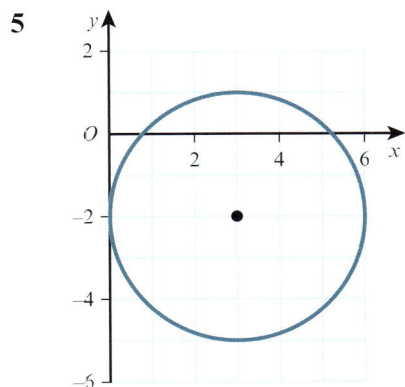

**6** $(x - 6)^2 + (y + 5)^2 = 25$

**7** Proof

**8** $(x - 5)^2 + y^2 = 8$ and $(x - 5)^2 + (y - 4)^2 = 8$

**9** $(x - 4)^2 + (y - 2)^2 = 20$

**10** $(x - 3)^2 + (y + 1)^2 = 16$, $(3, -1)$, 4

**11** $y = \dfrac{3}{4}x - \dfrac{21}{2}$

**12** $(x - 5)^2 + (y - 2)^2 = 29$

**13** **a** Proof      **b** $(x + 1)^2 + (y - 4)^2 = 20$

**14** $(x - 5)^2 + (y + 3)^2 = 40$

**15** $(x - 9)^2 + (y - 2)^2 = 85$

**16** $(x + 3)^2 + (y + 10)^2 = 100$,
$(x - 13)^2 + (y + 10)^2 = 100$

**17** **a** **i** $1 + \sqrt{2}$      **ii** Student's own answer
     **b** **i** $3 + 2\sqrt{2}$      **ii** Student's own answer

## Exercise 3E

**1** $(-1, -4)$, $(5, 2)$

**2** $2\sqrt{5}$

**3** Proof

**4** $-\dfrac{2}{29} < m < 2$

**5** **a** $(0, 6)$, $(8, 10)$      **b** $y = -2x + 16$
     **c** $\left(5 - \sqrt{5}, 6 + 2\sqrt{5}\right)$, $\left(5 + \sqrt{5}, 6 - 2\sqrt{5}\right)$
     **d** $20\sqrt{5}$

**6** $(4, 3)$

**7** **a** $(x - 12)^2 + (y - 5)^2 = 25$ and
$(x - 2)^2 + (y - 10)^2 = 100$
     **b** Proof

## End-of-chapter review exercise 3

**1** $2 < a < 26$

**2** **i** $\dfrac{4}{9}$ and $\dfrac{1}{4}$      **ii** $\dfrac{49}{24}$

**3** $a = -4$, $b = -1$ or $a = 12$, $b = 7$

**4** 10

**5** **a** $a = 5$, $b = -2$      **b** $(4, -5)$
     **c** $y = -\dfrac{2}{5}x - 3\frac{2}{5}$

**6** **i** $16t^2$      **ii** Proof

**7** $(13, -7)$

**8** **a** $(-2, 2)$, $(4, 5)$      **b** $y = -2x + 5\frac{1}{2}$

**9**   **a**  $(-2, -3)$          **b**  $y = -\dfrac{1}{2}x + 4\dfrac{3}{4}$

   **c**  $\dfrac{19}{2} - \sqrt{113},\ \dfrac{19}{2} + \sqrt{113}$

**10**  **i**  $2, m = 1$          **ii**  $(-1, 6)$
   **iii**  $(5, 12)$

**11**  **i**  $y = 2x - 2$          **ii**  $(0, -2), \left(\dfrac{8}{5}, \dfrac{6}{5}\right)$

**12**  **i**  $y = -2x + 6,\ (3, 0)$   **ii**  Proof
   **iii**  $(-1, 8),\ 2\sqrt{10}$

**13**  **a**  $y = -\dfrac{2}{3}x + 3$          **b**  $p = -1$

   **c**  $(x - 6)^2 + (y + 1)^2 = 26$

**14**  **a**  $(19, 13)$          **b**  $104$

**15**  **a**  $\left(\dfrac{10}{3}, 10\right)$          **b**  $k < -12, k > 12$

**16**  **a**  $y = -\dfrac{4}{3}x + 2$          **b**  Proof

   **c**  $(x - 15)^2 + (y - 7)^2 = 325$

**17**  **a**  $4, (4, -2)$          **b**  $4 - 2\sqrt{3},\ 4 + 2\sqrt{3}$
   **c**  Proof          **d**  Proof

## Cross-topic review exercise 1

**1**   $x = \pm \dfrac{2}{3},\ x = \pm \dfrac{\sqrt{2}}{2}$

**2**   **a**

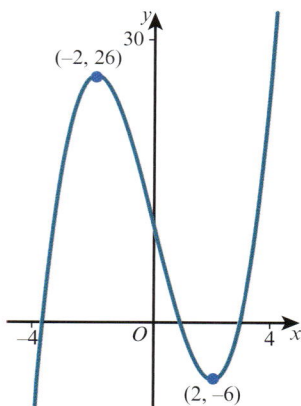

   **b**

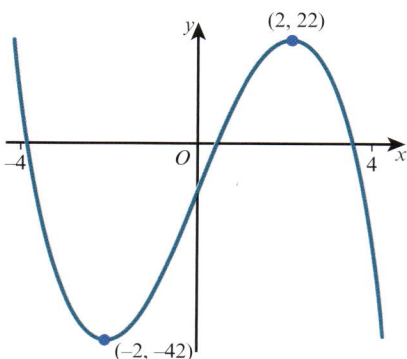

**3**   $a = 5, b = -2$

**4**   Translation $\begin{pmatrix} 5 \\ 0 \end{pmatrix}$, vertical stretch with stretch factor 2

**5**   $y = -x^2 + 6x - 8$

**6**

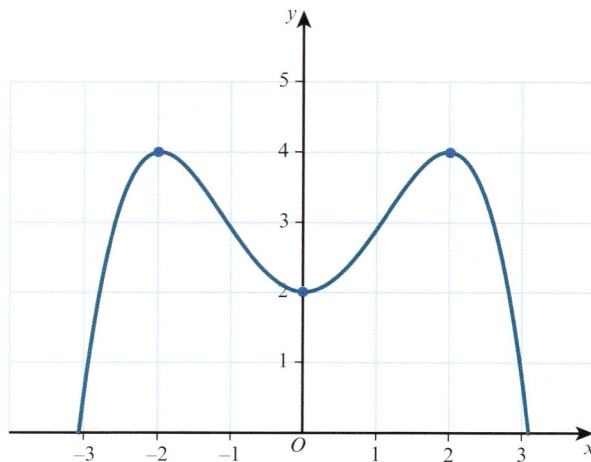

**7**   **a**  $1 \leqslant x \leqslant 5$          **b**  $-13, 3$

**8**   $(k^2, -2k)$

**9**   $6\sqrt{5}$

**10**  **a**  $k = 14$          **b**  $y = \dfrac{1}{3}x + 4$

**11**  **a**  $(-1, -11), (6, 3)$          **b**  $k < -\dfrac{25}{12}$

**12**  **a**  $k = -\dfrac{1}{2}$          **b**  $x > 5$

**13**  **i**  $fg(x) = 5x$, range is $fg(x) \geqslant 0$
   **ii**  $g^{-1}(x) = \dfrac{4 - 2x}{5x}$, domain is $0 < x \leqslant 2$

**14**  **a**  $b = -5, c = -14$
   **b**  **i**  $(2.5, -20.25)$   **ii**  $-3 < x < 8$

**15**  **a**  $2y = 3x + 25$          **b**  $(-3, 8)$

**16**  **a**  $36 - (x - 6)^2$          **b**  $36$
   **c**  $x \leqslant 36, g^{-1}(x) \geqslant 6$   **d**  $g^{-1}(x) = 6 + \sqrt{36 - x}$

**17**  **a**  $3(x + 2)^2 - 13$          **b**  $(-2, -13)$
   **c**  $6 < x < 18$

**18**  **a**  $a = 12, b = 2$          **b**  $-3$
   **c**  $g^{-1}(x) = -3 + \sqrt{\dfrac{26 - x}{2}}$

**19**  **a**  $(x - 8)^2 + (y - 3)^2 = 29$          **b**  $5x + 2y = 75$

**20** **a** $x = \dfrac{1}{2}$

**b** $f^{-1}(x) = \dfrac{x+7}{3}$, $g^{-1}(x) = 5 - \dfrac{18}{x}$

**c** Proof

**21** **a** $\dfrac{17}{4} - \left(x + \dfrac{3}{2}\right)^2$ **b** $\left(-\dfrac{3}{2}, \dfrac{17}{4}\right)$

**c** $-5$ and $-1$ **d** $(1, -2), (-1, 4)$

**22** **a** $(8, 0)$ **b** $10$

**c** $(-2, 0), (18, 0)$ **d** $y = -\dfrac{3}{4}x + 6$

**23** **a** $k = -2$

**b** **i** $fg(x) \geqslant 28$

**ii** $(fg)^{-1}(x) = -\sqrt{\dfrac{x+26}{6}}$, domain is $x \geqslant 28$,

range is $(fg)^{-1}(x) \leqslant -3$

**24** **a** **i** $(4, 5), (10, 2)$ **ii** $4x - 2y = 21$

**b** $k = \pm 4\sqrt{10}$

# 4 Circular measure

## Prerequisite knowledge

**1** $(12 + \pi)\,\text{cm}$, $3\pi\,\text{cm}^2$

**2** $13, 67.4$

**3** $5.14, 15.4\,\text{cm}^2$

## Exercise 4A

**1** **a** $\dfrac{\pi}{9}$ **b** $\dfrac{2\pi}{9}$

**c** $\dfrac{5\pi}{36}$ **d** $\dfrac{5\pi}{18}$

**e** $\dfrac{\pi}{36}$ **f** $\dfrac{5\pi}{6}$

**g** $\dfrac{3\pi}{4}$ **h** $\dfrac{7\pi}{6}$

**i** $\dfrac{5\pi}{4}$ **j** $\dfrac{5\pi}{3}$

**k** $\dfrac{13\pi}{36}$ **l** $3\pi$

**m** $\dfrac{\pi}{20}$ **n** $\dfrac{7\pi}{36}$

**o** $\dfrac{10\pi}{3}$

**2** **a** $90°$ **b** $60°$ **c** $30°$ **d** $15°$

**e** $240°$ **f** $80°$ **g** $54°$ **h** $105°$

**i** $81°$ **j** $810°$ **k** $252°$ **l** $48°$

**m** $225°$ **n** $420°$ **o** $202.5°$

**3** **a** $0.489$ **b** $0.559$

**c** $0.820$ **d** $3.49$

**e** $5.59$

**4** **a** $68.8°$ **b** $45.8°$

**c** $76.8°$ **d** $87.1°$

**e** $45.3°$

**5** **a.**

| Degrees | 0 | 45 | 90 | 135 | 180 | 225 | 270 | 315 | 360 |
|---|---|---|---|---|---|---|---|---|---|
| Radians | 0 | $\dfrac{\pi}{4}$ | $\dfrac{\pi}{2}$ | $\dfrac{3\pi}{4}$ | $\pi$ | $\dfrac{5\pi}{4}$ | $\dfrac{3\pi}{2}$ | $\dfrac{7\pi}{4}$ | $2\pi$ |

**b.**

| Degrees | 0 | 30 | 60 | 90 | 120 | 150 | 180 | 210 |
|---|---|---|---|---|---|---|---|---|
| Radians | 0 | $\dfrac{\pi}{6}$ | $\dfrac{\pi}{3}$ | $\dfrac{\pi}{2}$ | $\dfrac{2\pi}{3}$ | $\dfrac{5\pi}{6}$ | $\pi$ | $\dfrac{7\pi}{6}$ |

| Degrees | 240 | 270 | 300 | 330 | 360 |
|---|---|---|---|---|---|
| Radians | $\dfrac{4\pi}{3}$ | $\dfrac{3\pi}{2}$ | $\dfrac{5\pi}{3}$ | $\dfrac{11\pi}{6}$ | $2\pi$ |

**6** **a** $0.644$ **b** $14.1$

**c** $0.622$ **d** $0$

**e** $\dfrac{\sqrt{3}}{2}$ **f** $0.727$

**7** $7.79\,\text{cm}$

**8** $12.79°$

## Exercise 4B

**1** **a** $2\pi\,\text{cm}$ **b** $3\pi\,\text{cm}$

**c** $6\pi\,\text{cm}$ **d** $28\pi\,\text{cm}$

**2** **a** $13\,\text{cm}$ **b** $2.275\,\text{cm}$

**3** **a** $0.5\,\text{rad}$ **b** $0.8\,\text{rad}$

**4** $15.6\,\text{m}$

**5** **a** $19.2\,\text{cm}$ **b** $20.5\,\text{cm}$

**c** $50.4\,\text{cm}$

**6** **a** $0.927\,\text{rad}$ **b** $4\,\text{cm}$

**c** $17.6\,\text{cm}$

**7** **a** $14\,\text{cm}$ **b** $11.8\,\text{cm}$

**c** $25.8\,\text{cm}$

**8** **a** $13\,\text{cm}$ **b** $2.35\,\text{rad}$

**c** $56.6\,\text{cm}$

**9** **a** Proof **b** $43.4\,\text{cm}$

**10** Proof

## Exercise 4C

**1**  **a** $12\pi\,\text{cm}^2$  **b** $20\pi\,\text{cm}^2$

    **c** $\dfrac{9\pi}{4}\,\text{cm}^2$  **d** $54\pi\,\text{cm}^2$

**2**  **a** $867\,\text{cm}^2$  **b** $3.042\,\text{cm}^2$

**3**  **a** $1.125\,\text{rad}$  **b** $1.5\,\text{rad}$

**4**  **a** $1.25\,\text{rad}$  **b** $40\,\text{cm}^2$

**5**  **a** $1.75\,\text{rad}$  **b** $4.79\,\text{cm}$

    **c** $5.16\,\text{cm}^2$

**6**  $\left(32\sqrt{3} - \dfrac{32\pi}{3}\right)\text{cm}^2$

**7**  **a** $\dfrac{5\sqrt{3}}{3}\,\text{cm}$  **b** $\dfrac{25}{6}\left(2\sqrt{3} - \pi\right)\text{cm}^2$

**8**  $1.86\,\text{cm}^2$

**9**  **a** $29.1\,\text{cm}^2$  **b** $36.5\,\text{cm}^2$

    **c** $51.7\,\text{cm}^2$  **d** $13.9\,\text{cm}^2$

**10**  **a** Proof  **b** Proof

**11**  $\left(\dfrac{2\pi r^2}{3} - \dfrac{\sqrt{3}r^2}{2}\right)\text{cm}^2$

**12**  $100\left(1 + \dfrac{\pi}{3} - \sqrt{3}\right)\text{cm}^2$

**13**  **a** $\left(\tan x + \dfrac{1}{\tan x}\right)\text{cm}$  **b** $0.219\,\text{rad}$

**14**  **a** Proof  **b** Proof

## End-of-chapter review exercise 4

**1**  **a** $15 - 5\sqrt{3} + \dfrac{5\sqrt{3}\pi}{6}$  **b** $\dfrac{25\sqrt{3}}{4} - \dfrac{25\pi}{8}$

**2**  **i** $\alpha = \dfrac{\pi}{8}$  **ii** $8 + 5\pi$

**3**  **i** $8\tan\alpha - 2\alpha$  **ii** $\dfrac{4}{\cos\alpha} + 4\tan\alpha + 2\alpha$

**4**  **i** $r(1 + \theta + \cos\theta + \sin\theta)$  **ii** $55.2$

**5**  **i** $AC = r - r\cos\theta$  **ii** $\dfrac{7\pi}{3} + 2\sqrt{3} - 2$

**6**  **i** Proof  **ii** $36 - (r - 6)^2$

    **iii** $A = 36,\ \theta = 2$

**7**  **i** Proof  **ii** $r^2$

**8**  **i** $r\left(\cos\theta + 1 - \sin\theta + \dfrac{\pi}{2} - \theta\right)$  **ii** $6.31$

**9**  **i** $4\alpha\cos\alpha + 4\alpha + 8 - 8\cos\alpha$  **ii** $\dfrac{\pi}{3}$

**10**  **i** $2\pi r + r\alpha + 2r$  **ii** $\dfrac{3r^2\alpha}{2} + \pi r^2$

    **iii** $\alpha = \dfrac{2}{5}\pi$

# 5 Trigonometry

## Prerequisite knowledge

**1**  **a** $\sqrt{1 + r^2}$  **b** $\dfrac{r}{\sqrt{1 + r^2}}$

    **c** $\dfrac{1}{\sqrt{1 + r^2}}$  **d** $r$

**2**  **a** **i** $\dfrac{\pi}{4}$  **ii** $4\pi$  **iii** $\dfrac{5\pi}{6}$

    **b** **i** $30°$  **ii** $630°$  **iii** $195°$

**3**  **a** $0,\ 5$  **b** $-5,\ \dfrac{3}{2}$

## Exercise 5A

**1**  **a** $\dfrac{3}{5}$  **b** $\dfrac{3}{4}$  **c** $\dfrac{24}{25}$

    **d** $\dfrac{20}{3}$  **e** $\dfrac{4}{5}$  **f** $\dfrac{12}{19}$

**2**  **a** $\dfrac{2}{3}$  **b** $\dfrac{\sqrt{5}}{3}$

    **c** $1$  **d** $\dfrac{\sqrt{5}}{2}$

    **e** $\dfrac{6}{5}$  **f** $\dfrac{15(3 - \sqrt{5})}{4}$

**3**  **a** $\dfrac{\sqrt{15}}{4}$  **b** $\dfrac{\sqrt{15}}{15}$

    **c** $\dfrac{15}{16}$  **d** $\dfrac{15}{16}$

    **e** $4 + \sqrt{15}$  **f** $\dfrac{75 - 4\sqrt{15}}{15}$

**4**  **a** $\dfrac{1}{4}$  **b** $\dfrac{1}{2}$

    **c** $\dfrac{\sqrt{3} + \sqrt{2}}{2}$  **d** $\sqrt{3}$

    **e** $\dfrac{2 - \sqrt{3}}{2}$  **f** $1$

**5**  **a** $\dfrac{1}{2}$  **b** $\dfrac{1}{4}$  **c** $\dfrac{1}{2}$

    **d** $\dfrac{\sqrt{2} - 2\sqrt{6}}{2}$  **e** $-1$  **f** $\dfrac{2 + \sqrt{3}}{3}$

**6**

| | $\theta = \dfrac{\pi}{4}$ | $\theta = \dfrac{\pi}{3}$ | $\theta = \dfrac{\pi}{6}$ |
|---|---|---|---|
| $\tan\theta$ | $1$ | $\sqrt{3}$ | $\dfrac{1}{\sqrt{3}}$ |
| $\cos\theta$ | $\dfrac{1}{\sqrt{2}}$ | $\dfrac{1}{2}$ | $\dfrac{\sqrt{3}}{2}$ |
| $\dfrac{1}{\sin\theta}$ | $\sqrt{2}$ | $\dfrac{2}{\sqrt{3}}$ | $2$ |

## Exercise 5B

1. **a** 70°  **b** 40°
   **c** 20°  **d** 40°
2. **a** 2nd quadrant, 80°  **b** 3rd quadrant, 80°
   **c** 4th quadrant, 50°  **d** 3rd quadrant, 30°
   **e** 1st quadrant, 40°  **f** 2nd quadrant, $\dfrac{\pi}{3}$
   **g** 3rd quadrant, $\dfrac{\pi}{6}$  **h** 1st quadrant, $\dfrac{\pi}{3}$
   **i** 3rd quadrant, $\dfrac{4\pi}{9}$  **j** 4th quadrant, $\dfrac{\pi}{8}$
3. **a** 125°  **b** −160°
   **c** 688°  **d** $\dfrac{5\pi}{4}$
   **e** $\dfrac{8\pi}{3}$  **f** $-\dfrac{13\pi}{6}$

## Exercise 5C

1. **a** $-\sin 10°$  **b** $\cos 55°$
   **c** $-\tan 55°$  **d** $-\cos 65°$
   **e** $-\cos \dfrac{\pi}{5}$  **f** $-\sin \dfrac{\pi}{8}$
   **g** $-\cos \dfrac{3\pi}{10}$  **h** $\tan \dfrac{2\pi}{9}$
2. **a** $-\dfrac{1}{2}$  **b** $-\dfrac{\sqrt{3}}{3}$
   **c** $-\dfrac{\sqrt{2}}{2}$  **d** $\sqrt{3}$
   **e** $-\dfrac{\sqrt{3}}{2}$  **f** $\dfrac{1}{2}$
   **g** $-\dfrac{\sqrt{3}}{3}$  **h** $-\dfrac{1}{2}$
3. 4th quadrant
4. **a** $-\dfrac{\sqrt{21}}{5}$  **b** $-\dfrac{2}{\sqrt{21}}$
5. **a** $-\sqrt{\dfrac{2}{3}}$  **b** $\sqrt{2}$
6. **a** $-\dfrac{5}{13}$  **b** $\dfrac{12}{13}$
7. **a** $a$  **b** $\dfrac{a}{\sqrt{1+a^2}}$
   **c** $\dfrac{a}{\sqrt{1+a^2}}$  **d** $-\dfrac{a}{\sqrt{1+a^2}}$
8. **a** $\sqrt{1-b^2}$  **b** $\dfrac{b}{\sqrt{1-b^2}}$
   **c** $-\sqrt{1-b^2}$  **d** $\sqrt{1-b^2}$

9. **a** $-\dfrac{12}{13}$  **b** $-\dfrac{5}{12}$
   **c** $\dfrac{3}{5}$  **d** $-\dfrac{3}{4}$
10. **a** $-\dfrac{2}{\sqrt{13}}$  **b** $\dfrac{3}{\sqrt{13}}$
    **c** $-\dfrac{\sqrt{7}}{4}$  **d** $-\dfrac{\sqrt{7}}{3}$

11. 

|  | $\theta = 120°$ | $\theta = 135°$ | $\theta = 210°$ |
|---|---|---|---|
| $\tan\theta$ | $-\sqrt{3}$ | $-1$ | $\dfrac{1}{\sqrt{3}}$ |
| $\sin\theta$ | $\dfrac{\sqrt{3}}{2}$ | $\dfrac{1}{\sqrt{2}}$ | $-\dfrac{1}{2}$ |
| $\dfrac{1}{\cos\theta}$ | $-2$ | $-\sqrt{2}$ | $-\dfrac{2}{\sqrt{3}}$ |

## Exercise 5D

1. **a** 360°  **b** 180°
   **c** 360°  **d** 120°
   **e** 180°  **f** 180°
2. **a** 1  **b** 5
   **c** 7  **d** 3
   **e** 4  **f** 2
3. **a**

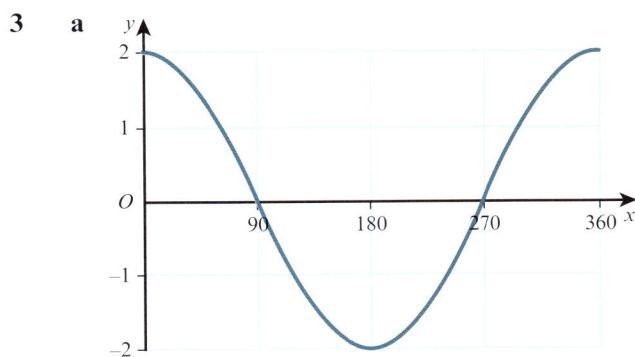

**b**

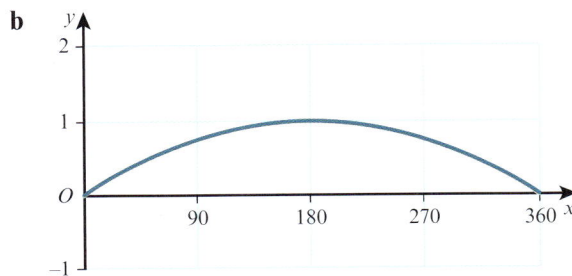

301

**c**

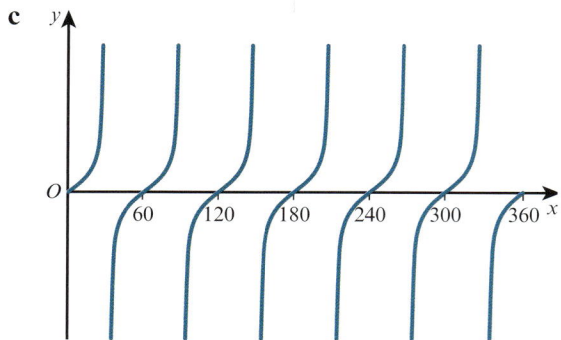

**d**

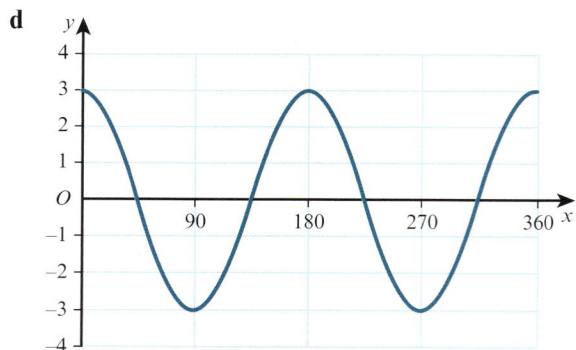

**e**

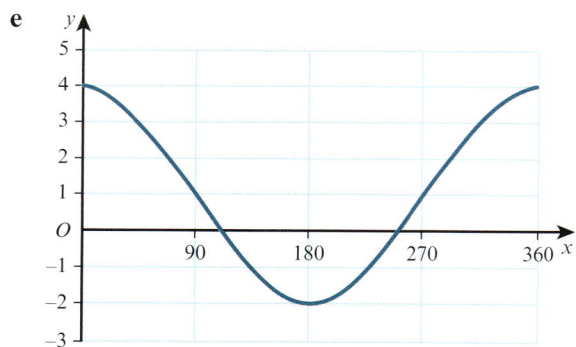

**f**

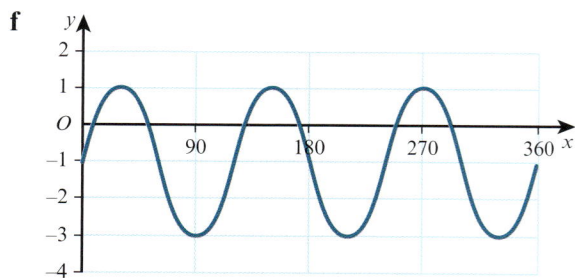

**g**

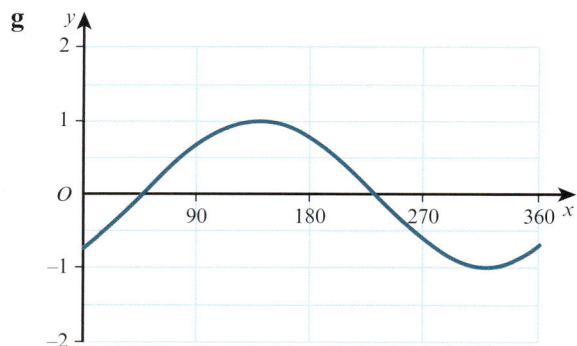

**h**

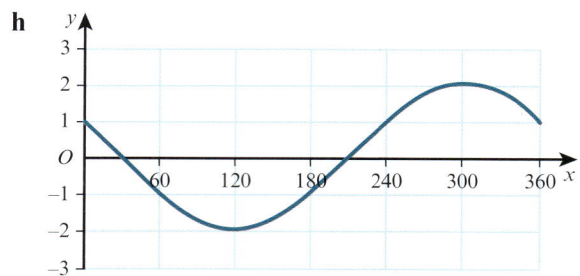

**i**

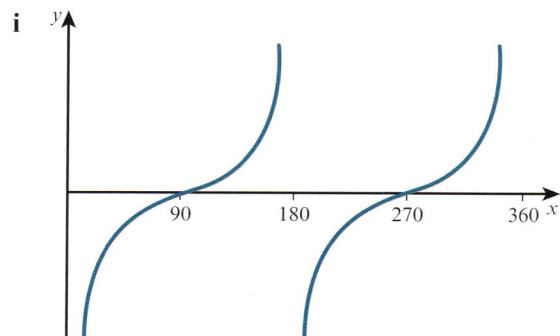

**4   a   i**

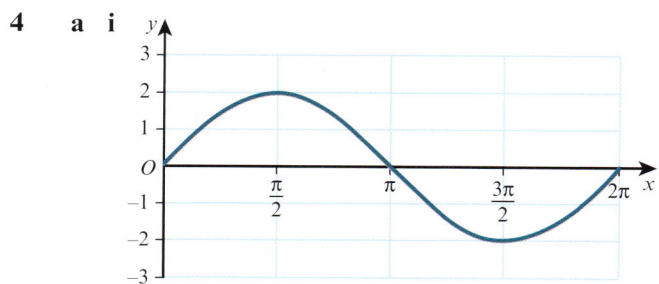

**ii**

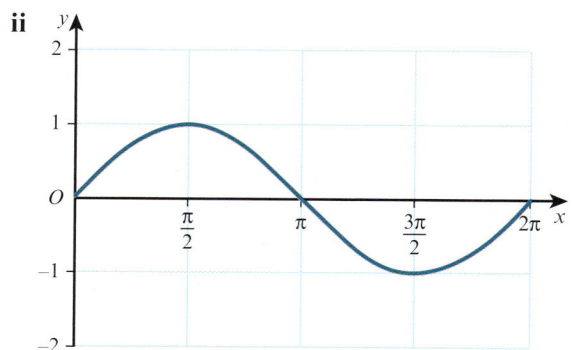

**iii**

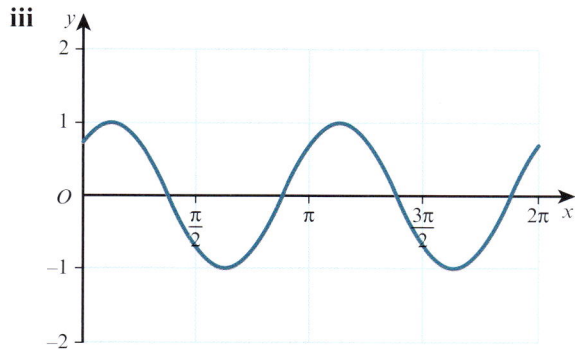

**b** $\left(\dfrac{\pi}{8},1\right),\left(\dfrac{5\pi}{8},-1\right),\left(\dfrac{9\pi}{8},1\right),\left(\dfrac{13\pi}{8},-1\right)$

**5 a**

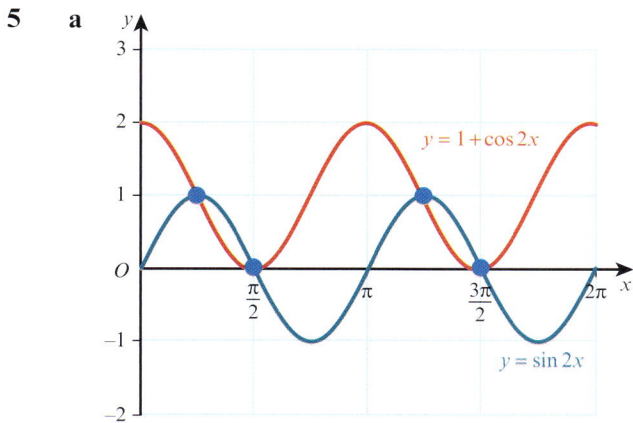

**b** 4

**6 a**

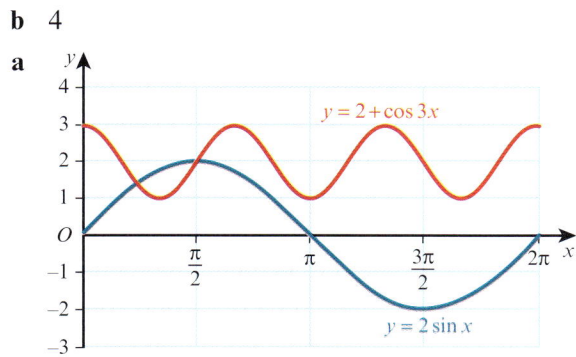

**b** 2

**7 a**

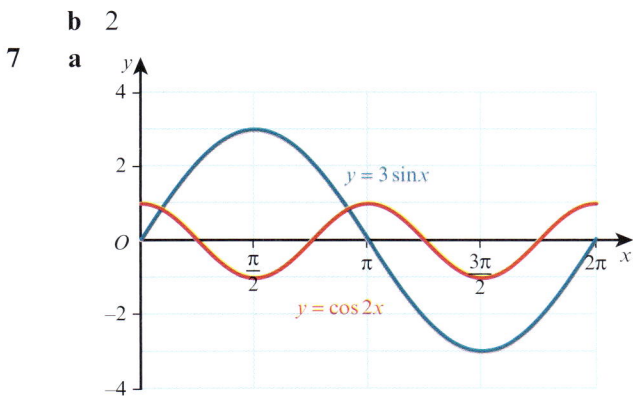

**b** 2

**8** $a = 4, b = 2, c = 5$

**9** $a = 3, b = 2, c = 3$

**10 a**

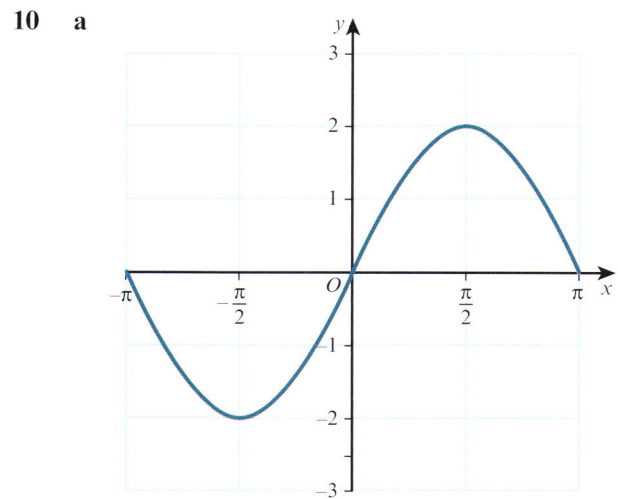

**b** $k = \dfrac{4}{\pi}$

**c** $(0, 0), \left(-\dfrac{\pi}{2}, -2\right)$

**11** $a = 3, b = 1, c = 5$

**12 a** $a = 3, b = 2$      **b** $1 \leqslant f(x) \leqslant 5$

**13 a** $a = 3, b = 5$

**b**

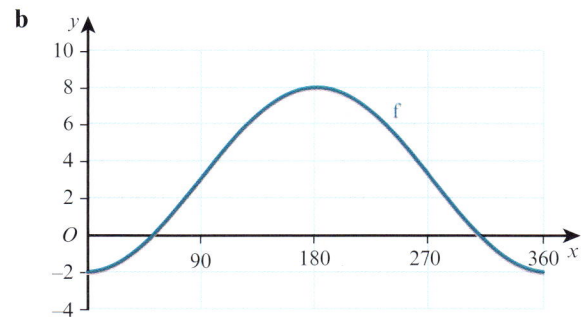

**14** $a = 5, b = 4, c = 3$

**15 a** $A = 2, B = 6$      **b** 5

**c**

**16**  $y = 2 + \sin x$

**17**  $y = 6 + \cos x$

## Exercise 5E

**1**  **a**  $0°$  **b**  $30°$
 **c**  $60°$  **d**  $-90°$
 **e**  $-60°$  **f**  $135°$

**2**  **a**  $0$  **b**  $\dfrac{\pi}{4}$
 **c**  $\dfrac{\pi}{4}$  **d**  $-\dfrac{\pi}{6}$
 **e**  $\dfrac{2\pi}{3}$  **f**  $-\dfrac{\pi}{3}$

**3**  **a**  $\dfrac{16}{25}$  **b**  $\dfrac{16}{9}$

**4**  **a**  $-7 \leqslant f(x) \leqslant -1$  **b**  $f^{-1}(x) = \sin^{-1}\left(\dfrac{x+4}{3}\right)$

**5**  **a**  $2 \leqslant f(x) \leqslant 6$

 **b**  f is one-one, $f^{-1}(x) = \cos^{-1}\left(\dfrac{4-x}{2}\right)$

**6**  **a**  $\dfrac{3\pi}{2}$

 **b**  $f^{-1}(x) = \sin^{-1}\left(\dfrac{5-x}{2}\right)$, $3 \leqslant x \leqslant 7$

**7**  **a**  $-9 \leqslant f(x) \leqslant -1$

 **b**  $f^{-1}(x) = 2\cos^{-1}\left(\dfrac{x+5}{4}\right)$, $0 \leqslant f^{-1}(x) \leqslant 2\pi$

## Exercise 5F

**1**  **a**  $56.3°, 236.3°$  **b**  $23.6°, 156.4°$
 **c**  $45.6°, 314.4°$  **d**  $197.5°, 342.5°$
 **e**  $126.9°, 233.1°$  **f**  $116.6°, 296.6°$
 **g**  $60°, 300°$  **h**  $216.9°, 323.1°$

**2**  **a**  $0.305, 2.84$  **b**  $\dfrac{\pi}{3}, \dfrac{5\pi}{3}$
 **c**  $1.25, 4.39$  **d**  $3.92, 5.51$
 **e**  $1.89, 5.03$  **f**  $\dfrac{2\pi}{3}, \dfrac{4\pi}{3}$
 **g**  $0.848, 2.29$  **h**  $2.19, 5.33$

**3**  **a**  $26.6°, 153.4°$
 **b**  $17.7°, 42.3°, 137.7°, 162.3°$
 **c**  $38.0°, 128.0°$  **d**  $105°, 165°$
 **e**  $24.1°, 155.9°$  **f**  $116.6°, 153.4°$
 **g**  $58.3°, 148.3°$  **h**  $5.77°, 84.2°$

**4**  **a**  $90°, 210°$  **b**  $\dfrac{\pi}{2}, \dfrac{7\pi}{6}$
 **c**  $139.1°, 175.9°$  **d**  $0.0643, 2.36, 3.21, 5.51$
 **e**  $278.2°$  **f**  $0, \dfrac{3\pi}{2}$

**5**  **a**  $26.6°, 206.6°$  **b**  $56.3°, 236.3°$
 **c**  $119.7°, 299.7°$
 **d**  $18.4°, 108.4°, 198.4°, 288.4°$

**6**  $0.298, 1.87$

**7**  **a**  $0°, 150°, 180°, 330°, 360°$
 **b**  $0°, 36.9°, 143.1°, 180°, 360°$
 **c**  $0°, 78.7°, 180°, 258.7°, 360°$
 **d**  $0°, 116.6°, 180°, 296.6°, 360°$
 **e**  $0°, 60°, 180°, 300°, 360°$
 **f**  $0°, 76.0°, 180°, 256.0°, 360°$

**8**  **a**  $60°, 120°, 240°, 300°$
 **b**  $56.3°, 123.7°, 236.3°, 303.7°$

**9**  **a**  $30°, 150°, 270°$
 **b**  $45°, 108.4°, 225°, 288.4°$
 **c**  $0°, 109.5°, 250.5°, 360°$
 **d**  $60°, 180°, 300°$
 **e**  $0°, 180°, 199.5°, 340.5°, 360°$
 **f**  $70.5°, 120°, 240°, 289.5°$
 **g**  $19.5°, 160.5°, 270°$
 **h**  $30°, 150°, 270°$

**10**  **a**  0.565, 2.58  **b**  $\dfrac{\pi}{6}, \dfrac{5\pi}{6}$

**11**  2.03, $\dfrac{3\pi}{4}$, 5.18, $\dfrac{7\pi}{4}$

## Exercise 5G

**1**  $9 \sin^2 x - 3$

**2**  **a**  Proof  **b**  Proof
 **c**  Proof  **d**  Proof
 **e**  Proof  **f**  Proof

**3**  **a**  Proof  **b**  Proof
 **c**  Proof  **d**  Proof

**4**  **a**  Proof  **b**  Proof
 **c**  Proof  **d**  Proof

**5**  **a**  Proof  **b**  Proof
 **c**  Proof  **d**  Proof
 **e**  Proof  **f**  Proof
 **g**  Proof  **h**  Proof

**6**  **a**  Proof  **b**  Proof
 **c**  Proof  **d**  Proof
 **e**  Proof  **f**  Proof

**7**  4

**8**  **a**  $4 + 3 \sin^2 x$  **b**  $4 \leqslant f(x) \leqslant 7$

**9**  **a**  $(\sin \theta + 2)^2 - 5$  **b**  4, −4

**10**  **a**  Proof

 **b**  $\sin \theta = \dfrac{1 - 4a^2}{1 + 4a^2}$ ,  $\cos \theta = \dfrac{4a}{1 + 4a^2}$

## Exercise 5H

**1**  **a**  Proof  **b**  76.0°, 256.0°
**2**  **a**  Proof  **b**  18.4°, 116.6°
**3**  **a**  Proof
 **b**  60°, 131.8°, 228.2°, 300°
**4**  **a**  Proof  **b**  30°, 150°, 210°, 330°
**5**  **a**  Proof  **b**  72.4°, 287.6°
**6**  **a**  Proof  **b**  65.2°, 245.2°
**7**  **a**  Proof  **b**  41.8°, 138.2°, 270°
**8**  **a**  Proof  **b**  30°, 150°
**9**  **a**  Proof  **b**  66.4°, 293.6°
**10**  **a**  Proof  **b**  70.5°, 289.5°
**11**  **a**  Proof  **b**  30°, 150°, 210°, 330°

## End-of-chapter review exercise 5

**1**  $a = 1, b = 2$

**2**  1.95

**3**  **a**  $\sqrt{1 - k^2}$  **b**  $\dfrac{\sqrt{1 - k^2}}{k}$
 **c**  $-k$

**4**  $x = \pm \dfrac{\sqrt{3}}{2}$

**5**  39.3° or 129.3°

**6**  30° or 150°

**7**  30° or 150°

**8**  **i**

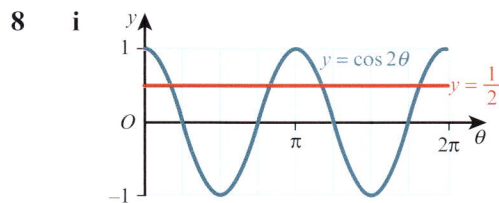

 **ii**  4  **iii**  20
**9**  **i**  Proof  **ii**  45°, 135°, 225°, 315°
**10**  **i**  60° or 300°  **ii**  120°
**11**  **i**  Proof  **ii**  109.5° or 250.5°
**12**  **a**  $\dfrac{\pi}{6}, \dfrac{5\pi}{6}$  **b**  −2.21, 0.927
**13**  **i**  $f(x) \leqslant 3$  **ii**  $3 - 2\sqrt{3}$
 **iii**

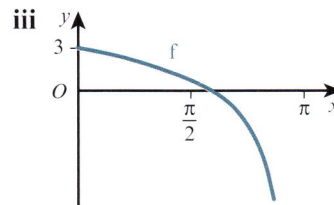

 **iv**  $f^{-1}(x) = 2 \tan^{-1}\left(\dfrac{3 - x}{2}\right)$
**14**  **i**  30° or 150°  **ii**  $n = 3, \theta = 290°$
**15**  **i**  Proof
 **ii**  54.7°, 125.3°, 234.7°, 305.3°
**16**  **i**  Proof  **ii**  194.5° or 345.5°
**17**  **i**  1.68

**ii**

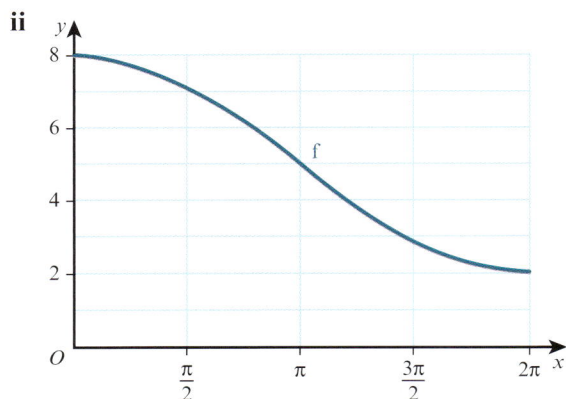

**iii** f is one-one

**iv** $f^{-1}(x) = 2\cos^{-1}\left(\dfrac{x-5}{3}\right)$

# 6 Series

## Prerequisite knowledge

**1**  **a** $4x^2 + 12x + 9$  **b** $9x^3 - 9x^2 - x + 1$

**2**  **a** $125x^6$  **b** $-32x^{15}$

**3**  **a** $2n + 3$  **b** $11 - 3n$

## Exercise 6A

**1**  **a** $x^3 + 6x^2 + 12x + 8$

   **b** $1 - 4x + 6x^2 - 4x^3 + x^4$

   **c** $x^3 + 3x^2y + 3xy^2 + y^3$

   **d** $8 - 12x + 6x^2 - x^3$

   **e** $x^4 - 4x^3y + 6x^2y^2 - 4xy^3 + y^4$

   **f** $8x^3 + 36x^2y + 54xy^2 + 27y^3$

   **g** $16x^4 - 96x^3 + 216x^2 - 216x + 81$

   **h** $x^6 + \dfrac{9x}{2} + \dfrac{27}{4x^4} + \dfrac{27}{8x^9}$

**2**  **a** 12  **b** 10

   **c** −90  **d** 16

   **e** 40  **f** −32

   **g** 768  **h** $-\dfrac{5}{2}$

**3**  $A = 486$, $B = 540$, $C = 30$

**4**  ±2

**5**  **a** $16 + 32x + 24x^2 + 8x^3 + x^4$

   **b** $97 + 56\sqrt{3}$

**6**  **a** $1 + 3x + 3x^2 + x^3$

   **b  i** $16 + 8\sqrt{5}$  **ii** $16 - 8\sqrt{5}$

   **c** 32

**7**  $16 + 112x + 312x^2 + 432x^3 + 297x^4 + 81x^5$

**8**  **a** $x^8 - 4x^6 + 6x^4 - 4x^2 + 1$  **b** −16

**9**  −216

**10**  54

**11**  **a** $1 + 4y + 6y^2$  **b** 142

**12**  5

**13**  $x^{11}$

**14**  **a** $x^5 + 5x^4y + 10x^3y^2 + 10x^2y^3 + 5xy^4 + y^5$

   **b** 113 100

**15**  **a** $p = 8$, $q = 8$  **b** $36\sqrt{2}$

**16**  **a** $y^3 - 3y$  **b** $y^5 - 5y^3 + 5y$

## Exercise 6B

**1**  **a** 35  **b** 84

   **c** 495  **d** 5005

**2**  **a** $\dfrac{n(n-1)}{2}$  **b** $n$

   **c** $\dfrac{n(n-1)(n-2)}{6}$

**3**  **a** 45  **b** 56

   **c** 364  **d** 792

**4**  **a** $1 + 16x + 112x^2$  **b** $1 - 30x + 405x^2$

   **c** $1 + \dfrac{7}{2}x + \dfrac{21}{4}x^2$  **d** $1 + 12x^2 + 66x^4$

   **e** $2187 + \dfrac{5103}{2}x + \dfrac{5103}{4}x^2$

   **f** $8192 - 53248x + 159744x^2$

   **g** $256 + 1024x^2 + 1792x^4$

   **h** $512 + 1152x^2 + 1152x^4$

**5**  **a** −84  **b** 5940

   **c** $\dfrac{35}{4}$  **d** −9720

**6**  7920

**7**  −224 000

**8**  −41 184

**9**  40 095

**10**  **a** $128 + 320x + 224x^2$  **b** $1 - 28x + 345x^2$

   **c** $1 - 3x + 3x^2$

**11**  **a** $1024 + 5120x + 11520x^2$

   **b** $1024 + 10240y + 30720y^2$

**12** **a** $1 - 4x + 7x^2$    **b** 1

**13** $16 + 224x + 1176x^2$

**14** $a = -2$, $b = 1$, $p = -364$

**15** $n = 8$, $p = 256$, $q = -144$

## Exercise 6C

**1** $a + 6d$, $a + 18d$

**2** **a** 22, 1210    **b** 35, 3535

**3** **a** 1037    **b** $-1957$

   **c** $38\frac{1}{3}$    **d** $-3160x$

**4** 7

**5** **a** 7, 29    **b** 2059

**6** 1442

**7** 1817

**8** 31

**9** 5586

**10** 25

**11** $360

**12** **a** 17, $-4$    **b** 20

**13** 7, 8

**14** 10, $-4$

**15** $\frac{1}{2}(5n - 11)$

**16** 9°

**17** **a** $a = 8d$    **b** $9a$

**18** Proof

**19** **a** $4 - 3\sin^2 x$    **b** Proof

**20** **a** Proof    **b** 900

## Exercise 6D

**1** **a** No    **b** 3, 15 309

   **c** $-\frac{1}{3}, -\frac{1}{27}$    **d** No

   **e** No    **f** $-1, -1$

**2** $ar^5$, $ar^{14}$

**3** $\frac{2}{3}$

**4** $-10.8$

**5** $\frac{3}{2}, 8$

**6** 64

**7** $-8, 2$

**8** **a** 765    **b** 255

   **c** $-85$    **d** $700\frac{5}{9}$

**9** 21

**10** **a** $8\left(\dfrac{3}{4}\right)^n$    **b** 48.8125 m

**11** **a** $\dfrac{48}{x+1}$    **b** $2, \dfrac{1}{3}$

**12** 40, $-20$

**13** **a** $17 715.61    **b** $94 871.71

**14** Proof

**15** Proof

**16** Proof

## Exercise 6E

**1** **a** 3    **b** $1\frac{1}{9}$

   **c** $26\frac{2}{3}$    **d** $-36\frac{4}{7}$

**2** $\dfrac{4}{3}$

**3** 32

**4** $\dfrac{2}{3}$, 810

**5** **a** $0.\dot{5}\dot{7} = \dfrac{57}{100} + \dfrac{57}{10\,000} + \dfrac{57}{1\,000\,000} + \ldots$

   **b** Proof

**6** 0.25, 199.21875

**7** 0.5, 9

**8** $\dfrac{52}{165}$

**9** **a** $\dfrac{2}{3}$, 13.5    **b** 40.5

**10** **a** $-0.25$, 256    **b** 204.8

**11** **a** 90    **b** 405

**12** **a** 36    **b** 192

**13** 93.75

**14** $a = 2$, $r = \dfrac{3}{5}$

**15** $\dfrac{\pi}{3} < x < \dfrac{\pi}{2}$

**16** **a** $5\pi$    **b** $\dfrac{11\pi}{8}$

**17** **a** Proof    **b** Proof

   **c** Proof

## Exercise 6F

**1**  **a**  352  **b**  788.125

**2**  **a**  100  **b**  16

**3**  **a**  2  **b**  384, 32

**4**  −2.5, 22.5

**5**  **a**  $\dfrac{3}{5}$  **b**  12.96, 68

**6**  **a**  4  **b**  $\dfrac{3}{2}$, $n = 6$

**7**  **a**  $x = -3$ or 5, 3rd term = 24 or 40

  **b**  $-\dfrac{4}{5}$

## End-of-chapter review exercise 6

**1**  240

**2**  5

**3**  2

**4**  $-\dfrac{864}{25}$

**5**  16 800

**6**  40

**7**  $\dfrac{135}{2}$

**8**  **a**  $6561x^{16} - 17\,496x^{15} + 20\,412x^{14}$

  **b**  −37 908

**9**  **a**  $1 + 8px + 28p^2x^2$  **b**  $-\dfrac{17}{7}$, 3

**10**  **a**  **i**  $1 + 10x + 40x^2$  **ii**  $243 - 405x + 270x^2$

  **b**  5940

**11**  23

**12**  **a**  $-\dfrac{2}{3}$  **b**  2187

  **c**  1312.2

**13**  **a**  $d = 2a$  **b**  $99a$

**14**  **a**  $a = 44$, $d = -3$  **b**  $n = 22$

**15**  **a**  $a = 60$, $d = -10.5$  **b**  $42\frac{2}{3}$

**16**  **a**  17  **b**  $r = -\dfrac{5}{7}$, $S = \dfrac{7}{4}$

**17**  **a**  $\dfrac{1}{5}$  **b**  16

**18**  **i**  41 000  **ii**  22 100

**19**  **a**  $a = 10$, $b = 45$

  **b**  **i**  $0 < \theta < \dfrac{\pi}{3}$  **ii**  1.125

**20**  **i**  $x = -2$ or 6, 3rd term = 16 or 48  **ii**  $\dfrac{16}{27}$

**21**  **a**  $2\frac{1}{4}$  **b**  115.2°

**22**  **a**  $d = 6$, $a = 13$  **b**  $a = \dfrac{12}{7}$, $r = \dfrac{5}{7}$

## Cross-topic review exercise 2

**1**  $x^{180}$

**2**  3840

**3**  **a**  $729x^6 - 2916x^3 + 4860$  **b**  −5832

**4**  **a**  $1 - 10x + 40x^2$  **b**  12

**5**  **a**  −25.6  **b**  $27\frac{7}{9}$

**6**  **a**  14  **b**  112

**7**  $\dfrac{625}{8}$

**8**  **i**  Proof

  **ii**  35.3°, 144.7°, 215.3°, 324.7°

**9**  **a**  Proof  **b**  $225 - (r - 15)^2$

  **c**  15  **d**  225, maximum

**10**  **a**  Proof  **b**  $625 - (r - 25)^2$

  **c**  25  **d**  625, maximum

**11**  **a**  Proof  **b**  $\dfrac{40\,000}{\pi} - \pi\left(r - \dfrac{200}{\pi}\right)^2$

  **c**  Proof  **d**  $\dfrac{40\,000}{\pi}$, maximum

**12**  **i**  Proof  **ii**  0.9273

  **iii**  5.90 cm$^2$

**13**  **i**  $r^2(\tan\theta - \theta)$  **ii**  $12 + 12\sqrt{3} + 4\pi$

**14**  **i**  $2 - 5\cos^2 x$  **ii**  −3 and 2

  **iii**  0.685, 2.46

**15**  **i**  Proof  **ii**  26.6°, 153.4°

**16**  **i**  $-5 \leqslant \text{f}(x) \leqslant 3$  **ii**  (0.253, 0), (0, −1)

  **iii**

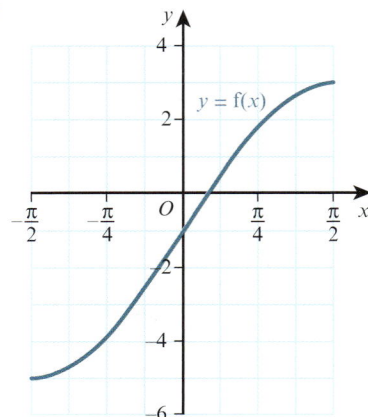

**iv** $f^{-1}(x) = \sin^{-1}\left(\dfrac{x+1}{4}\right)$, domain is $-5 \leqslant x \leqslant 3$,

range is $-\dfrac{1}{2}\pi \leqslant f^{-1}(x) \leqslant \dfrac{1}{2}\pi$

**17 a** 250

**b i** Proof **ii** 70

# 7 Differentiation

## Prerequisite knowledge

**1 a** $3x^{\frac{3}{2}}$ **b** $5x^{\frac{2}{3}}$

**c** $\dfrac{1}{2}x^{\frac{1}{2}}$ **d** $\dfrac{1}{2}x^{-1}$

**e** $3x^{-2}$ **f** $-\dfrac{2}{5}x^{\frac{5}{3}}$

**2 a** $4(x-2)^{-3}$ **b** $2(3x+1)^{-5}$

**3** $-\dfrac{3}{2}$

**4** $y = 2x + 1$

## Exercise 7A

**1 a**

| Chord | $AF$ | $BF$ | $CF$ | $DF$ | $EF$ |
|---|---|---|---|---|---|
| Gradient | 2 | 2.5 | 2.8 | 2.95 | 2.99 |

**b** 3

**2 a** 4 **b** $-2$

**c** $\dfrac{1}{2}$ **d** $-3$

**3 a** $5x^4$ **b** $9x^8$

**c** $-\dfrac{4}{x^5}$ **d** $-\dfrac{1}{x^2}$

**e** 0 **f** $\dfrac{2}{3\sqrt[3]{x}}$

**g** $5x^4$ **h** $3x^2$

**4 a** $8x^3$ **b** $15x^4$

**c** $3x^5$ **d** $-\dfrac{3}{x^2}$

**e** $-\dfrac{10}{3x^3}$ **f** 0

**g** $\dfrac{2}{\sqrt{x}}$ **h** $-\dfrac{1}{\sqrt{x^5}}$

**5 a** $10x - 1$ **b** $6x^2 + 8$

**c** $10x - 3$ **d** $2x + 1$

**e** $16x^3 - 24x$ **f** $\dfrac{10}{x^3} - \dfrac{2}{x^2}$

**g** $14x + \dfrac{3}{x^2} - \dfrac{4}{x^3}$ **h** $3 - \dfrac{5}{x^2} + \dfrac{1}{4\sqrt{x^3}}$

**i** $6\sqrt{x} + \dfrac{3}{2\sqrt{x}} + \dfrac{1}{\sqrt{x^3}}$

**6 a** 3 **b** 0.5

**c** $-\dfrac{5}{4}$

**7** 15

**8** $-3$

**9** $-8$

**10** $(-2, -10), (2, -6)$

**11** $\dfrac{5}{4}$

**12 a** $(-2, 7), (3, -8)$ **b** $-8, 2$

**13** $a = 2, b = -7$

**14** $a = -5, b = 2$

**15** $a = 4, b = -6$

**16** $a = -10.5, b = 18$

**17** $-2 < x < 3$

**18** $x \leqslant -1$ and $x \geqslant \dfrac{1}{2}$

**19** Proof

## Exercise 7B

**1 a** $6(x+4)^5$ **b** $16(2x+3)^7$

**c** $-20(3-4x)^4$ **d** $\dfrac{9}{2}\left(\dfrac{1}{2}x+1\right)^8$

**e** $10(5x-2)^7$ **f** $50(2x-1)^4$

**g** $-56(4-7x)^3$ **h** $\dfrac{21}{5}(3x-1)^6$

**i** $10x(x^2+3)^4$ **j** $-16x(2-x^2)^7$

**k** $6x^2(x+2)(x+4)^2$ **l** $\dfrac{5(x^3-5)^4(2x^3+5)}{x^6}$

**2 a** $-\dfrac{1}{(x+2)^2}$ **b** $-\dfrac{3}{(x-5)^2}$

**c** $\dfrac{16}{(3-2x)^2}$ **d** $-\dfrac{32x}{(x^2+2)^2}$

**e** $-\dfrac{72}{(3x+1)^7}$ **f** $-\dfrac{45}{2(3x+1)^6}$

**g** $-\dfrac{16(x+1)}{x^2(x+2)^2}$ **h** $-\dfrac{49(4x-5)}{(2x-5)^8 x^8}$

**3 a** $\dfrac{1}{2\sqrt{x-5}}$ **b** $\dfrac{1}{\sqrt{2x+3}}$

**c** $\dfrac{2x}{\sqrt{2x^2-1}}$ **d** $\dfrac{3x^2-5}{2\sqrt{x^3-5x}}$

**e** $-\dfrac{2}{3\sqrt[3]{(5-2x)^2}}$ **f** $\dfrac{3}{\sqrt{3x+1}}$

**g** $-\dfrac{1}{\sqrt{(2x-5)^3}}$ **h** $\dfrac{6}{\sqrt[3]{(2-3x)^4}}$

**4** 10

**5** 12

**6** $4, \dfrac{4}{3}$

**7** $(5,1)$

**8** $a=5, b=3$

## Exercise 7C

**1 a** $y=3x-7$  **b** $8x+y=17$
**c** $y=3x+9$  **d** $2y=x-1$

**2 a** $4y=x+4$  **b** $y=x-1$
**c** $x-6y=9$  **d** $5x-6y=3$

**3 a** $x+4y=4$  **b** $y=4x-7.5$

**4 a** Proof  **b** $(0.6, 2.48)$

**5** $(0, 7.5)$

**6** $(-2.5, 8.5)$

**7 a** $y=4x-68$  **b** $(17, 0)$

**8 a** $2+\dfrac{20}{x^3}$  **b** Proof

**9** $(-7.5, 2.5)$

**10** $(-6.2, 6.6)$

**11 a** $(-32.6, 28.4)$  **b** 317.2 units$^2$
**12 a** $(2\sqrt{3}, 8\sqrt{3}), (-2\sqrt{3}, -8\sqrt{3})$
**b** $x+4y=0$

**13** $(4, -1)$

**14** $y=0.6x+1.6, 30.96°$

**15** $\sqrt{73}$

**16 a** $-7$  **b** $(0.4, -5.48)$

## Exercise 7D

**1 a** 2  **b** $30x-14$

**c** $-\dfrac{36}{x^4}$  **d** $48(2x-3)^2$

**e** $-\dfrac{4}{\sqrt{(4x-9)^3}}$  **f** $\dfrac{27}{2\sqrt{(3x+1)^5}}$

**g** $\dfrac{4x-30}{x^4}$  **h** $24x^2-36x+20$

**i** $-\dfrac{5x+12}{4\sqrt{x^5}}$

**2 a** $\dfrac{30}{x^4}-\dfrac{45}{x^7}$  **b** $-\dfrac{3}{x^3}$

**c** $\dfrac{4}{x^3}-\dfrac{45}{4\sqrt{x^7}}$  **d** $-\dfrac{9}{4\sqrt{(1-3x)^3}}$

**e** $\dfrac{15\sqrt{x}}{4}-6$  **f** $\dfrac{80}{3\sqrt[3]{(2x+1)^7}}$

**3** $4-8(2x-1)^3, -48(2x-1)^2$

**4 a** $-1$  **b** 4
**c** 10

**5** $-\dfrac{48}{(2x-1)^9}$

**6** $\dfrac{2}{81}$

**7**

| $x$ | 0 | 1 | 2 | 3 | 4 | 5 | 6 | 7 |
|---|---|---|---|---|---|---|---|---|
| $\dfrac{dy}{dx}$ | + | + | 0 | − | − | 0 | + | + |
| $\dfrac{d^2y}{dx^2}$ | − | − | − | − | + | + | + | + |

**8** $x>5$

**9** Proof

**10** Proof

**11 a** Proof  **b** $-8, 8$

**12** $a=-4, b=4$

## End-of-chapter review exercise 7

**1** $3x^3+\dfrac{7}{4x^2}$

**2** $-3\frac{5}{9}$

**3** Proof

**4** $-15(3-5x)^2-2, 150(3-5x)$

**5** $-\dfrac{8}{15}$

**6 i** $4x+5y=66$  **ii** $(16.5, 0)$

**7 i** $5-\dfrac{24}{x^3}$  **ii** Proof

**8** 37.4

**9** $(4.25, -7.5)$

**10 a** $y=-4x+18$  **b** $y=\dfrac{1}{4}x+1$

**11 a** Proof
**b i** $(-0.8, 15.5)$  **ii** $(2.1, 8.25)$

**12** $y=-\dfrac{1}{2}x$

310

**13**   **i**   Proof       **ii**   $10\frac{3}{4}$

**14**   **i**   Proof       **ii**   $\left(-\dfrac{9}{2}, \dfrac{9}{4}\right)$

    **iii**   $\left(\dfrac{1}{2}, 4\frac{3}{4}\right)$

**15**   **i**   $y - 2 = \dfrac{1}{2}(x - 1),\ y - 2 = -2(x - 1)$

    **ii**   $2\frac{1}{2}$

    **iii**   $\left(\dfrac{6}{11}, \dfrac{12}{11}\right)$, $E$ not midpoint of $OA$

# 8 Further differentiation

## Prerequisite knowledge

**1**   **a**   $x < -1$ and $x > 3$    **b**   $-2 < x < 3$

**2**   **a**   $6x - 1, 6$           **b**   $-\dfrac{3}{x^3}, \dfrac{9}{x^4}$

    **c**   $\dfrac{9\sqrt{x}}{2}, \dfrac{9}{4\sqrt{x}}$

**3**   **a**   $10(2x - 1)^4$      **b**   $\dfrac{18}{(1 - 3x)^3}$

## Exercise 8A

**1**   **a**   $x > 4$                **b**   $x > 1$

    **c**   $x < -1.75$        **d**   $x < 0$ and $x > 8$

    **e**   $x < 1$ and $x > 4$   **f**   $-2\frac{2}{3} < x < 2$

**2**   **a**   $x < 1\frac{1}{3}$         **b**   $x > 4.5$

    **c**   $2 < x < 5$       **d**   $-1 < x < 3$

    **e**   $x < 2$ and $x > 6\frac{2}{3}$   **f**   $x < -4$ and $x > 2$

**3**   $1.5 < x < 3.5$

**4**   $\dfrac{8}{(1 - 2x)^2}$, increasing

**5**   $\dfrac{2x - 6}{(x + 2)^3}$, neither

**6**   Proof

**7**   $8x + 20,\ 8x + 20 \geqslant 0$, if $x \geqslant 0$

**8**   Proof

**9**   $7 < x < 20$

## Exercise 8B

**1**   **a**   $(2, 4)$ minimum

    **b**   $(-0.5, 6.25)$ maximum

    **c**   $(-2, 22)$ maximum; $(2, -10)$ minimum

    **d**   $(-3, -17)$ minimum; $(1, 15)$ maximum

    **e**   $(-1, -4)$ minimum

    **f**   $(1, -7)$ maximum; $(2, -11)$ minimum

**2**   **a**   $(9, 6)$ minimum      **b**   $(1, 12)$ minimum

    **c**   $(-3, -12)$ maximum; $(3, 0)$ minimum

    **d**   $(-2, -28)$ maximum; $(2, 36)$ minimum

    **e**   $(4, 4)$ maximum      **f**   $(2, 6)$ minimum

**3**   $\dfrac{18}{x^3}, \dfrac{18}{x^3} \neq 0$

**4**   **a**   $-2, 3$            **b**   $-44, 81$

**5**   **a**   $a = 3$            **b**   $-3 < x < 1$

**6**   **a**   $a = -15, b = 36$    **b**   $(2, -2)$, maximum

**7**   Proof

**8**   $x = \dfrac{3 + k}{2}$, minimum; $x = \dfrac{3 - k}{2}$, maximum

**9**   $(0, 1)$, minimum; $(1, 2)$, maximum;

    $(2, 1)$, minimum

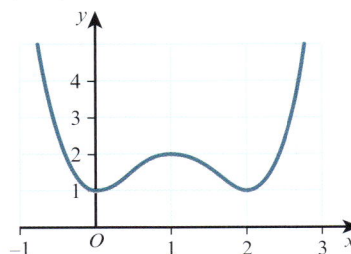

**10**   **a**   $a = -6, b = 5$     **b**   minimum

    **c**   $(0, 5)$, maximum    **d**   $(2, -11), -12$

**11**   **a**   $a = 4,\ b = 16$     **b**   minimum

    **c**   $x < 0$ and $x > 2$

**12**   **a**   $a = 54, b = -22$   **b**   minimum

    **c**   $x < 0$ and $0 < x < 3$

**13**   **a**   $a = -3, b = -12$   **b**   $(-1, 14)$

    **c**   $(2, -13) = $ minimum, $(-1, 14) = $ maximum

    **d**   $(0.5, 0.5), -13.5$

## Exercise 8C

**1**   **a**   $y = 9 - x$

    **b**   **i**   $P = 9x^2 - x^3$      **ii**   $108$

    **c**   **i**   $Q = 5x^2 - 36x + 162$   **ii**   $97.2$

**2**   **a**   $\theta = \dfrac{40 - 2r}{r}$      **b**   Proof

    **c**   $r = 10$           **d**   $A = 100$, maximum

**3**   **a**   $y = \dfrac{50 - x}{2}$      **b**   Proof

    **c**   $A = 312.5, x = 25$

**4**   **a**   $3x^2 - 10x + 160$   **b**   $x = 1\frac{2}{3}, 151\frac{2}{3}$ cm$^2$

**5**   **a**   Proof           **b**   $A = 37.5$, maximum

**6**   **a**   $QR = 9 - p^2$     **b**   Proof

    **c**   $p = \sqrt{3}$          **d**   $A = 12\sqrt{3}$, maximum

**7** **a** Proof   **b** $V = 486$, $x = 3$
**c** Maximum

**8** **a** $y = \dfrac{288}{x^2}$   **b** Proof
**c** 432, 12 cm by 6 cm by 8 cm

**9** **a** $y = 1 - \dfrac{1}{2}x - \dfrac{1}{4}\pi x$   **b** Proof
**c** $\dfrac{dA}{dx} = 1 - x - \dfrac{1}{4}\pi x,\ \dfrac{d^2A}{dx^2} = -1 - \dfrac{1}{4}\pi$
**d** $\dfrac{4}{4+\pi}$   **e** $A = \dfrac{2}{4+\pi}$, maximum

**10** **a** $h = \dfrac{5 - 2r - \pi r}{2}$   **b** Proof
**c** $\dfrac{dA}{dr} = 5 - 4r - \pi r,\ \dfrac{d^2A}{dr^2} = -4 - \pi$
**d** $\dfrac{5}{4+\pi}$   **e** $\dfrac{25}{8+2\pi}$, maximum

**11** **a** $r = \dfrac{50 - 2x}{\pi}$   **b** Proof
**c** $\dfrac{100}{(\pi + 4)} = 14.0,\ \dfrac{2500}{(4+\pi)} = 350$, minimum

**12** **a** $h = \dfrac{432}{r^2}$   **b** Proof
**c** $r = 6$   **d** $A = 216\pi$, minimum

**13** **a** $y = \dfrac{500 - 24x^2}{10x}$   **b** Proof
**c** $\dfrac{5\sqrt{10}}{6}$   **d** Proof

**14** **a** $h = \dfrac{160}{r} - \dfrac{3}{2}r$   **b** Proof
**c** $r = 8$

**15** **a** $r = \sqrt{20h - h^2}$   **b** Proof
**c** $13\frac{1}{3}$   **d** 1241, maximum

## Exercise 8D

**1** −0.315 units per second, decreasing
**2** $\dfrac{1}{300}$ units per second
**3** −0.04 units per second
**4** 0.08 units per second
**5** 1.25 units per second
**6** 0.09 units per second, increasing
**7** 1, 6
**8** 0.016 units per second
**9** $\dfrac{1}{3}$, 3

## Exercise 8E

**1** $\dfrac{4}{5}\pi$ cm$^2$ s$^{-1}$
**2** 18 cm$^3$ s$^{-1}$
**3** $\pi$ cm$^3$ s$^{-1}$
**4** 0.003 cm s$^{-1}$
**5** 0.125 cm s$^{-1}$
**6** $\dfrac{1}{320}$ cm s$^{-1}$
**7** $9\pi$ cm$^2$ s$^{-1}$
**8** **a** Proof   **b** $\dfrac{\sqrt{3}}{120}$ cm s$^{-1}$
**9** **a** $\dfrac{1}{3}$ cm s$^{-1}$   **b** $\dfrac{1}{7}$ cm s$^{-1}$
**10** **a** Proof   **b** $-\dfrac{9}{100\pi}$ cm s$^{-1}$
**11** $32\pi$ cm$^2$ s$^{-1}$
**12** **a** $2\sqrt{\dfrac{10}{\pi}}$ cm   **b** $\dfrac{5}{4\pi}\sqrt{\dfrac{\pi}{10}}$ cm s$^{-1}$
**13** **a** $40\pi$ cm$^3$ s$^{-1}$
**b** **i** 1.024 cm s$^{-1}$   **ii** $8\pi$ cm$^2$ s$^{-1}$

## End-of-chapter review exercise 8

**1** $\dfrac{2}{45\pi}$ cm s$^{-1}$
**2** $300\pi$ m$^2$ hr$^{-1}$
**3** Maximum
**4** $k = 0.0032$ kg cm$^3$, 0.096 kg day$^{-1}$
**5** **i** $A = 2p^2 + p^3$   **ii** 0.4
**6** **i** Proof   **ii** 20 m by 24 m
**7** **i** Proof   **ii** 120, minimum
**8** **i** $y = \dfrac{4(6-x)}{3}$   **ii** Proof
**iii** $A = 72$
**9** **i** $\dfrac{dy}{dx} = -\dfrac{8}{x^2} + 2,\ \dfrac{d^2y}{dx^2} = \dfrac{16}{x^3}$
**ii** (2, 8), minimum since $\dfrac{d^2y}{dx^2} > 0$ when $x = 2$
(−2, −8), maximum since $\dfrac{d^2y}{dx^2} < 0$ when $x = -2$
**10** **i** $y = 30 - x - \dfrac{\pi x}{4}$   **ii** Proof
**iii** $x = 15$   **iv** Maximum

**11** **a** $-2 < x < \dfrac{4}{3}$

**b** Maximum at $\left(-\dfrac{5}{3}, \dfrac{364}{27}\right)$, minimum at $(1, 4)$

**12** **i** Proof **ii** Maximum

**13** **i** $\left(-\dfrac{2p}{3}, \dfrac{4p^3}{27}\right)$

**ii** $(0, 0)$ minimum, $\left(-\dfrac{2p}{3}, \dfrac{4p^3}{27}\right)$ maximum

**iii** $0 < p < 3$

# 9 Integration

## Prerequisite knowledge

**1** **a** $-19$ **b** $-1$

**2** **a** $\left(-\dfrac{2}{3}, 5\right)$ **b** $(0, 9)$

**3** **a** $24x^7 - 13$ **b** $10x - 4 + \dfrac{5}{\sqrt{x}}$

## Exercise 9A

**1** **a** $y = 5x^3 + c$ **b** $y = 2x^7 + c$

**c** $y = 3x^4 + c$ **d** $y = -\dfrac{3}{x} + c$

**e** $y = -\dfrac{1}{4x^2} + c$ **f** $y = 8\sqrt{x} + c$

**2** **a** $f(x) = x^5 - \dfrac{x^4}{2} + 2x + c$

**b** $f(x) = \dfrac{x^6}{2} + \dfrac{x^3}{3} - x^2 + c$

**c** $f(x) = 3x^2 - \dfrac{1}{x^2} - \dfrac{8}{x} + c$

**d** $f(x) = -\dfrac{3}{2x^6} + \dfrac{3}{x} - 4x + c$

**3** **a** $y = \dfrac{x^3}{3} + \dfrac{5x^2}{2} + c$

**b** $y = \dfrac{3x^4}{2} + \dfrac{2x^3}{3} + c$

**c** $y = \dfrac{x^4}{4} - 2x^3 - 8x^2 + c$

**d** $y = \dfrac{x^2}{4} - \dfrac{5}{4x^2} + \dfrac{1}{x} + c$

**e** $y = \dfrac{2x^{\frac{7}{2}}}{7} - \dfrac{12x^{\frac{5}{2}}}{5} + 6x^{\frac{3}{2}} + c$

**f** $y = 2x^{\frac{5}{2}} + 2x^{\frac{3}{2}} + 2\sqrt{x} + c$

**4** **a** $2x^6 + c$ **b** $5x^4 + c$

**c** $-\dfrac{3}{x} + c$ **d** $-\dfrac{2}{x^2} + c$

**e** $\dfrac{4\sqrt{x}}{3} + c$ **f** $-\dfrac{10}{\sqrt{x}} + c$

**5** **a** $\dfrac{x^3}{3} + \dfrac{5x^2}{2} + 4x + c$ **b** $\dfrac{x^3}{3} - 3x^2 + 9x + c$

**c** $-\dfrac{8x^{\frac{3}{2}}}{3} + 2x^2 + x + c$ **d** $\dfrac{3x^{\frac{10}{3}}}{10} + \dfrac{3x^{\frac{4}{3}}}{4} + c$

**e** $\dfrac{x}{2} + \dfrac{1}{2x} + c$ **f** $\dfrac{x}{2} - \dfrac{3}{2x^2} + c$

**g** $\dfrac{x^2}{6} + \dfrac{4\sqrt{x}}{3} + c$ **h** $\dfrac{2x^{\frac{7}{2}}}{7} + \dfrac{20}{\sqrt{x}} + c$

**i** $2x^2 + \dfrac{12}{x} - \dfrac{9}{4x^4} + c$

## Exercise 9B

**1** **a** $y = x^3 + x + 2$ **b** $y = 2x^3 - x^2 + 5$

**c** $y = 10 - \dfrac{4}{x}$ **d** $y = x^2 + \dfrac{6}{x} - 4$

**e** $y = 4\sqrt{x} - x + 2$

**f** $y = 2\sqrt{x} - 2x + \dfrac{2}{3}x^{\frac{3}{2}} - 1$

**2** $y = \dfrac{3}{x} + 2$

**3** $y = 2x^3 - 6x^2 + 5x - 4$

**4** $y = 5x^3 + \dfrac{3}{x^2} - 2$

**5** **a** $y = 2x^2\sqrt{x} + 2x - 1$ **b** $y = 42x - 97$

**6** $y = 2x^2 + 3x - 7$

**7** $f(x) = 4 + 8x - x^2$

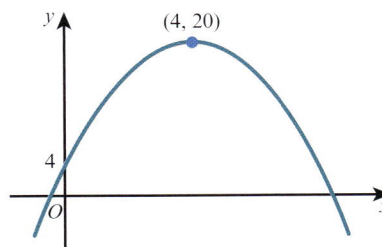

**8** **a** $y = x^3 + \dfrac{1}{2}x^2 - 10x + 3$

**b** $x < -2$ and $x > 1\frac{2}{3}$

**9** **a** $y = 2x^3 + 6x^2 + 10x + 4$ **b** Proof

**10** $y = 2 + 4x - 2x^2 - x^3$

**11** **a** $-3.5$

**b** $P = $ minimum, $Q = $ maximum

**12** **a** $-6$ **b** $y = \dfrac{1}{2}x^2 - 6x + 2$

**13** $f'(x) = 2x - \dfrac{2}{x^2}$, $f(x) = x^2 + \dfrac{2}{x} - 4$

313

**14** $\left(-11, 408\frac{1}{3}\right)$

**15** **a** $y = 9 + 3x - x^2$  **b** $5y = x + 21$

**16** **a** $y = 2x\sqrt{x} - 6x + 10$  **b** $(4, 2)$, minimum

**17** $(1, 7)$, maximum

**18** **a** $y = x^2 - 5x + 2$  **b** $x + y = -1$
  **c** $(1, -2)$

## Exercise 9C

**1** **a** $\dfrac{1}{18}(2x - 7)^9 + c$  **b** $\dfrac{1}{18}(3x + 1)^6 + c$

  **c** $\dfrac{2}{45}(5x - 2)^9 + c$  **d** $-\dfrac{1}{4}(1 - 2x)^6 + c$

  **e** $-\dfrac{3}{16}(5 - 4x)^{\frac{4}{3}} + c$  **f** $\dfrac{1}{5}(2x + 1)^{\frac{5}{2}} + c$

  **g** $\dfrac{4}{3}\sqrt{3x - 2} + c$  **h** $-\dfrac{2}{(2x + 1)^2} + c$

  **i** $\dfrac{5}{32(7 - 2x)^4} + c$

**2** **a** $y = \dfrac{1}{8}(2x - 1)^4 + 2$  **b** $y = \dfrac{1}{3}(2x + 5)^{\frac{3}{2}} - 7$

  **c** $y = 2\sqrt{x - 2} + 5$  **d** $y = \dfrac{2}{3 - 2x} + 6$

**3** $y = 3(x - 5)^4 - 1$

**4** **a** $x + 5y = 7$  **b** $y = 5\sqrt{2x - 3} - 4$

**5** **a** Maximum
  **b** $y = 8\sqrt{3x + 1} - 2x^2 - 2x + 5$

**6** $y = 4\sqrt{2x - 5} - 2$

## Exercise 9D

**1** **a** $8x(x^2 + 2)^3$  **b** $\dfrac{1}{8}(x^2 + 2)^4 + c$

**2** **a** $20x(2x^2 - 1)^4$  **b** $\dfrac{1}{20}(2x^2 - 1)^5 + c$

**3** **a** $k = -2$  **b** $-\dfrac{2}{x^2 - 5} + c$

**4** **a** $\dfrac{6x}{(4 - 3x^2)^2}$  **b** $\dfrac{1}{8 - 6x^2} + c$

**5** **a** $6(2x - 3)(x^2 - 3x + 5)^5$

  **b** $\dfrac{(x^2 - 3x + 5)^6}{3} + c$

**6** **a** $\dfrac{4(\sqrt{x} + 3)^7}{\sqrt{x}}$  **b** $\dfrac{1}{4}(\sqrt{x} + 3)^8 + c$

**7** **a** $15\sqrt{x}(2x\sqrt{x} - 1)^4$  **b** $\dfrac{1}{5}(2x\sqrt{x} - 1)^5 + c$

## Exercise 9E

**1** **a** $7$  **b** $\dfrac{16}{9}$

  **c** $-6$  **d** $21$

  **e** $9$  **f** $\dfrac{5}{2}$

**2** **a** $\dfrac{11}{2}$  **b** $3$

  **c** $\dfrac{107}{6}$  **d** $\dfrac{4}{15}$

  **e** $\dfrac{37}{8}$  **f** $18$

**3** **a** $10$  **b** $\dfrac{26}{3}$

  **c** $\dfrac{2}{5}$  **d** $4$

  **e** $2$  **f** $8$

**4** **a** $-\dfrac{4x}{(x^2 + 5)^2}$  **b** $\dfrac{4}{45}$

**5** **a** $15x^2(x^3 - 2)^4$  **b** $2\frac{1}{15}$

**6** **a** $\dfrac{(\sqrt{x} + 1)^4}{4\sqrt{x}}$  **b** $84\frac{2}{5}$

## Exercise 9F

**1** **a** $21\frac{1}{3}$  **b** $8$
  **c** $20\frac{5}{6}$  **d** $5\frac{1}{3}$

**2** Proof

**3** **a** $11\frac{5}{6}$  **b** $40\frac{1}{2}$
  **c** $4\frac{3}{32}$  **d** $21\frac{1}{12}$

**4** **a** $7\frac{1}{5}$  **b** $5\frac{1}{2}$
  **c** $15.84$  **d** $14$
  **e** $16$  **f** $9$

**5** **a** $48\frac{3}{4}$  **b** $6$

**6** $3\frac{1}{3}$

**7** $10\frac{2}{3}$

**8** **a** $9$  **b** $1.6$

**9** **a** Proof  **b** $3 - \sqrt{5}$

**10** $18\frac{2}{3}$

**11** **a** $(-1, 0)$  **b** $90$

**12** $10\frac{1}{2}$

**13** $34$

**14** $26$

## Exercise 9G

**1** $26\frac{2}{3}$

**2** $10\frac{2}{3}$

**3** $57\frac{1}{6}$

**4** **a** 36      **b** $10\frac{2}{3}$

     **c** 36

**5** $\dfrac{1}{3}$

**6** $1\frac{1}{3}$

**7** **a** $y = \dfrac{1}{3}x + 2$      **b** $\dfrac{1}{2}(2\sqrt{3} - 3)$

**8** **a** $y = -3x + 46$      **b** 64

**9** **a** $y = -8x + 16$      **b** 108

**10** **a** $2y = x - 1$      **b** 8.83

## Exercise 9H

**1** **a** 2      **b** $\dfrac{1}{256}$

     **c** $-\dfrac{5}{4}$      **d** 4

     **e** 50      **f** 16

     **g** $6\sqrt{3}$      **h** 1

     **i** $\dfrac{20}{9}$

**2** Proof

**3** **a** Proof      **b** Proof

     **c** Proof      **d** Proof

     **e** Proof      **f** Proof

## Exercise 9I

**1** **a** $\dfrac{71\pi}{5}$      **b** $\dfrac{16\pi}{3}$

     **c** $\dfrac{15\pi}{8}$      **d** $\dfrac{25\pi}{4}$

**2** **a** $\dfrac{81\pi}{2}$      **b** $\dfrac{124\pi}{15}$

**3** 6

**4** $\dfrac{39\pi}{4}$

**5** **a** $24\pi$      **b** $24\pi$

**6** **a** $\cup$-shaped curve, $y$-intercept = 4,
     vertex = (2, 0)

     **b** $\dfrac{32\pi}{5}$

**7** **a** (25, 0)      **b** $\dfrac{3125\pi}{6}$

**8** **a** (0, 3)      **b** $16\pi$

**9** **a** $\left(\dfrac{1}{3}, 7\right)$, (2, 7)      **b** $\dfrac{250\pi}{9}$

**10** Proof

**11** **a** $\dfrac{52\pi}{3}$      **b** $\dfrac{128\pi}{3}$

**12** **a** $\dfrac{8\pi}{3}$      **b** $\dfrac{32\pi}{5}$

**13** **a** $\dfrac{1888\pi}{3}$ cm$^3$      **b** $171\pi$ cm$^3$

**14** Proof

## End-of-chapter review exercise 9

**1** $f(x) = 3x^4 + 5x^2 - 7$

**2** $\dfrac{25}{3}x^3 - 20x - \dfrac{4}{x} + c$

**3** $y = 30 - \dfrac{6}{x} - \dfrac{5}{2}x^2$

**4** $f(x) = 6\sqrt{x+2} + \dfrac{4}{x^2} - 10$

**5** $\dfrac{194\pi}{9}$

**6** **a** 1      **b** $f(x) = 3x^2 - 6x + 8$

**7** $10\frac{2}{3}$

**8** **a** $\cup$-shaped curve, $y$-intercept (0, 11),
     vertex at (3, 2)

     **b** $\dfrac{483\pi}{5}$

**9** **i** Proof      **ii** $\dfrac{9}{4}$

**10** **i** $B(0, 1)$, $C(4, 3)$      **ii** $y = -3x + 15$

     **iii** $\dfrac{2\pi}{15}$

**11** **i** Proof      **ii** 1      **iii** $\dfrac{5\pi}{3}$

**12** **i** $\dfrac{1}{9}$ or 9

     **ii** $f''(x) = \dfrac{3}{2}x^{-\frac{1}{2}} - \dfrac{3}{2}x^{-\frac{3}{2}}$ at $x = \dfrac{1}{9}$ max, at
     $x = 9$ min

     **iii** $f(x) = 2x^{\frac{3}{2}} + 6x^{\frac{1}{2}} - 10x + 5$

**13** **i** $\left(4, \dfrac{20}{3}\right)$      **ii** $P$ and $Q$ are both $\dfrac{32}{3}$

**14** **i** $y = -24x + 20$      **ii** $\dfrac{9}{8}$

**15** **i** 29.7°      **ii** 1

## Cross-topic review exercise 3

**1**   $y = \dfrac{2}{3}x^3 - 3x + 7$

**2**   **i**  $-0.6$

  **ii**  $y = 2x - \dfrac{16}{3}(3x+4)^{\frac{1}{2}} + 12$

**3**   **i**  $y = 2x^{\frac{3}{2}} - 6x + 2$    **ii**  $x = 4$, minimum

**4**   **i**  $y = -\dfrac{3}{2(1+2x)} + 1$  **ii**  $x > 1, x < -2$

**5**   **i**  $\left(\dfrac{1}{2}, 0\right)$    **ii**  $\dfrac{1}{6}$

**6**   **i**  $\dfrac{4}{9}$    **ii**  $-\dfrac{2}{27}$

**7**   **a**  Proof    **b**  Proof

  **c**  $\dfrac{1250}{(\pi+4)} = 175\,(3\text{ s.f.})$, maximum

**8**   **i**  12    **ii**  $x = -1$ or $x = -3, 1\frac{1}{3}$

**9**   **i**  $\dfrac{dy}{dx} = \dfrac{9}{(2-x)^2}$, no turning points since $\dfrac{dy}{dx} \neq 0$

  **ii**  $\dfrac{81\pi}{2}$    **iii**  $k < -8, k > 4$

**10**  **a**  $f'(x) = -\dfrac{8}{(2x+1)^2}$, $\dfrac{8}{(2x+1)^2} < 0$

  **b**  $f^{-1}(x) = \dfrac{4-x}{2x}, 0 < x \leqslant 4$

  **c**

**11**  **i**  $y = \dfrac{2}{3}x^{\frac{3}{2}} - 2x^{\frac{1}{2}} - \dfrac{2}{3}$  **ii**  $\dfrac{1}{2}x^{-\frac{1}{2}} + \dfrac{1}{2}x^{-\frac{3}{2}}$

  **iii**  $(1, -2)$, minimum

**12**  **i**  $y - 6 = -\dfrac{2}{7}(x-2)$  **ii**  $y = x^2 + \dfrac{2}{x} + 1$

  **iii**  $x = 1$, minimum since $f''(1) > 0$

**13**  **i**  13

  **ii**  $x = -1$ (max), $x = 2$ (min)

**14**  **i**  $B\left(\dfrac{5}{4}, 0\right), C\left(0, \dfrac{3}{4}\right)$  **ii**  $\dfrac{\sqrt{17}}{4}$

  **iii**  $\dfrac{3}{40}$

**15**  **i**  Proof    **ii**  $\dfrac{284\pi}{3}$

## Practice exam-style paper

**1**   $f'(x) = 2 + \dfrac{15}{x^4} > 0$ for all $x$

**2**   $y = -x^3 + 6x^2 - 12x + 11$

**3**   Proof

**4**   **a**  $2187 - 10\,206x + 20\,412x^2$    **b**  $-30\,618$

**5**   **a**  $\dfrac{3}{2}(4\pi - 3\sqrt{3})$ cm$^2$

  **b**  $(2\pi + 3 + 3\sqrt{3})$ cm

**6**   **a**  $(x-3)^2 + (y+2)^2 = 20$    **b**  $x - 2y = 17$

**7**   **a**  $7, -8$

  **b**  **i**  $\dfrac{1}{3}$    **ii**  $\dfrac{27}{8}$

**8**   **a**  $21 - 2(x-3)^2$    **b**  $(3, 21)$

  **c**  $x \leqslant 3 - \sqrt{13}, x \geqslant 3 + \sqrt{13}$

**9**   **a**  $1 \leqslant f(x) \leqslant 11$

  **b**

  **c**  0.927 rad, 5.36 rad

  **d**  $g^{-1}(x) = \cos^{-1}\left(\dfrac{6-x}{5}\right)$

**10**  **a**  $3x + 4y = 17$    **b**  $\dfrac{1}{240}$ units per second

**11**  **a**  $-\dfrac{16}{x^2} - 2x, \dfrac{32}{x^3} - 2$  **b**  $(-2, -12)$, maximum

  **c**  $\dfrac{431\pi}{5}$

316

# Glossary

## A

**Amplitude:** the distance between a maximum (or minimum) point and the principal axis of a sinusoidal function

**Arithmetic progression:** each term in the progression differs from the term before by a constant

**Asymptote:** a straight line such that the distance between a curve and the line approaches zero as they tend to infinity

## B

**Basic angle:** the acute angle made with the $x$-axis

**Binomial:** a polynomial with two terms

**Binomial coefficients:** the coefficients in a binomial expansion

**Binomial theorem:** the rule for expanding $(1 + x)^n$ or $(a + b)^n$

## C

**Chain rule:** the rule for computing the derivative of the composition of two functions

**Common difference:** the difference between successive terms in an arithmetic progression

**Common ratio:** the constant ratio of successive terms in a geometric progression

**Completed square form:** the equation $(x - a)^2 + (y - b)^2 = r^2$, where $(a, b)$ is the centre and $r$ is the radius of the circle

**Completing the square:** writing the expression $ax^2 + bx + c$ in the form $d(x + e)^2 + f$

**Composite function:** a function obtained from two given functions by applying first one function and then applying the second function to the result

**Convergent series:** a sequence that tends to a finite number

## D

**Decreasing function:** a function whose value decreases as $x$ increases

**Definite integral:** an integral between limits whose result does not contain a constant of integration

**Derivative:** denoted by $\dfrac{dy}{dx}$ of $f(x)$; gives the gradient of a curve

**Differentiation:** the process of finding the gradient of a curve

**Differentiation from first principles:** the process of finding the gradient of a curve using small increments

**Discriminant:** the part of the quadratic formula underneath the square root sign

**Domain:** the set of input values for a function

## F

**Factorial:** $6 \times 5 \times 4 \times 3 \times 2 \times 1 = 6!$ (read as '6 factorial')

**First derivative:** see Derivative

**Function:** a rule that maps each $x$ value to just one $y$ value for a defined set of input $(x)$ values

## G

**General form of a circle:** the equation $x^2 + y^2 + 2gx + 2fy + c = 0$, where $(-g, -f)$ is the centre and $\sqrt{g^2 + f^2 - c}$ is the radius of the circle

**Geometric progression:** each term in the progression is a constant multiple of the preceding term

**Gradient function:** the derivative $f'(x)$ is also known as the gradient function of the curve $y = f(x)$

## I

**Identity:** a mathematical relationship equating one quantity to another

**Improper integrals:** a definite integral that has either one limit or both limits are infinite, or a definite integral where the function to be integrated approaches an infinite value at either or both endpoints in the interval (of integration)

**Increasing function:** a function whose value increases as $x$ increases

**Indefinite integral:** an integral without limits whose result contains a constant of integration

**Integration:** the reverse process of differentiation

**Inverse function:** the inverse of a function, $f^{-1}(x)$, is the function that undoes what $f(x)$ has done

## M

**Many-one:** a function which has one output value for each input value but each output value can have more than one input value

**Mapping:** a diagram to show how the numbers in the domain and range are paired

**Maximum point:** a point, $P$, on a curve where the value of $y$ at this point is greater than the value of $y$ at other points close to $P$

**Minimum point:** a point, $Q$, on a curve where the value of $y$ at this point is less than the value of $y$ at other points close to $Q$

## N

**Normal:** the line perpendicular to the tangent at a point on a curve

## O

**One-one:** a function where exactly one input value gives rise to each value in the range

## P

**Parabola:** the graph of a quadratic function

**Pascal's triangle:** a triangular array of the binomial coefficients, where each number is the sum of the two numbers above

**Period:** the length of one repetition or cycle of a periodic function

**Periodic functions:** a function that repeats its values in regular intervals or periods

**Point of inflexion:** a point on a curve at which the direction of curvature changes

**Principal angle:** the angle that the calculator gives is called the principal angle

## Q

**Quadrant:** the Cartesian plane is divided into four quadrants

**Quadratic formula:** the formula $x = \dfrac{-b \pm \sqrt{b^2 - 4ac}}{2a}$, which is used to solve the equation $ax^2 + bx + c = 0$

## R

**Radian:** one radian is the angle subtended at the centre of a circle by an arc that is equal in length to the radius of a circle

**Range:** the set of output values for a function

**Reference angle:** the acute angle made with the $x$-axis

**Roots:** if $f(x)$ is a function, then the solutions to the equation $f(x) = 0$ are called the roots of the equation

## S

**Second derivative:** denoted by $\dfrac{d^2 y}{dx^2}$ or $f''(x)$ and is used to determine the nature of stationary points on a curve

**Series:** the sum of the terms in a sequence

**Self-inverse function:** a function f where $f^{-1}(x) = f(x)$ for all $x$

**Solid of revolution:** the solid formed when an area is rotated through 360° about an axis

**Stationary point:** a point on a curve where the gradient is zero, also known as a turning point

## T

**Tangent:** a straight line that touches a curve at a point

**Term:** a number in a sequence

**Turning point:** a point on a curve where the gradient is zero, also known as a stationary point

## V

**Vertex:** the vertex of a parabola is the maximum or minimum point

**Volume of revolution:** the volume of the solid formed when an area is rotated through 360° about an axis

# Index

319